Molecular biology of development

Volume I
MOLECULAR EVENTS

A.A.Neyfakh, D.Sc. (Biol.)

Institute of Developmental Biology,
USSR Academy of Sciences

M.Ya.Timofeeva, Cand.Sc. (Biol.)

Institute of Molecular Biology,
USSR Academy of Sciences

Translated from Russian
by M.G.Zimmerman

Springer Science+Business Media, LLC

Originally published in Russian as *Molekulyarnaya biologiya protsessov razvitiya* by Nauka Publishers, USSR, in 1977, under the general editorship of **G. P. Georgiev**, Corresponding Member of the USSR Academy of Sciences

Translated into English
from the revised and updated Russian version
English Editor:
A. M. Kolchinsky, Cand. Sc. (Biol.)

Originally published by Plenum Publishing Corporation in 1984
Softcover reprint of the hardcover 1st edition 1984

Library of Congress Catalog Card Number

ISBN 978-1-4899-5372-8 ISBN 978-1-4899-5370-4 (eBook)
DOI 10.1007/978-1-4899-5370-4

PREFACE TO THE ENGLISH TRANSLATION

Molecular biology of development is a discipline shared by molecular and developmental biology. It endows molecular biology with its specific range of objects, and developmental biology, with its more profound level of research. Such an interpenetration of two sciences is quite natural. At this juncture the traditional path of investigation from the organism to the cells and molecules, i.e. from the effect to the cause, no longer appears promising. It seems more natural to view development as the realization of genetic information and to try and trace this process step by step, from DNA via RNA to proteins, and thence to the properties of cells and the formation of organs.

This approach promises no easy solutions. Even having overcome the molecular stage and answered its two principal questions — what is the mechanism of differential expression of genes and why are different sets of proteins synthesized in different parts of the embryo — we are bound to come up against two other challenges: finding the way for passing on, first, from a set of proteins to the properties and behavior of cells, and, second, from the properties of cells to the formation of tissues and organs. But it is perhaps the unfathomable depths of developmental biology that are now pushing it to the foreground among the main biological problems and attracting new effort to it.

The present book follows the logic of molecular, rather than developmental, biology. Therefore it reviews successively the available data on DNA, RNA, and proteins. Within each respective area the changes in the content, composition, and synthesis rate of macromolecules which occur in oogenesis and embryogenesis are considered. This mode of presentation seems to be the simplest alternative. Volume I is dedicated to the description of molecular events. Therefore it hardly mentions the synthesis control mechanisms — the subject plaguing the minds of scientists today. This subject is covered in Volume II, which is a logical sequel to Volume I and will appear in English in the near future.

Writing books on molecular biology is an ungrateful task, since the first chapter becomes obsolete before the last one is completed: each year and even each month brings new information, which must necessarily be included in the text. In the course of translation of the present book we did our best to update the text with all the essential data published up to 1981, inclusive. Besides, this particular book focuses on figures and facts which will hardly change substantially with time, rather than on speculations. We would deem our goal achieved if the book proved useful for an appreciable period of time.

A. A. Neyfakh
M. Ya. Timofeeva

CONTENTS

Chapter One

DNA

One of the principal postulates of modern genetics, molecular biology, and developmental biology is the conservation of genetic information, i.e. of DNA, in the series of generations of cells and organisms, and hence maintaining its constancy throughout the individual development. This must be true in any case for the germ-line cells, i.e. for the series of cell generations from gamete to gamete.

The actual situation, however, which is already revealed when determining the quantity of DNA in oocytes and embryonic cells, is much more complicated. The quantity of DNA in oocytes increases from tens to millions of times owing to the increasing number of mitochondria and, accordingly, to the increasing quantity of mitochondrial DNA. In the oocyte nucleus of many animals the total quantity of DNA is increased by amplification of ribosomal genes — the nucleolar organizer. In some species, at the onset of embryogenesis, the somatic cells experience diminution of chromatin or elimination of chromosomes and, hence, a decrease in nuclear DNA. The question of amplification of structural genes as a means of regulating their function during differentiation is still being discussed. Finally, the classical picture of the mitotic cycle during egg cleavage has changed substantially because of the almost complete disappearance of the G_1 and G_2 phases. The organization of chromosomes and nuclei suffers substantial changes in oogenesis and embryogenesis: lampbrush loops are formed, the structure of the nucleoli in the oocyte changes, there appear karyomeric nuclei during early cleavage and polytene chromosomes in the provisory organs of dipteran larvae. The changes in DNA content and chromosome structure are naturally associated with the peculiarities of their function in development and are a means of regulation.

DNA IN THE EGG

The determination of the DNA content in the egg is very complicated as regards the techniques used. The egg is much larger than the normal cell and contains a considerable number of reserve substances, including such chemically similar compounds as RNA, nucleotidepeptides, low-molecular-weight deoxyribonucleotides, and their derivatives. Many animals have no nucleus as a structural entity by the completion of egg maturation, because the nuclear envelope is dissolved and the chromosomes are in the cytoplasm. Therefore, cytophotometric determination of DNA in the nucleus by Feulgen's staining method is possible only at the oogenesis stage, and also in the egg after fertilization (or after parthenogenetic activation), when the female pronucleus has been formed.

For chemical determination of DNA in a whole egg, highly specific methods had to be elaborated. But even these methods do not provide exact values of the DNA content in the egg for many species.

Paradoxical results were obtained in the early investigations carried out in the 1950s, when the biochemistry and chemistry of nucleic acids were only beginning to develop. Thus, Marshak and Marshak (1952, 1953, 1954, 1955, 1956) failed to detect DNA in the nucleus of mature eggs of the sea urchin *Arbacia* by applying the Feulgen reaction. This even enabled them to postulate that DNA is not the only component of the nucleus which is responsible for the transmission of genetic information, because at the crucial stage — the formation of gametes — the nucleus has no DNA. This idea did not gain much popularity, because exact proofs of the genetic role of DNA had been produced by then, but the real causes for the wrong measurements of the DNA content in the egg were established much later.

It was not until 1955 that the improvement of Feulgen's method helped detect DNA also in the nuclei of sea urchin eggs (Burgos, 1955; Brachet and Ficq, 1956). At a later date, Hinegardner (1961) isolated nuclei from eggs of the sea urchin *Echinometria mathei* and demonstrated the presence of DNA in them. A DNA content of 4 C was detected in the nucleus of the grasshopper oocyte by Swift and Kleinfeld (1953) using the cytophotometric method.

In contrast to Marshak and Marshak's data, enormous DNA contents were concurrently reported in whole eggs of various animals: amphibians (Hoff-Jørgensen and Zeuthen, 1952; Sze, 1953), chickens (Hoff-Jørgensen, 1954; Solomon, 1957a, b), various species of sea urchin (Elson and Chargaff, 1952; Mazia, 1949; Vendrely and Vendrely, 1949; Zeuthen, 1951), and insects (Nigon and Dailler, 1958; Durante, 1961; Schultz, 1956). The quantities of DNA in the eggs, as determined by different methods, were found to be equal to many hundreds and thousands of haploid DNA equivalents (Table 1). This contradicted the generally accepted notions of the constancy of DNA content in chromosomes, according to which the oocyte must have no more than four haploid equivalents (in accordance with the number of chromosomes appearing during the preparation for meiosis). Calculations of the volume which can be occupied by the total DNA found in the egg showed that it greatly exceeds the volume of the female pronucleus (Hoff-Jørgensen and Zeuthen, 1952; Sze, 1953). Hence Hoff-Jørgensen and Zeuthen concluded that at least part of the DNA is concentrated in the cytoplasm. This assumption was confirmed by them in experiments, where DNA was found in enucleated frog oocytes (Hoff-Jørgensen and Zeuthen, 1952).

The "redundant" DNA, which was detected in the egg cytoplasm, was

Table 1

DNA Content in the Eggs of Various Animals

Species	Methods of isolation and determination of DNA in the egg	DNA in egg, μg	DNA in egg, diploid equivalents	Reference
Sea urchins:				
Paracentrotus lividus	Determination by microbiol. method of thymine in DNA fraction isolated by Schmidt-Tannhauser method	2.6×10^{-5}	13	Elson and Chargaff, 1952
Strongylocentrotus purpuratus	Determination by UV absorption and by deoxyribose in DNA fraction isolated by centrifuging in CsCl and treated with amylase	3.3×10^{-6}	2.15	Pikó *et al.*, 1967
Lytechinus pictus	ditto	8.3×10^{-6}	4.6	Pikó *et al.*, 1967
Arbacia acquituberculata	Determination by deoxyribose in DNA fraction isolated by Schmidt-Tannhauser method	2.2×10^{-4}	170	Vendrely and Vendrely, 1949
Arbacia lixula	Determination by microfluorometric method in DNA fraction	—	12	Baltus *et al.*, 1965
Insects:				
Drosophila melanogaster	Determination by fluorometric method in DNA fraction isolated by Schmidt-Tannhauser method	4.9×10^{-3}	1.5×10^{4}	Tsien and Wattiaux, 1971
	Determination by microbiological method		5×10^{2}	Shultz, 1956

Table 1 (*continued*)

Species	Methods of isolation and determination of DNA in the egg	DNA in egg, μg	DNA in egg, diploid equivalents	Reference
Echiuroids:				
Urechis caupo		1×10^{-5}	5	Schwartz, 1970
Fishes:				
Acipenser güldenstädti	Determination by deoxyribose in acid-insoluble fraction isolated by Schmidt-Tannhauser method	1.8×10^{-1}	4×10^{4}	Shmerling, 1965
Misgurnus fossilis	Determination by microbiological method of thymine	5.8×10^{-3}	1×10^{3}	Timofeeva, 1967
Salmo irideus	Determination by absorption at 260 nm in DNA fraction isolated by Ogur-Rosen method	2.5		Hagenmaier, 1969
Amphibians (Urodela):				
Triturus alpestris	Determination by absorption at 260 nm in DNA fraction isolated by Ogur-Rosen method	2.4	—	Chen, 1960 (for *T. alpestris*, *T. cristatus* and *T. palmatus*)
Triturus cristatus	ditto	1.1	—	
Triturus palmatus	ditto	1.5	—	
Triturus pyrrhogaster	Determination by microbiological method	2.7×10^{-2}	—	Kuriki and Okazaki, 1959
Amphibians (Anura):				
Bufo vulgaris	ditto	3.2×10^{-2}	4×10^{3}	Grant, 1958 (for *B. vulgaris* and *R. pipiens*)
Rana pipiens	ditto	2.2×10^{-2}	1.5×10^{3}	

Rana pipiens	Determination by deoxyribose in DNA fraction isolated by Schmidt-Tannhauser method	9.6×10^{-1}	7×10^{4}	Sze, 1953
Rana pipiens	Determination by deoxyribose in acid-insoluble fraction;	2.2×10^{-1}	2.5×10^{4}	Bieber *et al.*, 1959
	by the nucleotides sum;	7.2×10^{-2}	8×10^{3}	
	by microbiological method of thymine;	4.0×10^{-2}	4×10^{3}	
	by absorption at 260 nm in NaCl extract of acid-insoluble fraction	3.5×10^{-2}	3.5×10^{3}	
Rana pipiens	Determination by deoxyribose in acid-insoluble fraction;	8.4×10^{-3}	5.6×10^{2}	Dawid, 1965
	by absorption at 260 nm in DNA fraction isolated by centrifuging in CsCl	3.6×10^{-3}	2.4×10^{2}	
Xenopus laevis	Determination by deoxyribose in acid-insoluble fraction:	6.6×10^{-3}	1×10^{3}	
	by absorption at 260 nm in DNA fraction isolated by centrifuging in CsCl	3.1×10^{-3}	5×10^{2}	
Birds:				
Gallus domesticus	Determination by means of isotopic dilution in fraction isolated by Ogur-Rosen method	3.6×10^{2}	1.8×10^{8}	Solomon, 1957

assumed to have purely accessory functions of the stockpile, the "building material," which is utilized in the process of development for building up nuclear DNA in an increasing number of embryonic cells. This was supported by data which seemed to indicate the absence of DNA increase in the embryo at the early stages of development despite the intensive *de novo* formation of nuclei during cleavage (Gregg and Løvtrup, 1955; Steinert, 1951; Hoff-Jørgensen and Zeuthen, 1952; Agrell and Persson, 1956; Chen, 1960). It was even suggested that the DNA stockpiles could be utilized by inserting whole blocks of cytoplasmic DNA into the nuclear chromosomes (Moore, 1959).

These hypotheses are of no more than historic interest now, but a whole decade was required for this experimental material to be revised, and it was only the development of new methods for isolation, fractionation, and analysis of DNA that resolved the question of the content, localization, and function of the egg DNA from an entirely different angle.

At the same time the pioneering investigations of the 1950s played an essential role in arousing an interest in cytoplasmic DNA and also in the development of studies into cytoplasmic heredity, the genetic independence of cytoplasmic structures, etc.

DNA Content in the Egg

The quantity of DNA in the eggs of different animals correlates with the size of these eggs. In the small sea urchin eggs it is equal to several pg. According to Pikó *et al.* (1967), the egg of *Strongylocentrotus lividus* contains 3.3 ± 0.3 pg, and that of *Lytechinus*, 8.3 ± 0.3 pg of DNA. Higher values were obtained by other authors who used various chemical, microbiological, and microfluorometric determination methods (Vendrely and Vendrely, 1949; Elson and Chargaff, 1952; Sugino *et al.*, 1960; Baltus *et al.*, 1965; Bibring *et al.*, 1965). The measurement of the DNA content is impeded by polysaccharide impurities present in the eggs of echinoderms. Pikó *et al.* (1967) worked out a special procedure for removing them from DNA by treatment of the homogenate with 4% sodium dodecylsulfate followed by the separation of DNA by centrifugation in CsCl, digestion of the polysaccharides in these DNA preparations by α-amylase, and isolation of the pure DNA by centrifugation in CsCl. With this purification technique, the chemical diphenylamine method and that of DNA determination by the optical density at 260 nm yielded similar values. No DNA was lost in the course of isolation. This was checked by the isotopic dilution technique: labeled thymidine was injected to female *Lytechinus*, and eggs were isolated two months after the isotope was introduced. The DNA from such eggs was labeled, and if such labeled DNA was added prior to purification, no label was lost. Pikó *et al.* believed that the obtained values for DNA — 3 and 8 pg per egg for two sea urchin species — were quite reliable.

Quantities of DNA as small as in sea urchins were determined with the use of the same techniques in the oocytes of mammals and worms. In the mouse, for instance, it is equal to 8 pg (Pikó, 1975), i.e. it exceeds two- to three-fold the haploid equivalent in the nucleus. The mature oocyte of the sea echiuroid worm *Urechis caupo* contains 10 pg of DNA (Schwartz, 1970).

In insect eggs, which are much larger than those of sea urchins, the DNA content is also several orders of magnitude higher (Schultz, 1956; Tsien and Wattiaux, 1971). In *Drosophila* it is 4.9 ng, varying with the female's age: it is higher (6 ng) in young and aged specimens and lower (3 to 4 ng) in medium-aged females (Tsien and Wattiaux, 1971). Still more DNA is contained in the large eggs of fishes and amphibians (from 0.003 μg to several micrograms). The values obtained depend to a great extent on the determination method, and therefore they vary within a wide range even for one and the same species: for instance, from 0.22 to 0.004 μg in the frog *R. pipiens*. As in sea urchins, the large stockpiles of proteins and polysaccharides, which are hard to remove during material fractionation, substantially affect the DNA determination in the eggs of fishes and amphibians. Schmidt and Tannhauser's and Ogur and Rosen's methods of DNA isolation (see Table 1) with subsequent DNA determination by the optical density at 260 nm or by the diphenylamine reaction can hardly be applied to these objects. The possibility of bacterial contamination should also be taken into account. For instance, according to Shmerling (1965), the treatment of sturgeon eggs with trypsin and a sodium dodecylsulfate solution, which removes bacteria from the egg surface, reduces the DNA amount detectable in the eggs by a factor of five.

A more specific technique is the microbiological method with the use of a thymine-deficient strain of lactic-acid bacteria *Lactobacillus*. This method can be utilized for DNA determinations in crude material. It is, however, necessary to introduce a correction for the content of thymine nucleotides, which may be considerable in the egg. The value for the DNA content in the egg, as determined microbiologically, is smaller, as a rule, than that determined chemically and is probably closest of all to the true DNA content.

To determine DNA in the eggs of two frog species — *Xenopus laevis* and *Rana pipiens* — Dawid (1965) carried out preliminary isolation of high-molecular-weight DNA preparations and their purification by centrifuging in a CsCl gradient. The resulting values were much lower than those reported by other authors (see Table 1). It is possible that with this method of DNA determination in the egg part of the hard-to-extract or low-molecular-weight material, for instance, the DNA of yolk platelets (Hanocq-Quertier *et al.*, 1968; Baltus *et al.*, 1968), was neglected. In birds with very large eggs the DNA amount runs into hundreds of micrograms. Chicken eggs, for instance, contain 360 μg DNA according to Solomon (1957a, b).

Nuclear DNA

The quantity of DNA found in the eggs of various animals exceeds many times over the diploid equivalent, i.e. the DNA content in the chromosome set of the nuclei of somatic cells (see Table 1). Calculations show that this excess DNA amounts to 0.5 to 5 diploid equivalents in the eggs of mammals, worms, and sea urchins, hundreds of equivalents in insects, thousands in amphibians and fishes, and millions in birds (see Table 1).

The content of chromosomal nuclear DNA in male gametes — spermatozoa — is equal to the haploid amount. In the eggs of various animals the content of nuclear DNA varies from 1 C to 4 C, depending on the meiosis stage of the nucleus. In the nuclei of mature sea urchin eggs which have completed meiotic divisions, the DNA content is equal to the haploid equivalent (Hinegardner, 1961). Isolated nuclei of mature frog oocytes contain a diploid amount of DNA (England and Mayer, 1957). An unfertilized mouse egg is in the metaphase of the second maturation division and also contains 4 C of DNA (Alfert, 1950).

A mature egg of fishes and amphibians has no morphologically defined nucleus. The dissolution of the germinal vesicle during ovulation is followed by meiotic divisions; in the mature egg one part of the nuclear DNA forms the condensed first polar body, while the other part (2 C) is contained in the metaphase chromosomes of the second meiotic division.

Thus, in the small eggs of mammals, worms, and echinoderms the chromosome DNA constitutes 1/3 to 2/3 DNA of the eggs, whereas in the large eggs of fishes, amphibians, and birds it is a negligible part of the total DNA of the eggs. The remaining part of the egg DNA is located in the cytoplasm. The localization of the DNA in the cytoplasm was first shown by Hoff-Jørgensen and Zeuthen (1952) in frog oocytes and by Baltus *et al.* (1965) in the eggs of sea urchins. Hoff-Jørgensen and Zeuthen used enucleated frog oocytes. Baltus *et al.* divided a sea urchin egg into nucleate and anucleate halves by centrifugation in a discontinuous sucrose gradient and showed that both halves contain DNA. The DNA isolated from the anucleate halves did not differ in its sedimentation characteristics when centrifuged in CsCl (i.e. in buoyant density) from the DNA taken from the nucleate halves (Baltus *et al.*, 1965). As shown in subsequent works, cytoplasmic DNA is concentrated in the mitochondria and in yolk granules.

Mitochondrial DNA

Mitochondrial DNA is one of the main components of the DNA of oocytes and mature eggs. Its function is associated with the maintenance of hereditary continuity of mitochondria and thus with the maintenance of the energy metabolism of the cell. It plays the principal, if not the only, role in determining the cytoplasmic heredity.

The quantity of mitochondrial DNA (mtDNA) in the eggs of different animals varies from one to many hundreds or even thousands of haploid equivalents (Dawid, 1966; Pikó and Matsumoto, 1976), i.e. it is either equal to or many times greater than that of nuclear DNA. Thus, the ratio of nuclear and mitochondrial DNA in the egg differs sharply from that in the somatic cell, where the mitochondrial DNA usually does not exceed 1% of the nuclear DNA (Nass, 1969a). This is due to the formation, in oogenesis, of considerable stockpiles of mitochondria, whose quantity would be sufficient for many hundreds and thousands of cells. While somatic cells contain 2 to 5 $\times$ 10^4 mitochondria (Nass *et al.*, 1965), it has been shown by electron microscopy (EM) that the loach oocyte at the large-growth stage contains 1.2×10^7 mitochondria (Ozernyuk and Palmbakh, 1975). Even small mammalian eggs have 9.2×10^4 mitochondria (Pikó and Matsumoto, 1976), and sea urchin eggs, 2×10^5 mitochondria (Pikó *et al.*, 1967).

Structure and nucleotide composition. Mitochondrial DNA's are circular supercoiled molecules (Nass, 1969b; Borst, 1972; see Gause, 1977). They are not bound with histones and do not form chromosome-type structures in mitochondria. EM investigations of mitochondrial DNA's showed that mtDNA preparations isolated from eggs contain different forms of circular structures: open, covalently closed (supercoiled), monomeric, and oligomeric forms. In amphibians, each mtDNA molecule has only one ring (monomer) (Dawid and Wolstenholme, 1967). Concatenates, or oligomeric forms (Pikó *et al.*, 1968; Pikó and Matsumoto, 1976), have been discovered in the mtDNA of sea-urchin and mammalian eggs. They are mostly dimers (11% in sea urchins and 7.5% in mice), but there are also trimers (2 to 3%), tetramers (0.3 to 0.4%), and even penta- and hexamers (0.2% in mice). The content of such oligomeric forms of mtDNA in normal somatic cells depends on the cell state: their amount increases in the period of active growth, and in some varieties of cancer (Nass, 1969b; Clayton *et al.*, 1970).

The mitochondrial DNA may differ substantially from the chromosomal nuclear DNA or be closely similar to it in total nucleotide composition. This is established by centrifuging in a CsCl gradient. In the sea worm *Urechis caupo* the buoyant density of mtDNA and nuclear DNA is the same and equals 1.699 g/cm^3 (Dawid and Brown, 1970). In the ascarid *Ascaris lumbricoides*, the buoyant density of the nuclear DNA is 1.697 g/cm^3 and that of the mtDNA, 1.685 g/cm^3 (Bielka *et al.*, 1968). Pikó found large variations in sea urchins (by 0.01 g/cm^3). Thus, the density of nuclear DNA of *Strongylocentrotus nudus* is 1.693 g/cm^3, whereas that of mtDNA is 1.703 g/cm^3 (Pikó *et al.*, 1967). In another species, *Lytechinus variegatus*, however, the nuclear and mitochondrial DNA has the same density of 1.693 g/cm^3 (Hartman *et al.*, 1971). Similar fluctuations are observed in amphibians and mammals. The nuclear and mtDNA of frogs *Xenopus laevis* and *Rana pipiens* are practically hard to separate by centrifugation in CsCl. The densities of these DNA's are 1.702 to 1.704 g/cm^3 (Dawid and Wolstenholme, 1967, 1968). In two other representatives of amphibians, axolotls *Siredon mexicanum* and *Necturus maculosus*, the difference between the densities of the mtDNA and the nuclear DNA is 0.007 g/cm^3 (Wolstenholme and Dawid, 1968). In man and birds the buoyant densities of the nuclear DNA's differ considerably from those of the mtDNA's (Borst, 1972).

The differences in the buoyant density of mitochondrial DNA are determined by the differences in its GC content. As can be seen from Table 2, the content of GC pairs in mtDNA is the lowest in insects and ascarids (21 to 25%). In other organisms, the GC content varies from 35 to 45%. On the whole, according to its nucleotide composition, mtDNA must belong to the AT type. It has been found to contain the minor base 5-methylcytosine. Its content, according to Vanyushin and Kirnos (1974) and Kirnos (1975), is about 2%, which corresponds to approximately 300 residues of m^{5}C per circular molecule of mtDNA.

The differences in the total nucleotide composition of mtDNA in various species do not indicate their complete difference in primary structure, i.e. nucleotide sequence. A comparison of mtDNA from different

Table 2

The Content of GC Pairs in mtDNA from Eggs and Embryos of Various Animals According to Buoyant Density (ρ) Data

Animal	ρ in CsCl, g/cm^3	GC pairs, %	Reference
Sea urchins:			
Strongylocentrotus nudus	1.703	43	Pikó *et al.*, 1967
Lytechinus variegatus	1.693	33	Hartmann *et al.*, 1971
Insects:			
Drosophila melanogaster	1.681	21	Sager and Kitchin, 1975
Worms:			
Ascaris lumbricoides	1.685	25	Bielka *et al.*, 1968
Urechis caupo	1.695	35	Dawid and Brown, 1970
Fishes:			
Misgurnus fossilis	1.707	47	Gause, 1977
Amphibians:			
Xenopus laevis	1.704	44	Wolstenholme and Dawid, 1968
Rana pipiens	1.702	42	
Siredon mexicanum	1.695	35	
Necturus maculosus	1.695	35	

sources by the hybridization method showed that even animals of different classes exhibit a definite homology in the sequence. Annealing of denatured mtDNA of worm and chick (Dawid and Brown, 1970; Dawid, 1972b), clawed frog and chick (Dawid and Wolstenholme, 1968), and two species of clawed frog *Xenopus mulleri** and *Xenopus laevis* (Dawid, 1972b) produces hybrid molecules. This is readily ascertained by centrifugation in a CsCl gradient, because denatured single-stranded DNA's have a higher density than double-stranded DNA's (native, renaturated, or hybrids). The degree of homology, however, is not high. Even in the related species of *Xenopus* (Dawid, 1972b) it does not exceed 20%. The partial homology between mtDNA of different animals evidently reflects the similarity in nucleotide sequences (conservatism) of ribosomal genes, and also of the genes of mitochondrial transfer RNA. The differences, on the contrary, are associated with the divergences in the sequences of "spacers," i.e. of mtDNA portions with no coding functions (Dawid, 1972b). The degree of divergence in the structural genes of mtDNA still remains obscure. A comparison by means of hybridization of mtDNA and nuclear DNA showed that these DNA's, even when taken from one and the same animal, practically never form hybrids. On annealing, only the mtDNA renatures, while the nuclear DNA remains in the form of single-stranded molecules (Dawid and Wolstenholme, 1968). This goes to show that the homology between mtDNA and nuclear DNA in the nucleotide sequence is practically very low or altogether absent. No homology between the mtDNA's of animals and yeast has been found (Dawid and Wolstenholme, 1968).

Genetic autonomy of mtDNA. Measurement of the contour length of the circular molecules of mtDNA on electron micrographs helps determine the molecular weight. A DNA length of 1 μm corresponds to about 2×10^6 daltons. The length of mtDNA molecules varies from 4.5 to 5.9 μm for different animals, and the fluctuations within the same species are very small (0.5 to 0.6 μm), while those between different species are sometimes considerable, so that the mixing of mtDNA's of two species produces a distribution curve with two clearly defined peaks corresponding to the length of the mtDNA molecules of these species. For instance, in amphibians the average length of mtDNA molecules of frogs *Xenopus* and *Rana pipiens* is 5.8 to 5.9 μm, and that of the axolotls *Siredon mexicanum* and *Necturus maculosus*, 4.8 to 4.9 μm (Wolstenholme and Dawid, 1968). In sea urchins *Strongylocentrotus nudus* and *Lytechinus variegatus* the length of mtDNA molecules is 4.5 and 5.4 μm, respectively (Pikó *et al*., 1967; Hartmann *et al*., 1971). Molecular weights show similar differences: from 8.8 to 11×10^6 daltons. The total length of the mtDNA oligomers is a multiple of the length of one monomeric circular molecule, which suggests the

* The authors of these papers have reported that the species which they designated as *Xenopus mulleri* here and in other papers is actually *Xenopus borealis* (Brown *et al*., 1977).

formation of such oligomeric, or concatenate, molecules by multiple replication without segregation into monomers.

The genetic potency of each mtDNA molecule with a molecular weight of 10^7 daltons (1.5×10^4 nucleotide pairs) is rather low, being equal to approximately 10 to 12 genes of size 1 to 1.5×10^3 b.p. each, or 40 to 50 genes of smaller size. The coding strand in mtDNA is predominantly the heavy H-chain (Dawid, 1970; Attardi *et al.*, 1970; Raste and Dawid, 1979).

Using *Xenopus* oocytes (Dawid, 1970, 1972a) and mammalian cells (Attardi *et al.*, 1965. 1970; Battey and Clayton, 1978) it was shown that mtDNA encodes the RNA of the protein-synthesizing apparatus of mitochondria. Each molecule contains one gene of 16S and one gene of 12S mitochondrial ribosomal RNA (the molecular weight of these rRNA's is 0.53×10^6 and 0.36×10^6 daltons, respectively. It has been established by rRNA saturation hybridization that 18 to 20% of the *Xenopus* mtDNA (Dawid, 1970, 1972a) and 20% of the HeLa mtDNA (Attardi *et al.*, 1970) code for them. Three percent of the total *Xenopus* mtDNA and 4% of the H-chain of the HeLa mtDNA hybridize with the fraction of low-molecular-weight mitochondrial RNA which includes 5S rRNA and tRNA (Hatlen and Attardi, 1971). Besides, six tRNA genes are localized in the L-chain of the mtDNA (Attardi *et al.*, 1970; Ohi *et al.*, 1978). The detected mitochondrial tRNA's are indeed transcripts from mtDNA, and not an impurity of cytoplasmic tRNA which have mtDNA-homologous sequences. This is evident from the sensitivity of tRNA synthesis to ethidium bromide (Zylber and Penman, 1969), which blocks mtDNA transcription *in vivo*.

The number of tRNA genes in an mtDNA molecule may possibly vary from one organism to another. *Xenopus* mitochondria have 20 tRNA genes; 14 of them are localized in the H-chain, and 6 in the L-chain of mtDNA (Ohi *et al.*, 1976). The mitochondria possess a specific set of tRNA's. These tRNA's differ from those of the cytoplasm in acceptor properties and primary structure. Among the mitochondrial tRNA's, experimentors have detected formylmethionine tRNA involved in the initiation of protein synthesis only in mitochondria. Despite the limited content of tRNA genes, the multiplicity phenomenon, i.e. the presence of isoacceptor tRNA's, has been described for some tRNA's (see Gause, 1977).

Apart from the RNA genes of the protein-synthesizing apparatus, mtDNA has sequences complementary to the mitochondrial nonribosomal RNA. In *Xenopus*, for instance, these sequences constitute 16% of the mtDNA (Dawid, 1970). They are probably the mRNA's of certain subunits of specific mitochondrial enzymes, such as cytochrome oxidase and oligomycin-sensitive adenosinetriphosphatase. It has been established by now that mtDNA encodes three of the six subunits of cytochrome *c* oxidase and three of the ten subunits of ATPase.

An analysis by means of hybridization of poly(A)-containing

RNA's from HeLa mitochondria with mtDNA showed that eight distinct fractions of such RNA's have homology with mtDNA, i. e. are encoded by the mitochondrial genome (Ojala and Attardi, 1974).

Twelve most abundant transcripts of mtDNA have been found in mitochondria of *X. laevis*. Ten of them contained poly(A) and the other two were ribosomal RNA's. All of them were transcribed from the H-chain and constituted 80% of the coding capabilities of mtDNA. For seven of them, it has been definitely established that the sequences encoding these RNA's contain no introns (Raste and Dawid, 1979). The presence of introns has been detected in the oxi-3 locus of yeast mtDNA which contains the gene for the subunit I of cytochrome *c* oxidase (Van Ommen *et al.*, 1979).

According to Attardi and coworkers (Attardi *et al.*, 1975; Attardi and Attardi, 1971), both the H- and L-chains of the mtDNA are transcribed; the high-molecular-weight transcripts are first synthesized and then subjected to processing with the formation of rRNA, tRNA, and other functionally significant mitochondrial RNA's. Thus, it has been shown that symmetrical mtDNA transcription occurs in HeLa cells. In the course of processing, part of the sequences corresponding to "spacers" is eliminated. According to Dawid (1972a), about 1/3 of the mtDNA sequences are not represented in the RNA's functioning in the mitochondria.

The mtDNA sequences are sufficiently conservative within one species, so the DNA of the whole mitochondrial population of one cell, even as large as an oocyte, or of cells of different tissues and even different animals of this species, are practically the same (Dawid, 1972b). This was demonstrated in experiments on mtDNA renaturation: the hybrids formed had no mismatched regions. Conversely, although the mtDNA's of different animal species are capable of renaturing, the renatured molecules melt at low temperatures, and one can see with an electron microscope that a considerable part of the DNA sequences are not involved in the hybrids. Annealing of the mtDNA of the oocytes of two closely related species of *Xenopus*, *X. laevis* and *X. mulleri*, results in renaturation of only 70% of DNA sequences. Of these, in 50% of the molecules 1/4 of all the bases is unpaired and only 20% of the sequences form heteroduplexes with a small number of unpaired bases (6%) (Dawid, 1972b). All this suggested that in *Xenopus* the divergence of the mtDNA in the course of evolution of the species proceeds faster than that of nuclear DNA (Dawid, 1972b), and this results from the fact that the mitochondria, as multiple structures of the cells, undergo a greater than in nuclear DNA number of generations (replication cycles), which promotes the accumulation of mutations. The foregoing, however, explains neither the mechanism by which the similarity of mtDNA in a huge population of individuals within the same species is maintained, nor the origin of the species-specific differences of mtDNA in the course of evolution.

Biogenesis of mtDNA. It has been shown for the simplest eukaryotic organisms — such as yeast (Bolotin *et al.*, 1971) and neurospora (Diacumakos *et al.*, 1965; Reich and Luck, 1966) — that many hereditary properties of mitochondria are transferred through the cytoplasm rather than through the nucleus. No rigorous proof in favor of the cytoplasmic route has been obtained for the higher organisms. On the other hand, there has been evidence for the involvement of chromosomal DNA in mitochondrial biogenesis. Hybridization of mtDNA with nuclear DNA revealed the presence of a small number of sequences of homologous mtDNA in nuclear chromosomal DNA. On the basis of the molecular weight of mtDNA and the degree of hybridization of nuclear DNA with it one can calculate the number of loci of the mitochondrial genome present in the nuclear genome. According to the data of Borst and Kroon (1969), the chick genome, for instance, contains five mtDNA loci. The hypothesis about the involvement of the nucleus in the heredity of mtDNA is attractive in that it explains the intraspecific conservatism of the nucleotide sequence of mtDNA, because in this case the whole population of mtDNA molecules are copies of several loci of nuclear DNA (the master-slave principle). However, the hybridization experiments are not convincing enough, since it is practically impossible to obtain nuclear preparations absolutely free of mitochondrial impurity. Therefore, in hybridization with nuclear DNA, binding may occur not so much with the nuclear DNA's themselves as with those of the mitochondrial impurities.

A new approach to the study of the mtDNA heredity was applied in the work of Dawid and Blackler (1972). The authors used sexual hybrids of two *Xenopus* species: *X. laevis* and *X. mulleri*. The mitochondrial DNA's of these species differ considerably in primary structure, and this can be revealed by hybridization, even if they are in a mixture (Dawid, 1972b). First, it has been shown, that F_1 hybrids inherit only the maternal-type mtDNA. Then the nuclear DNA of F_1 hybrid was annealed with mtDNA (more precisely, with the RNA obtained by *in vitro* transcription of mtDNA) of the parental species of *X. mulleri* or *X. laevis*. In this case, if the mitochondrial genome is inherited only via the cytoplasm, preparations of the nuclear DNA of *X. mulleri*♀/*X. laevis* ♂ F_1 hybrid must be contaminated only by the maternal mtDNA of *X. mulleri*. Experiments have shown that the hybridization level of the F_1 hybrid nuclear DNA with the mtDNA of *X. laevis* was not higher than that in the control, i. e. with the nuclear DNA of *X. mulleri*. Thus, the presence of *X. laevis* paternal chromosomes in the hybrid nucleus did not affect the binding level, i. e. the hybridization with the mtDNA of *X. laevis*. At the same time, if mtDNA of *X. laevis* was hybridized with the nuclear DNA of *X. laevis*, which could have been contaminated with the cytoplasmic mtDNA of *X. laevis*, the binding was considerable and corresponded, according to calculations, to the presence of appoximately five mtDNA copies in the genome of *X. laevis*. The authors are right in concluding that these results favor the cytoplasmic hereditary route of mtDNA, and also the absence in chromosomal DNA of mtDNA-homologous loci.

Yolk DNA

The presence of DNA in the yolk has been shown for such animals as frogs *Rana pipiens* and *Xenopus* (Baltus and Brachet, 1962; Baltus *et al.*, 1968; Hanocq-Quertier *et al.*, 1968, 1972), newts (Ohno *et al.*, 1964; Steinert and Van Gansen, 1971), and sea urchins (Pikó *et al.*, 1967). According to Baltus *et al.* (1968), in *Xenopus* the yolk DNA constitutes 65 to 75% of the cytoplasmic DNA, while in the sea urchin it does not exceed 1/3 of the egg DNA (Pikó *et al.*, 1967). The presence of DNA in the yolk platelets of amphibians has been shown histochemically by the binding of labeled actinomycin D (Steinert and Van Gansen, 1971). This method was first used by Brachet and Ficq (1965) for detecting DNA in the cytoplasm of the newt oocyte. Besides, yolk DNA was determined biochemically after isolation from amphibian eggs (Hanocq *et al.*, 1972; Baltus *et al.*, 1968). The DNA in amphibian yolk platelets is bound very firmly with the phosphoprotein components and therefore its extraction from these structures as a high-molecular-weight preparation involves difficulties. Only the use of Kirby and Cook's method (1967) made it possible to isolate DNA and determine its characteristics (Baltus *et al.*, 1968; Hanocq *et al.*, 1972).

Electron micrographs have shown that yolk platelet DNA has linear structure, and this distinguishes it from circular mitochondrial DNA's. The length of the molecules varies from 0.67 to 22.6 μm, which corresponds to a molecular weight of 1.3 to 45×10^6 daltons. The nucleotide composition of this DNA in *Xenopus* and sea urchins practically coincides with that of nuclear and mitochondrial DNA. Therefore they cannot be separated by centrifugation in a CsCl density gradient, since the buoyant density of the mitochondrial DNA and that of the nuclei and yolk platelets is the same. At the same time the yolk DNA differs from mitochondrial DNA in sequence complexity, which is evident from a comparison of their renaturation rates. When denatured DNA preparations are annealed at low C_0t (DNA concentration per time multiple), the mitochondrial DNA renatures completely, whereas the yolk and nuclear DNA remain single-stranded. This has been confirmed by centrifugation in CsCl, which helps separate single-stranded denatured DNA from double-stranded renatured DNA (Hanocq *et al.*, 1972). In the axolotl *Ambystoma mexicanum*, the buoyant density of the yolk DNA (1.699 g/cm^3) is lower than that of the nuclear (1.704 g/cm^3) and higher than that of the mitochondrial DNA (1.696 g/cm^3); hence, these DNA's differ in their nucleotide composition (Hanocq *et al.*, 1972). This suggests that yolk DNA is a special type of DNA.

The origin and role of yolk DNA are unknown. Investigations along these lines have been carried out mainly in the laboratory of Brachet. These investigations have shown that at the vitellogenesis stage no labeled thymidine is incorporated into the DNA of the yolk platelets of amphibian oocytes (Baltus *et al.*, 1968) and thus it was suggested that yolk DNA is of exogenous origin. Since vitellogenin synthesized in the liver is then transported with the blood to the ovary and is assimilated selectively by the oocytes (see Chapter 3), Brachet (1969) suggests a DNA involvement in this process. He describes considerable cytological changes in liver cells and erythrocytes in the region of blood sinuses of the liver, including degenerative changes in the nuclei during active vitellogenesis in the frog. Further, Brachet believes that DNA, in combination with lipoproteins, may play a part in the determination of the animal-vegetal polarity of eggs. According to him this is realized by the binding of the DNA-vitellogenin complex by specific receptors of the cortical layer of the oocyte, whose localization is strictly defined.

The amount of extractable yolk DNA in amphibian eggs is not always reproducible (Hanocq *et al.*, 1972). Considerable losses are occasionally observed. This is believed to depend on the change in the content of divalent cations (Mg^{2+}, Ca^{2+}) in the egg during ovulation and egg passage through the oviducts. Hanocq *et al.* refer to their unpublished data stating that incubation of ovulated oocytes in the Ringer solution with 10^{-2}M Mg^{2+} greatly reduces the DNA content in the yolk fraction (Hanocq *et al.*, 1972). It is possible that at the end of oogenesis part of the yolk DNA is hydrolyzed, which results in replenishment of the pool of low-molecular-weight DNA precursors, deoxyribonucleotides, necessary for intensive postfertilization DNA synthesis. Brachet, however, does not rule out an active role of DNA as carrier of genetic information for yolk resorption in the course of embryogenesis (Hanocq *et al.*, 1972).

DNA SYNTHESIS IN OOGENESIS

Nuclear DNA of the Oocyte

Replication of chromosomal DNA occurs at the very outset of oogenesis, in the premeiotic S phase (Ficq, 1968; Gall, 1968; MacGregor, 1968). The further course of meiosis is blocked at the diplotene stage of the first-division prophase, and throughout the oogenesis the nucleus remains diploid as regards the number of chromosomes, but each chromosome contains a doubled quantity of DNA, so that its total content must be equal to four haploid equivalents (4 C). Indeed, 4 C of DNA have been found in the nucleus of the oocyte of mammals (Van de Kerckhove, 1959) and of some species of worms (Govaert, 1957) and insects (Mulnard, 1954).

Even in the early investigations of amphibians it was found that the oocyte nucleus — germinal vesicle — greatly exceeds in size the somatic cell nuclei and contains a larger quantity of DNA (Finamore *et al.*, 1960). In the oocytes of the frog *Rana pipiens*, the detected quantity of nuclear DNA was hundreds of times as large (Haggis, 1966), and in the newt *Triturus viridescens*, twice as large (Izawa *et al.*, 1963b) as 4 C. As in determination of the total DNA content in the egg, the magnitude of this "excess" of nuclear DNA strongly depended on the specificity of the method used. According to the observations of Izawa *et al.* (1963b) the fluorometric method yields values about ten times those obtained by the diphenylamine method. Even after correcting for the nonspecific reaction the resulting values were too high. The DNA content in the oocyte chromosomes in the newt was four times as large as in the erythrocytes (0.39×10^{-9} g in the oocyte and 0.089×10^{-9} g in the erythrocytes). Besides, DNA was detected in the nucleoplasm, and its content was the same as in the chromosomes, 0.8×10^{-9} g. Thus, the total quantity of DNA found in the nucleus was 1.4 to 1.5×10^{-9} g (Izawa *et al.*, 1963b).

The presence of excess DNA in the oocyte nucleus, as compared with the somatic nuclei, has been confirmed in subsequent investigations, which used more advanced methods of analysis. Brown and Dawid (1968) determined the DNA content after the isolation of high-molecular-weight DNA preparations from oocyte nuclei and their centrifuging in CsCl. It was found that the actual amount of DNA in the oocyte nucleus of the axolotls *Siredon mexicanum* and *Necturus maculosus* exceeds that in the erythrocyte nuclei by 20%. In the frog *Xenopus* the excess is five times as large, but the authors believe that it partly consists of an mtDNA impurity due to the sticking of mitochondria to the surface of the nuclear membrane. The main result of this work was the discovery of an additional fraction in the nuclear DNA, which was revealed as a heavy satellite on centrifugation in a CsCl gradient. The absolute quantity of this satellite DNA in the amphibians under investigation is about the same. Hybridization showed that it contains sequences complementary to rRNA (Brown and Dawid, 1968). This suggests that the excess of nuclear DNA of the oocyte consists of numerous copies of ribosomal genes.

DNA of Free Oocyte Nucleoli

In earlier studies it was established that the germinal vesicle of the oocyte of various amphibian species contains numerous nucleoli (Wilson, 1928; Painter and Taylor, 1942). Similar observations were made on fish oocytes (Jamamoto, 1956; Sakun, 1961). An electron-microscopic study of these nucleoli showed that they were not bound with the chromosomes but located on the nuclear membrane (Miller, 1964, 1966). The nucleoli structure resembles beads consisting of RNase-sensitive material with a thread sensitive to DNase. Thus, the components of free nucleoli are similar to those of the chromosome nucleoli of the somatic cells. Besides, incorporation of labeled uridine and the effect of actinomycin D showed that free nucleoli experience an active synthesis of RNA on a DNA template. Since the nucleoli of somatic cells are the site of synthesis of ribosomal RNA (McMaster-Kaye and Taylor, 1958; Woods and Taylor, 1958) and their structure includes a part of a chromosome — the so-called locus of the nucleolar organizer containing ribosomal genes — rDNA (Ritossa and Spiegelman, 1965; Birnstiel *et al.*, 1966), it could be thought that the same functions were performed by the free oocyte nucleoli. This was corroborated by the works of Gall and Pardue (1969), and John *et al.* (1969). Using *in situ* hybridization on fixed cytological preparations, it was shown that free many-segmented nucleoli contain DNA complementary to ribosomal RNA.

The number and size of free nucleoli in the oocyte germinal vesicle may be different. They are numerous, for instance, in amphibians, 600 in the newt *Triturus viridescens* (MacGregor, 1965), 1000 in the axolotl *S. mexicanum* (Callan, 1966), and 1500 in the frog *Xenopus* (Miller, 1966; Perkowska *et al.*, 1968). They are as numerous in fish oocytes (Sakun, 1961; Chmilevsky, 1971). The diameter of these nucleoli varies between 0.4 and 1 μm.

Some marine invertebrates, such as *Spisula solidissima* and *Urechis caupo*, have a single but greatly enlarged nucleus (Brown and Dawid, 1968). In insects, for instance in crickets, the oocyte nuclei have one large (primary) nucleolus surrounded by numerous small (secondary) nucleoli (Lima-de-Faria *et al.*, 1969). In amphibians, the nucleoli contain one or several (up to four) circular DNA molecules, which are incomplete replicas from the nucleolar organizer locus (MacGregor, 1968). The number of nucleoli is equal to the number of the nucleolar organizer copies, while the number of circular molecules greatly exceeds it. Thus, the rDNA content in amphibian nuclei (Brown and Dawid, 1968) in three different species *N. maculosus*, *S. mexicanum*, and *X. laevis* is almost the same: 5.3 to 5.5 pg rDNA (Table 3). This corresponds to 7×10^5 copies of ribosomal genes. Since the nucleolar organizer in *Xenopus* has 450 genes of rRNA (Brown and Weber, 1968), the total number of the organizer copies must be 1.5×10^3. Precisely such a number of nucleoli has been found in *Xenopus*. The number of circular rDNA's in one nucleus is, however, 5×10^3, i. e. much greater (Brown and Dawid, 1968). The nucleolus contains from 8 to 1000 ribosomal cistrons (see Hough and Davidson, 1976).

More accurate measurements carried out recently showed that the

Table 3

The Content of Chromosomal and Nucleolar DNA in Somatic Nuclei and Germinal Vesicles of Four Amphibians (in pg) (Brown and Dawid, 1968)

Species	Chromosomal DNA			Nucleolar DNA		
	4 C[1]	Germinal vesicle[2]	4 C Total	4 C rDNA[3]	Germinal vesicle[2] total	Germinal vesicle[4] rDNA
Xenopus laevis	12.6	70	0.02[5]	0.014	25	5.3
Siredon mexicanum	140	170	—	0.16	13	5.5
Necturus maculosus	380	500	—	0.08	30	5.3
Triturus viridescens	178	—	—	0.16	—	8.5

[1] The value for *N. maculosus* was determined by the diphenylamine reaction in a sample of counted erythrocytes. Literature citations for the other values are: *X. laevis* (Dawid, 1965), *S. mexicanum* and *T. viridescens* (Edström, 1964; Gall, 1963).

[2] Calculated from the band areas in the experiment with CsCl centrifugation and corrected for losses during purification by comparison with the band area of a known amount of dAT (see also Perkowska *et al.*, 1968).

[3] These values have been calculated from the 4 C complement of DNA for each species (see first column) and from the percentage of the genome homologous to *X. laevis* ^{3}H-rRNA. The percentage of the genome that is homologous to *X. laevis* rRNA in *X. laevis, S. mexicanum, N. maculosus,* and *T. viridescens* is 0.057, 0.05, 0.01, and 0.05, respectively. The homology of *X. laevis* rRNA with these heterologous DNA's is extensive; if it is not 100 percent, then rDNA complements are underestimated here. In calculating the amount of rDNA it is assumed that only one strand of DNA is homologous to rRNA and therefore each value has been doubled.

[4] The amount of rRNA homologous to germinal-vesicle DNA was determined from the extent of hybridization found in the experiments with CsCl centrifugation. The hybridization in each sample was compared to that of a known amount of somatic DNA. Since there is a direct relationship between the amount of DNA on filters and the radioactive RNA hybridized, even at the subsaturating concentration of ^{3}H-rRNA used in these experiments, the amount of rDNA can be calculated for each unknown preparation (Brown and Weber, 1968). Since losses of rDNA may have occurred during the hybridization steps, the rDNA values are minimum. Once again it is assumed that only one strand of the DNA is homologous to rRNA and therefore each value has been doubled.

[5] Estimated from experiments of Birnstiel *el al.* (1966) in which the proportion of high-density DNA in somatic DNA was directly measured by CsCl centrifugation in an analytical ultracentrifuge and found to be between 0.15 and 0.2 percent of the total DNA

number of amplified nucleoli may vary considerably (from 0.5×10^3 to 2.5×10^3) in *Xenopus* both between oocytes of the same stage and from the same female and between different oocyte stages and different females (Thieband, 1979). The quantity of rDNA per oocyte, however, remains constant (35 pg of rDNA per oocyte) since the quantity of DNA per nucleolus is inversely proportional to their quantity in the oocyte nucleus. The author believes the variations in the number of nucleoli to be due to their fusions or deletions.

The DNA of free oocyte nucleoli differs in its buoyant density from that of somatic cell nucleoli (Evans and Birnstiel, 1968; Birnstiel *et al.*, 1968; Brown and Dawid, 1968): in normal somatic nucleoli this density is 1.723 g/cm^3, while in the DNA of oocyte nucleoli it is 1.729 g/cm^3 (Fig. 1). The total content of the GC pairs is, however, equal in the two DNA: 67%. It was found that the differences were due to the different degree of methylation of the cytosine residue: while in the DNA of somatic nucleoli 13% of the cytosine residues were methylated, there were no methyl groups whatsoever in the DNA of the free nucleoli (Dawid *et al.*, 1970).

The structure peculiarity of the nucleolar DNA (absence of methylation) and its localization outside the chromosomes enabled Wallace to advance the hypothesis of the episomal route of nucleoli heredity in the oocyte (Wallace *et al.*, 1971). This was also supported by other data. First, the relative number of arising nucleoli in the oocyte nucleus is independent of the genome size and of the number of nucleolar organizers. For instance, in amphibians *S. mexicanum*, *N. maculosus*, and *X. laevis*, whose genome sizes differ by 10 and more times (see Table 2), the number of amplified nucleoli is very similar: from 10^3 to 1.5×10^3. The number of nucleoli is the same in species containing one or two nucleoli (Perkowska *et al.*, 1968; Brown and Dawid, 1968). All this presupposes the existence of a special mechanism for regulating the formation of nucleoli. A second important factor was the discovery of nucleoli in the nucleus of spermatogonia at the initial stages of spermatogenesis. In the course of further differentiation during meiosis, the nucleoli disappear from spermatozoa, and in the DNA of a mature spermatozoon hybridization determines only the haploid content of rDNA (Brown and Blackler, 1972).

According to Wallace's hypothesis the nucleoli in the oocyte are formed from one or several episome-like structures preserved in the germ-line cells. During spermatogenesis the nucleoli disappear completely, while in oogenesis, on maturation, one or several nucleoli remain and, after fertilization, find their way into a special portion of the egg — the sexual plasm, and then into the nuclei of germ-line cells. As can be seen, the episomal heredity presumes the transfer of nucleoli only through the cytoplasm.

The hypothesis of the episomal heredity was not, however, corroborated by Brown and Blackler's data (1972). As in the study of mtDNA heredity, the authors used two species of *Xenopus*: *X. laevis* and *X. mulleri*, which, on crossing, produce viable offspring. Ribosomal genes of these species differ in the primary structure of their spacers (they have identical sequences encoding the ribosomal RNA) (Brown *et al.*, 1972). This is manifested in the small differences of the buoyant density of rDNA (1.729 g/cm^3 in *X. laevis* and 1.727 g/cm^3 in *X. mulleri*). Besides, these differences can be revealed by hybridizing rDNA in the presence of unlabeled ribosomal RNA, i. e. under conditions in which only the level of homology between the spacer

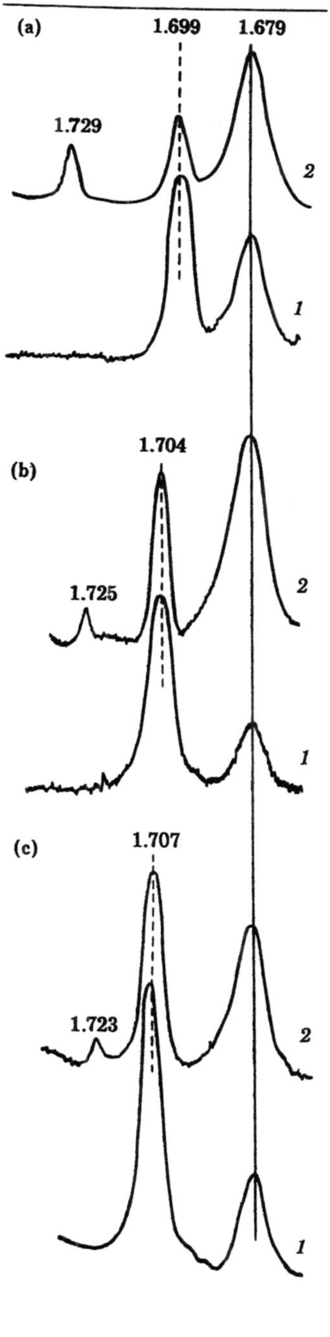

Fig. 1. Detection of amplified ribosomal genes in DNA of amphibian oocytes. The DNA was isolated from the nuclei of somatic cells (*1*) and from the germinal vesicles of oocytes (*2*) of three species of amphibians: *X. laevis* (*a*), *S. mexicanum* (*b*), and *N. maculosus* (*c*), and centrifuged in a CsCl density gradient. A poly dA-dT copolymer with a buoyant density of 1.679 served as a reference substance. It can be seen that the DNA from somatic cells (*1*) displays only one peak, its buoyant density being 1.699 for *X. laevis*, 1.704 for *S. mexicanum*, and 1.707 for *N. maculosus*. The DNA of the germinal vesicles (*2*) exhibits an additional satellite peak with a buoyant density of 1.729 to 1.723. As demonstrated by hybridization with rRNA, it contains sequences complementary to rRNA, i. e. corresponds to rDNA (Brown and Dawid, 1968).

Table 4

Investigation into the Nature of the rDNA of Free Nucleoli in Hybrids of *X. Laevis* and *X. Mulleri* (Brown and Blackler, 1972)

Hybridization of rDNA isolated from chromosomes or free nucleoli was carried out using a mixture of ^{14}C-RNA, transcribed from ribosomal genes of *X. laevis*, and ^{3}H-RNA, transcribed from ribosomal genes of *X. mulleri*, in the presence and absence of the competitor rRNA of *X. laevis*. The ^{3}H/^{14}C ratio reflects the degree of homology between the genes of *X. mulleri* and *X. laevis* and serves as a characteristic of the nucleolar DNA formed during oogenesis in the F_1 hybrid.

Genetic combinations of two *Xenopus* species (♀×♂)	^{3}H/^{14}C: Chromosomal rDNA		^{3}H/^{14}C: rDNA of free nucleoli		Buoyant density of rDNA of free nucleoli, g/cm^3
	Without competitor	In presence of rRNA of *X. laevis*	Without competitor	In presence of rRNA of *X. laevis*	
X. laevis and *X. laevis*	0.22	0.06	0.26	0.04	1.729
X. mulleri and *X. mulleri*	1.21	1.97	1.6	5.1	1.727
X. laevis and *X. mulleri*	0.61	0.67	0.34	0.06	1.729
X. mulleri and *X. laevis*	0.72	0.70	0.28	0.05	1.729

sequences is tested (Wellauer and Reeder, 1975). It was shown that in hybrids of *X. laevis* and *X. mulleri* the oocyte nuclei contained only *X. laevis* rDNA even when the inherited cytoplasm was that of *X. mulleri* and only half of the chromosomes in the nucleus was inherited from *X. laevis* (paternal genes), while the other half was taken from *X. mulleri* (maternal genes) (Table 4). These experiments show convincingly that it is chromosomes (in this case paternal chromosomes of *X. laevis*) that are involved in the formation of rDNA of the free nucleoli, and that nucleolar rDNA is inherited by the chromosomal route rather than via the cytoplasm (Brown and Blackler, 1972). Thus, the formation of numerous nucleoli in oogenesis reflects the multiplication of ribosomal genes by extracopying of the nucleolar organizer. This phenomenon was termed the amplification of ribosomal genes, while the free nucleoli were called supernumerary, or amplified.

Amplification of rDNA

Amplification of rDNA occurs in oogenesis, and more often than not this process terminates at the earliest stages of oogenesis. Two amplification phases have been described in amphibians: the slow phase and the active-amplification phase.

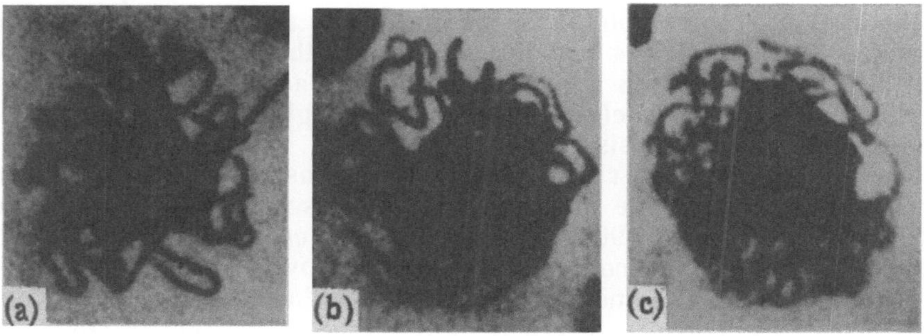

Fig. 2. Electron micrograph of nucleolar DNA amplification in oogenesis of the axolotl. Three stages of formation of amplified ribosomal genes are seen as dark patches. At each stage, rDNA was tested by *in situ* hybridization with labeled rRNA on a cytological preparation (Gall, 1968).

The first phase, in *Xenopus*, is amplification in oogonia and spermatogonia (Gall and Pardue, 1969; Wallace *et al.*, 1971). It has been shown by *in situ* hybridization that the nuclei of these cells bind more labeled rRNA than the nuclei of somatic cells (Gall and Pardue, 1969). The total duration of this period is at least three weeks (the premeiotic phase takes two weeks, the leptotene phase 3 to 7 days, and the zygotene phase 5 to 9 days (Watson-Coggins and Gall, 1972; Bird and Birnstiel, 1971).

The second phase, the active-amplification period, occurs at the pachytene stage of the first meiotic division, and precisely this process was described by most investigators (Gall, 1968; Birnstiel *et al.*, 1968; MacGregor, 1968). It lasts 15 days, after which amplification still continues for some time though at a very low rate (Fig. 2).

The formation of extra copies of rDNA (gene amplification) and the formation of the nucleoli themselves may not coincide in time (MacGregor, 1968, 1972). In *Xenopus*, in the second phase of amplification, at first a mass of rDNA extra copies is formed, which has the form of a cap and is intensively Feulgen-stained. Then, independent nucleoli tie off and migrate to the nuclear membrane. In the toad *Bufo bufo* the extra copies of rDNA in combination with protein immediately migrate to the nuclear membrane, where they take the form of nucleoli. In the tailed frog *Ascaphus truei* a few nucleoli appear at the early stages of amplification, and then several giant nucleoli 200 μm in diameter are formed, which contain several hundred circular extra copies of rDNA each (MacGregor, 1968). Similar data have been obtained for the amplification of nucleoli in fishes (Sakun, 1961; Chmilevsky, 1971). In the ruff (*Acerina cernua* L.), for instance, at the pachytene stage a large nucleolus (primary nucleolus) first appears, from which then small nucleoli (of diameter 1 to 3 μm) tie off and become attached to the nuclear membrane.

In distinction to the amphibians, the number of nucleoli in this species increases throughout the oocyte's small-growth phase. It is possible that rDNA amplification is completed earlier, and only the formation of the nucleoli themselves proceeds throughout the subsequent stages until vitellogenesis (Chmilevsky, 1971).

The accumulation of rDNA in oogenesis has also been described for a number of marine invertebrates: molluscs *Spisula solidissima* (Brown and Dawid, 1968) and *Mulinia lateralis* (Kidder, 1976), and the worm *Urechis caupo* (Dawid and Brown, 1970).

These animals have one enlarged nucleolus in the oocyte, but it has been shown by hybridization with rRNA that in the oocyte DNA the content of rDNA is higher than in the sperm DNA: five-fold in *Spisula solidissima*, two-fold in *Mulinia lateralis*, and 24-fold in *Urechis caupo*.

In invertebrates, such as the starfish *Asterias forbesii*, the rDNA is not amplified (Vincent *et al.*, 1968, 1969), apparently because oocyte formation is due to the polyploid nurse cells.

Among the insects, the amplification of ribosomal genes has been described for crickets *Acheta* (Lima-de-Faria *et al.*, 1969; Hansen-Delkeskamp, 1969; Cave, 1972, 1973; Lima-de-Faria, 1973). These organisms have two nucleolar organizers: on the 6th and the 11th chromosome (Jaworska and Lima-de-Faria, 1973a; Cave, 1973). These two loci differ in the number of ribosomal genes (cistrons), the structure of the gene, and the organization of the chromosomal site subjected to amplification. The 6th chromosome has three units of amplification (amplificons) organized in the form of chromomeres. In each of the ribosomal genes of this chromosome the sequence encoding 28S and 18S rRNA makes up 7%, and the spacer portion, 93%. These genes form clusters of different sizes: two clusters (chromomeres) of 40 cistrons each and one smaller cluster of 18 cistrons. The 11th chromosome has two similar amplificons which contain clusters of 50 and 22 ribosomal cistrons, respectively. In each gene, 10% corresponds to rDNA, and 90%, to the spacer portions (Pero *et al.*, 1973; Lima-de-Faria *et al.*, 1973a, b; Lima-de-Faria, 1973). The total number of ribosomal genes is 171 per haploid set (Pero *et al.*, 1973). The 6th and the 11th chromosome begin amplification at different times of the meiotic cycle. This is manifested in the formation of puff-like thickenings. The DNA content in these puffs is 100 times higher than that in the locus of the nucleolar organizer of the somatic chromosomes (Lima-de-Faria *et al.*, 1973a, b): 1.8×10^9 b. p. in the 6th chromosome, and 1.4×10^9 b. p. in the 11th chromosome as compared to 2×10^7 b. p. and 1×10^7 b. p. in the respective chromosomes of somatic cells.

The release of extra copies into the nucleoplasm also occurs at different times. At the pachytene stage after active amplification, extra copies in the form of separate nucleoli are released only from the 11th chromosome. The release of the nucleoli from the 6th chromosome is delayed until the onset of the diplotene stage. At this stage one can distinguish the main large nucleolus (primary nucleolus) and a number of secondary small nucleoli both free in the nucleoplasm and attached to the nuclear membrane (Lima-de-Faria *et al.*, 1973a, b). Proceeding from the total DNA content in the puffs of both chromosomes, the total number of extra copies in the germinal vesicle should be 100 haploid units (or about 1.6×10^4 genes) (Lima-de-Faria *et al.*, 1973a, b). However, Pero *et al.* (1973), using the hybridization method, found only a five-fold increase in rDNA content (i. e. not more than 800 genes) in the oocytes at the pachytene stage. But the centrifugation in a CsCl gradient showed that at the early pachytene stage the oocyte DNA contains 15 times as much rDNA's as does the sperm DNA (14% as compared with 0.8%) (Lima-de-Faria *et al.*, 1969). These differences are probably due to the fact that in the course of oogenesis the newly formed secondary nucleoli are released into the cytoplasm (Lima-de-Faria, 1973).

In the cricket, not even a temporary rDNA amplification in spermatogenesis has been detected: the same relative rDNA content has been found in spermatozoa and spermatogonia (Pero *et al.*, 1973; Lima-de-Faria *et al.*, 1969).

Like in molluscs, in the insects, whose oocyte development depends on the nurse cells (e. g. in the bug *Oncopeltus fasciatus*), no amplification of ribosomal genes takes place (Cave, 1974, 1976).

Finally, a small, but reliable quadruple amplification of rDNA was recently detected for the first time in mammals, in human oocytes (Wolgemuth *et al.*, 1979), although additional nucleoli in mammalian oocytes have been known for a long time. The same method — *in situ* hybridization with 18S and 28S rRNA's labeled to high specific activity with radioactive iodine (^{125}I) — was also used to demonstrate the quadruple amplification in baboon oocytes (Wolgemuth *et al.*, 1980).

rDNA Amplification Mechanism

The molecular mechanism involved in amplification of ribosomal genes still remains somewhat obscure. It is known that this process is of a cascading nature; this is manifested in the fact that the rate of incorporation of labeled thymidine into the rDNA is not constant but increases substantially in the course of amplification. This means that the initially formed copies of the ribosomal operon serve then as templates for subsequent copying, i. e. that the number of copying templates increases in geometrical progression (MacGregor, 1968, 1972). This is also confirmed by the fact that there are two amplification periods, and during the first period the rDNA accumulation rate is low but it increases considerably in the second period. Many data have been obtained suggesting that rDNA amplification proceeds by the "rolling circle" mechanism (Gilbert and Dressler, 1968; Bird *et al.*, 1973; Rochaix *et al.*, 1974; Hourcade *et al.*, 1973a, b), i. e. many copies are formed from one circular DNA molecule. The process involves the formation, along the circular molecule, of single-stranded copies which are then detached in the form of linear molecules. A complementary chain is then built on each replica, the 3′-end of the replica serving as a primer for further replication. Such circles with long attached DNA segments (tails) were detected by EM in DNA preparations from early *Xenopus* oocytes at the rDNA amplification stage. The ratio of the tail length to the contour size of the circle varied from 1 to 15 (Rochaix *et al.*, 1974). It has been shown that the size of the circle-forming DNA molecules may de different, from 8×10^6 to 140×10^6 daltons, most of the circles having about 40×10^6 daltons. Since the size of one ribosomal gene is 8×10^6 daltons, the smallest circles probably correspond to 1 cistron, and the largest, to 2-15 cistrons (Hourcade *et al.*, 1973a, b; Rochaix *et al.*, 1974).

The circular molecules and tailed circles (rolling circles) visible on the micrographs indeed represent rDNA. This was proved primarily by centrifugation in a CsCl density gradient: DNA preparations had the same buoyant density as rDNA. Besides, use was made of the EM-analysis of the distribution of single-stranded regions (loops) formed under mild conditions which cause denaturation only of AT-rich sequences (Inman, 1967; Inman and Schnos, 1970; Wolfson *et al.*, 1972). In a ribosomal operon, the spacers are relatively rich in GC pairs (Wensink and Brown, 1971). Hourcade *et al.* (1973a, b) achieved partial denaturation of circular DNA and rolling-circle DNA using heating to 39°C and high formamide concentrations and showed that the distribution and relative length of single- and double-stranded sections are the same in both cases as in chromosomal rDNA.

The functioning of this rolling-circle model was demonstrated in experiments on the kinetics of incorporation of labeled thymidine into circular and linear (tail) rDNA sequences at the amplification stage (Rochaix *et al.*, 1974). The method of pulse labeling was employed. The number of detected labeled rings was greater with short labeling times (1 labeled circle per 6 with a 2-hour label and per 12 with a 6-hour label). But the more time was used for incorporating the label, the longer was the labeled linear sequence attached to the circle (tail). If, however, a long-term cold thymidine chase was made after short-term labeling, the number of labeled circular structures decreased, and in the tail sequences the labeled section moved away from the ring to the distal end (Rochaix *et al.*, 1974). Thus, the experiments demonstrated that first the replication of the ring structure takes place, and then the DNA replica is displaced and transported to the end of the tail by subsequent replication acts. Many questions still remain obscure in this model. For instance, it is not clear what determines the number of replication acts and how the large circular molecules of the nucleolar DNA are created, which contain from 200 to 500 copies of ribosomal genes. It is also unknown how the first circles appear and how their multiplication occurs, which is responsible for the cascading nature of the amplification (MacGregor, 1968). It is also unclear whether all the nucleolar organizer loci are involved in the initial process of formation of circular molecules. According to the calculations of Bird *et al.* (Bird, 1974; Rochaix *et al.*, 1974), at an average replication rate of 10 μm/h, the whole amplification process can be provided by the total number of genes present in the four chromosomes of the tetraploid nuclei of *Xenopus* oocytes within 20 days (time of replication of nucleolar DNA). Besides, it has been shown for crickets that amplification involves both nucleolar organizers located in different chromosomes: they form puff-like structures (Lima-de-Faria, 1973). One of the most important points concerning the amplification mechanism is the finding of the factors responsible for the selectivity of replication of the nucleolar organizer locus at the moment when the replication of the entire chromosomal DNA has been completed (Gall, 1968; MacGregor, 1968, 1972).

In this respect the hypothesis of Crippa and Tocchini-Valentini on the participation in rDNA amplification of a special enzyme, reverse transcriptase, capable of synthesizing DNA molecules on RNA templates seems rather attractive (Tocchini-Valentini and Crippa, 1971; Crippa and Tocchini-Valentini, 1971; Ficq and Brachet, 1971; Brown and Tocchini-Valentini, 1972; Mahdavi and Crippa, 1972; Tocchini-Valentini *et al.*, 1973).

This enzyme — RNA-dependent DNA polymerase — was isolated from *Xenopus* ovaries (Brown and Tocchini-Valentini, 1972, 1973). It showed an ability for using polyribonucleotides as templates 30 times more efficiently than polydeoxynucleotides. It was shown at a later date that the enzyme isolated from the ovaries can synthesize DNA on the template of 9S globin mRNA (a test for the true activity of the reverse transcriptase possessed by an enzyme of carcinogenic viruses) (Tocchini-Valentini *et al.*, 1973).

Data on the isolation of RNA/DNA complexes from ovaries indicate that reciprocal transcription from RNA templates takes place in oocytes (Tocchini-Valentini and Crippa, 1971; Crippa and Tocchini-Valentini, 1971). The RNA from the complexes contained sequences identical to 28S and 18S of mature RNA and to spacer sections of the precursor of the ribosomal RNA. The size of RNA in the complexes was larger than the transcription unit of the ribosomal genes of *Xenopus* (45S versus 40S) (Tocchini-Valentini *et al.*, 1973). The complex contained ten times as much DNA as RNA, but the DNA sedimentation constant in the sucrose gradient was much lower

(27S) than the total molecular weight of the RNA and DNA molecules found in the complex. The authors attribute this to the loose structure of the complex (Tocchini-Valentini *et al.*, 1973). According to the most recent scheme proposed by Tocchini-Valentini *et al.* (1973) the initial event in amplification is the synthesis of 45S RNA read out from the whole sequence of the ribosomal gene, including the sections of spacers, which are not read in normal transcription. DNA's are synthesized from 45S RNA by means of reverse transcription. These DNA's form circles and then function as rDNA amplification machines. This scheme, however, has not been confirmed by other authors. Thus, they failed to detect RNA/DNA complexes in the ovaries of *Xenopus* tadpoles during metamorphosis (Bird *et al.*, 1973; Hourcade *et al.*, 1973a, b). According to Tocchini-Valentini, however, these authors used tadpoles at later stages of metamorphosis, when the process of reverse transcription had already been completed in the oocyte nuclei.

Tocchini-Valentini and Crippa's model can explain, to some extent, the main difficulty: the selectivity of rDNA replication during the amplification of ribosomal genes. Further investigations are, however, required, for final elucidation of the molecular mechanism of rDNA amplification.

Fate of Nucleolar DNA in Oogenesis

It has been shown in morphological investigations that amplified nucleoli persist throughout oogenesis. At the small-growth stage up to vitellogenesis the *Xenopus* nucleoli have only the fibrillar component (DNA), while the granular structure is absent, which presumably reflects the absence of rRNA synthesis (Thomas, 1972). At the large-growth stage, during active synthesis of rRNA, the nucleoli have a characteristic structure: they are elongated, the DNA fiber is shifted toward the site of attachment of the nucleoli to the nuclear membrane, and there are spherical ribonucleoprotein complexes arranged along the DNA filament (MacGregor, 1972).

In the cricket, the secondary small nucleoli migrate toward the nuclear membrane and are discharged into the cytoplasm (Jaworska and Lima-de-Faria, 1973a, b; Jaworska *et al.*, 1973; Lima-de-Faria *et al.*, 1973a, b), where they continue synthesizing rRNA.

On completion of the growth of the amphibian oocyte, when the chromosomes become packed together and the lampbrush-type structure begins to disappear, the nucleoli also become more compact and start moving toward the center of the nucleus (see Brown, 1966; MacGregor, 1972). Then, in the course of oocyte maturation, after the germinal vesicle rupture, the nucleoli disappear (Brown, 1966).

Changes in the nucleolar apparatus in fishes have been studied very thoroughly (Sakun, 1961; Chmilevsky, 1971). The migration and fusion of nucleoli occur when the nucleus shifts from the oocyte center to the micropyle region and a homogeneous yolk is formed from individual yolk grains. Further, with the onset of maturation and disappearance of the nuclear membrane the nucleoli transform into thick tangled-up filaments, forming a ball with a cavity. Later, the ball breaks down into separate fragments, whose remains, in the pike, white fish, and smelt, are preserved almost until the time the spindle is formed during the first meiotic division (Sakun, 1961). In the ruff, the nucleoli disappear in the cytoplasm after the germinal vesicle is dissolved (Chmilevsky, 1971).

In most of the animals new nucleoli appear rather late in the course of embryonic development: at the blastula-gastrula stage. It has been shown by the hybridization method that the *Xenopus* nucleus at the gastrula stage contains no more than the diploid number of ribosomal genes (Brown and Dawid, 1968). Thus, at the gastrula stage no extra copies of rDNA are present. However, at the cleavage stage some amount of rDNA is still there (Thomas *et al.*, 1976). There are no exact data on the time and the process of the rDNA extra copies elimination. According to our data, in the loach at the blastula stage the nuclear DNA contains an additional rapidly renaturing fraction constituting 5% of the genome DNA (Timofeeva *et al.*, 1975). At the same time, this fraction is not detected in the DNA of the liver and sperm. It is likely that this fraction represents some survived rDNA extra copies. In one of the nematode species, *Panagrellis silusiae*, cytophotometrical studies revealed a considerable DNA excess (23 C) in the nucleus by the time of oocyte maturation, which persisted (7 C) in the 8-cell embryo. It is only at the 32-blastomere stage that the quantity of DNA decreased to 2-4 C. Since a considerable increase in the size of nucleoli occurred in oogenesis, it is possible that the rDNA content in *Panagrellis silusiae* also increased and these copies remained after fertilization (Pasternak and Haight, 1975). This may be due to the later times of the ribosomal RNA synthesis at the end of oogenesis. It is interesting to note that the enlargement of the nucleolus occurs not only in the oocyte, but also toward the end of maturation of the spermatozoon. Then, during meiosis, the nucleolus is destroyed, but it is possible that the extra copies persist and even function. For instance, in another nematode species, the ascarid, an active rRNA synthesis has been described in a male pronucleus in the period between fertilization and cleavage although the nucleolus has not yet been formed in the nucleus and it will be completed only at the 4-blastomere stage (Kaulenas and Fairbairn, 1968; Kaulenas *et al.*, 1969).

As during the production of extra copies, when the replication process concerns only one section of the chromosome, during the degradation of rDNA extra copies there must be special mechanisms ensuring the selectivity and specificity of these processes. As for rDNA hydrolysis, it is possible that the different degrees of rDNA modification play a certain part: no cytosine methylation occurs in the extra copies, while in the chromosomal rDNA 13% of the cytosine residues are methylated (Dawid *et al.*, 1970). It is quite probable that the hydrolysis of the rDNA in the extra copies is related to the appearance of specific nucleases.

Synthesis of Mitochondrial DNA in Oogenesis

The store of mitochondrial DNA available in an unfertilized egg is built up in the oocyte by DNA replication in the mitochondria. The DNA synthesis in cytoplasmic DNA was first demonstrated in the works of Finamore and Volkin (1958), who detected it in amphibian oocytes by incorporation of labeled phosphate. Mezger-Freed (1963) also discovered the synthesis of cytoplasmic DNA in activated enucleated frog eggs. Direct evidence of DNA synthesis in insect mitochondria was further obtained by autoradiography (Muckenthaler and Mahowald, 1966).

Some idea of the dynamics of mtDNA synthesis in oogenesis can be obtained by investigating the accumulation of mitochondria in oocytes. The most detailed data have been obtained for loach oocytes (Ozernyuk, 1972; Ozernyuk and Palmbakh, 1975). In these investigations the number of mitochondria was determined by EM of thin sections of oocytes at different stages of oogenesis. It was shown that the number of mitochondria in oocytes increases 84-fold during small growth and 240-fold toward the beginning of vitellogenesis, reaching 1.5×10^6. This is 10^3 times more than in the somatic cell. Then, during large growth, they continue to multiply, so that in the mature egg their number increases by two more orders of magnitude. In the mature egg, the formation of mitochondria ceases.

The dynamics of accumulation of mitochondria in oogenesis also reflects the dynamics of accumulation of mtDNA. But little is known about the rate of mtDNA synthesis. The number of mitochondria in the loach oocyte doubles within 1.5 months. The time of DNA doubling is probably much shorter.

A direct determination of mtDNA accumulation was made in *Xenopus* oocytes by hybridization with labeled complementary RNA (Webb and Smith, 1977). Its quantity in previtellogenesis oocytes (less than 250 μm in diameter) is about 1 ng per oocyte and increases rapidly with oocyte growth from stage 1 to stage 4 (diameter 600 μm), reaching 4 to 4.5 ng. This value remains practically unaltered during the further growth up to stage 6 (1200 μm), although the oocyte volume increases by another factor of eight (Webb and Smith, 1977). The same quantity of mtDNA remains during maturation and early embryonic development (Chase and Dawid, 1972).

To compare mtDNA synthesis in early and late oocytes, the mitochondria were isolated from stage 3, when the quantity of mtDNA increases rapidly, and from stage 6, when it remains unchanged, and incubated with ^{3}H-TTP (Webb and Camp, 1979). It was found that the quantity of mtDNA synthesized *in vitro* per hour per oocyte was 2.3 pg at stage 3 and 0.28 pg at stage 6. In the mitochondria of early oocytes the activity of DNA polymerase was three times as high as in the late ones (Webb and Camp, 1979).

Using a system of isolated mitochondria Gaùse and Mikhailov (1974) compared the rate of mtDNA synthesis in oocytes, unfertilized eggs, and embryos of the loach at the 35th hour of development by incorporation of ^{3}H-dTTP into the mitochondrial DNA. It was found that the rate of mtDNA synthesis in the small- and large-growth oocytes was insignificant as compared with the embryos. Toward the end of oocyte maturation the rate of label incorporation into the mtDNA drops still further.

Replication of mtDNA in oogenesis occurs in the same way as in somatic cells. It is well known that the synthesis is effected with the aid of a special DNA polymerase which differs from nuclear DNA polymerases in a number of properties (see Shmerling, 1970; Weissbach, 1975). Its molecular weight is 2×10^{5} daltons, it shows an endonuclease activity, is capable of replicating circular double-stranded DNA molecules, and exhibits a high sensitivity to ethidium bromide (Weissbach, 1975). This DNA polymerase "works" inside the mitochondria, but its synthesis is controlled by the nuclear, rather than the mitochondrial, genome (Weissbach, 1975).

A DNA polymerase activity was detected in the mitochondria of oocytes and unfertilized eggs of the loach (Gause and Mikhailov, 1973; Mikhailov and Gause, 1974a, b). Pinon *et al.* (1975) have reported that DNA polymerase with the properties of a mitochondrial enzyme, i. e. capable of replicating double-stranded circular DNA molecules, accumulates in *Xenopus* oocytes. The replication of mitochondrial DNA occurs in a semiconservative way (Reich and Luck, 1966; Kirschner *et al.*, 1968). EM studies have shown that circular structures with loops of different sizes appear in the course of mtDNA replication (Fig. 3). The loop is a section of a single-stranded DNA sequence, and its complementary chain is associated with a short, newly synthesized fragment. Such structures were termed D-loops (displacement loops) (Kasamatsu *et al.*, 1971). Their studies ultimately yielded a model of mtDNA replication (Robberson *et al.*, 1972; Robberson and Clayton, 1972). According to this model, the process begins with untwisting of circular mtDNA due to the local break followed by rapid sealing. After that, the complementary chain is synthesized along one strand, while the other strand is discarded (the D-loop structure). The replication of this discarded chain begins later. As a result, two circular mtDNA molecules are produced

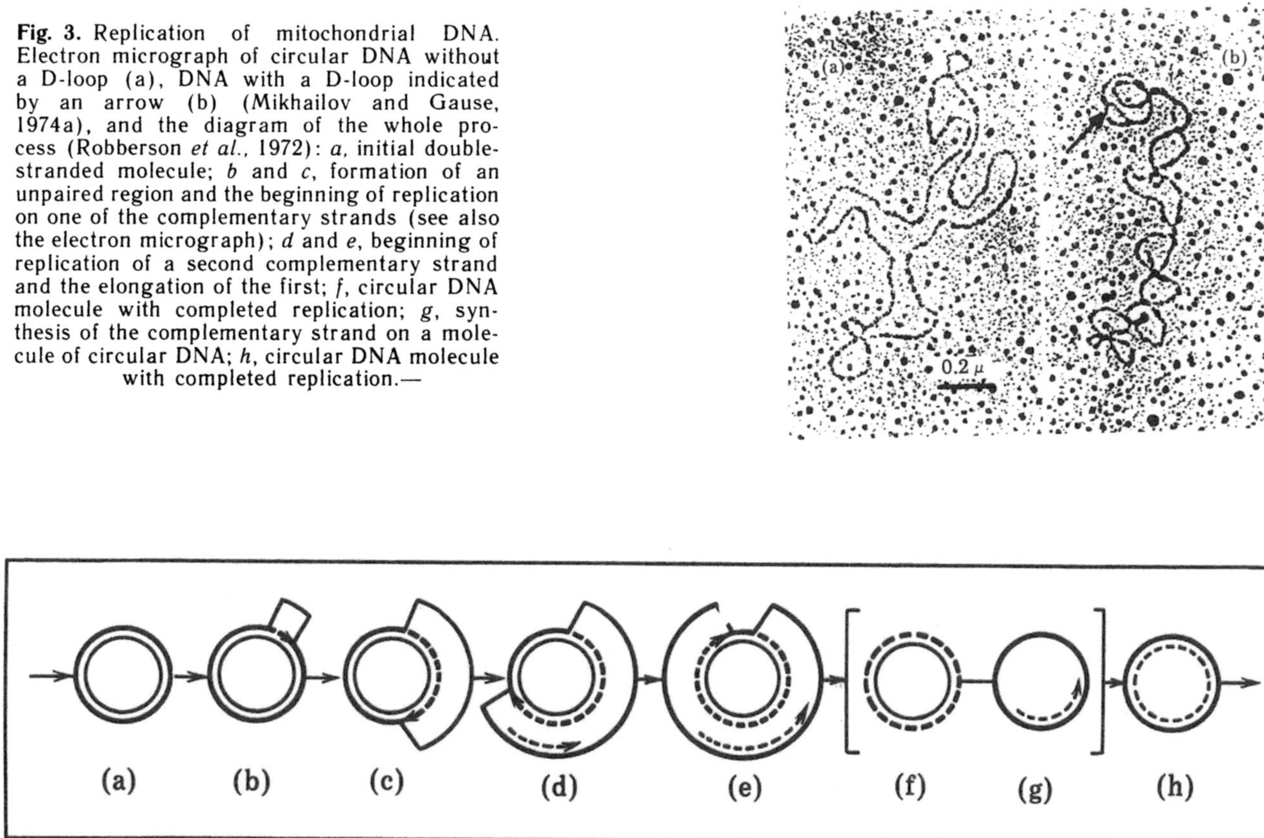

Fig. 3. Replication of mitochondrial DNA. Electron micrograph of circular DNA without a D-loop (a), DNA with a D-loop indicated by an arrow (b) (Mikhailov and Gause, 1974a), and the diagram of the whole process (Robberson *et al.*, 1972): *a*, initial double-stranded molecule; *b* and *c*, formation of an unpaired region and the beginning of replication on one of the complementary strands (see also the electron micrograph); *d* and *e*, beginning of replication of a second complementary strand and the elongation of the first; *f*, circular DNA molecule with completed replication; *g*, synthesis of the complementary strand on a molecule of circular DNA; *h*, circular DNA molecule with completed replication.—

(Robberson *et al.*, 1972; see Gause, 1977). According to Kasamatsu *et al.* (1971), molecules with D-loops constitute a considerable part of the mtDNA of L-cells in the period of exponential growth. D-mtDNA's are found in preparations of cytoplasmic DNA of the oocytes of different animals.

In unfertilized eggs, where the mtDNA synthesis practically ceases, the amount of D-mtDNA greatly decreases. Thus, in mtDNA preparations from loach eggs the quantity of D-mtDNA is less than 1% (Gause and Mikhailov, 1974). A low content of D-loop molecules was found in the DNA of unfertilized sea urchin eggs (Matsumoto *et al.*, 1973, 1974). It is possible that the blocking of mtDNA synthesis toward the end of oogenesis is due to the inactivation of the mechanism responsible for the generation of D-loops. But in the mouse (Pikó, 1975) and the frog (Hallberg, 1974), 70 to 80% of the mtDNA molecules of the unfertilized eggs contain D-loops although there is no mtDNA synthesis. This indicates that the conservation of the mtDNA synthesis in these objects occurs at a stage following the formation of D-loops.

DNA SYNTHESIS IN EMBRYOGENESIS

Synthesis of Nuclear DNA

Throughout the period of egg formation in oogenesis, which may last for many days or months, no synthesis of nuclear chromosomal DNA is observed. The synthesis begins after fertilization, which has been shown in many autoradiographic and biochemical investigations (see reviews by Grant, 1965; Gurdon, 1969; Giudice, 1973). According to Graham (1966) the DNA synthesis in the frog *Xenopus* is switched on 20 min after fertilization, simultaneously in both pronuclei. Until that moment, the nuclei swell considerably (Graham and Morgan, 1966; Gurdon, 1967). The factors determining these processes are localized in the cytoplasm. Thus, in Gurdon's classical experiments with injection of nuclei into the cytoplasm of a fertilized egg (Gurdon, 1967, 1969, 1974; Graham *et al.*, 1966) it was shown that any nucleus with a blocked DNA synthesis, for instance a nerve cell nucleus, undergoes the same changes as does the male pronucleus, i. e. it swells and begins to synthesize the DNA. At a later date, Anderson (1969) demonstrated by electron microscope autoradiographic study that in the case of polyspermy DNA synthesis begins simultaneously in all the nuclei. Besides, in parthenogenetic activation of eggs DNA synthesis is switched on in the female pronucleus.

DNA synthesis in sea urchins also begins a few minutes after fertilization simultaneously in both pronuclei (Simmel and Karnofsky,

1961; Hinegardner *et al.*, 1964). These data were obtained by using labeled thymidine. However, Zeitz *et al.* found that incorporation of bromodeoxyuridine and iododeoxyuridine into the DNA of *Strongylocentrotus purpuratus* and *Arbacia punctulata* begins immediately after fertilization (Zeitz *et al.*, 1969, 1970; see Giudice, 1973), whereas that of thymidine and deoxyuridine begins 20 min later. The authors believe that exogenous thymidine does not transform to thymidine triphosphate at this stage.

In the course of cleavage the rate of DNA synthesis is very high. In amphibians, it increases during the first divisions, reaching a maximum by the middle of the period of synchronous divisions. In this period the duration of synthesis of one haploid DNA equivalent is 10 min (Graham and Morgan, 1966). After the 12th division, with the beginning of desynchronization, the haploid equivalent synthesis time begins to increase, reaching 6 h at the neurula stage. The synthesis frequency during cleavage is also high (one cycle per 12 min) and it decreases 240-fold (one cycle per 48 h) toward the gastrula stage (Gurdon, 1969).

In sea urchins, the highest rate of DNA synthesis is maintained during several cleavage divisions, after which it starts to decline. A more detailed investigation into the DNA synthesis time during the first cleavage divisions up to the 16-cell stage was carried out by Nemer (1962). He showed that in that period the division does occur synchronously, and that half of the DNA has already been synthesized by the moment 50% of the cells have completed mitosis. When investigating the rate of DNA synthesis by the labeled thymidine method the thymidine permeability may vary in the course of the cell cycle. Therefore Nemer used the technique of preincubation (preloading) of embryos with labeled thymidine and corrected his figures for the change in the thymidine cell pool.

Nemer's data agree with earlier investigations, in which it was shown by microspectrometry and autoradiography (Lison and Pasteels, 1951; Ficq *et al.*, 1963; Zimmerman, 1963) that the synthesis of DNA at the first-divisions stage is synchronized and begins at each division even before the cells complete their mitosis. Using *Strongylocentrotus purpuratus*, Hinegardner *et al.* (1964) compared the DNA synthesis with the morphological changes in the nucleus and chromosomes at the first two cleavages. They showed that DNA synthesis begins at the end of the telophase, when the chromosomes are being decondensed, and continues for 13 min (at 15°C). In fishes, the pattern of changes in the DNA synthesis rate after fertilization is the same as in amphibians. In the loach, for instance, synchronous divisions continue for the first 6 h (10 to 11 divisions). The S period at this moment lasts 15 min (at 21°C). Then, after desynchronization, it increases to 110 min (Rott and Popov, 1969).

The description of the changes in DNA synthesis in the development of insects is given by Harris and Forrest (1967), who studied a milkweed bug (*Oncopeltus fasciatus*). The DNA synthesis rate increases within the first 50 h after fertilization, i. e. up to gastrulation, then it is maintained at a high level for the next 40 h, and begins to decrease only during organogenesis (differentiation of the nerve tissue).

DNA synthesis during cleavage involves most diverse low-molecular-weight precursors, including ribonucleotides, CO_2, glycine, etc. (see Grant, 1965). Apparently, early embryos have all their metabolic systems, producing deoxyribonucleotides, in an activated state to ensure a high rate of DNA synthesis during the synchronous-division period. The active DNA synthesis is attended by a high level of respiration (see Rossi *et al.*, 1975), which probably reflects the need for macroergic triphosphates.

Synthesis of Mitochondrial DNA

An unfertilized egg has a mitochondrial store, and it remains unchanged for a long developmental period after fertilization. In amphibians, an increase in mitochondrial protein is observed only after stage 38, i. e. in the period of active organogenesis.

The synthesis of mitochondrial DNA at the early stages of development is negligible or nil. According to Pikó (1970), even in the mouse, whose egg is not very rich in mitochondria, there is no mtDNA synthesis up to the blastocyst stage. In sea urchins, no mtDNA synthesis is detected throughout the cleavage and up to the free-blastula stage (Matsumoto *et al.*, 1973, 1974). In the course of further embryogenesis from the cleavage to the gastrula stage the rate of mtDNA replication increases by a factor of 2 to 3 (Bresch, 1973). DNA synthesis in mitochondria at different stages of development was investigated in the loach by Mikhailov and Gause (1974a, b). The experiments were run on isolated mitochondria and the synthesis rate was estimated by incorporation of ^{3}H-TTP into the mtDNA. The mitochondria of unfertilized eggs showed a very low level of ^{3}H-TTP incorporation (0.01 pmole per 1 mg protein per hour). It was shown by centrifugation in the CsCl/ethidium bromide density gradient that this incorporation rather reflected the processes of mtDNA reparation. Therefore there was apparently no mtDNA synthesis in unfertilized loach eggs (Mikhailov and Gause, 1974a). After fertilization the rate of incorporation of precursors into the mtDNA started to grow exponentially, doubling each 7 h. After 35 h, i. e. in the period of active organogenesis, the incorporation rate was 0.4 pmole TTP per 1 mg protein per hour, i. e. 40 times higher than before. To find out to what extent the increase in the rate of label incorporation into mtDNA reflects the changes in the rate of DNA synthesis and is not affected by the changes in intramitochondrial pool of the precursors, the authors investigated the content of TTP in the mitochondria. It turned out that the TTP pool in the mitochondria varied: it reduced 2.5 times in the first 10 h of development, then stabilized and remained constant throughout the subsequent development. The authors conclude that a considerable increase in mtDNA synthesis occurs after 10 h, i. e. with the onset of gastrulation (Mikhailov and Gause, 1974a; Gause, 1977). A peculiarity of mtDNA synthesis in the loach is that even at the moment of its considerable activation there is no increase in the number of D-loop structures of mtDNA (Mikhailov and Gause, 1974a). They constitute only 4% of the mtDNA. Gause and Mikhailov attribute the low content of D-loop molecules to the fact that in the embryo only a small part of the mtDNA is replicated or that in early embryogenesis the D-loop formation is blocked in some way or other. Indeed, at the 80th hour of development the content of D-mtDNA in a freely swimming larva reaches 10 to 15%.

In *Xenopus*, mtDNA synthesis was studied by Chase and Dawid (1972) by incorporation of labeled thymidine into the mtDNA of developing embryos. The authors took advantage of a very efficient technique of mtDNA separation from the nuclear DNA admixture by means of denaturing the DNA and its subsequent short-term annealing. The mitochondrial genome differs in size from that of nuclei by more than four orders of magnitude (10^5 versus 2×10^9 b. p.); therefore the rate of mtDNA renaturation is 10^4 times higher. If, however, nuclear DNA is present only in trace amount, only mtDNA will renature during annealing, while the nuclear DNA, even its repetitive fraction, will not. Double-stranded mtDNA's are readily further separated by centrifugation in CsCl (Chase and Dawid, 1972). Chase and Dawid showed that up to stage 11 (gastrula) the incorporation of labeled thymidine into mtDNA is not detected. Then the synthesis becomes detectable and increases by a factor of 2-3 each 10 h. Thus, by stage 35 the rate of thymidine incorporation into the mtDNA increases 10-fold. This is approximately the same order of magnitude of rate increase as in loach embryos.

A feature peculiar to mtDNA synthesis in animal embryogenesis is that during highly intensive replication of nuclear DNA, at the synchronous cleavage stage, it is actually absent, but its activity grows at the stages where the intensity of replication of nuclear DNA drops by hundreds of times. This strongly suggests a mutual independence of the syntheses of mtDNA and nuclear DNA. Such an independence of mitochondrial and nuclear DNA replication has also been described for other systems. Thus, in the slime mould *Phyzarum* and the infusorian *Tetrachymena*, no mtDNA synthesis takes place in the S period when the nuclear DNA is replicated (Parsons and Rustad, 1968; Holt and Gurney, 1969).

Change of DNA Content in Embryogenesis

The quantity of DNA in embryos increases throughout the development by multiplication of nuclei, i. e. by the synthesis of nuclear DNA. At the same time, in a number of objects — some fishes and amphibians — the content of total DNA at the early stages of development does not increase or increases very slightly, whereas the number of nuclei increases in this period in geometrical progression. This is due to the fact that the eggs of such animals contain a very large number of mitochondria, and, hence, huge stockpiles of cytoplasmic mtDNA, which exceed the content of nuclear DNA by one to three orders of magnitude. Against the background of such quantities of mtDNA the total DNA increase in the egg can be detected only when hundreds of nuclei have formed, i. e. at the blastula stage (Fig. 4).

In the course of normal development the ratio of nuclear DNA to cytoplasmic DNA, i. e. to mtDNA, changes. In the egg and at early division stages this ratio shifts strongly toward the mtDNA. However, because of the rapid increase in the number of nuclei and the absence of mtDNA synthesis this ratio levels off soon and then shifts toward the nuclear DNA. In amphibians at the gastrula stage the mtDNA constitutes 1% or less of the nuclear DNA, i. e. a nuclear-to-mtDNA ratio typical for somatic cells is established in the embryo.

In the development of most organisms, the DNA content per nucleus varies between 2 C and 4 C, reflecting its variation in the course of the mitotic cycle, i. e. its doubling within the S period. But in some animals (diptera) in the nuclei of definite specialized cells of the larva the amount of DNA greatly increases owing to the polytenization of the chromosomes. Many animals undergo nuclear polyploidization, sometimes by tens and hundreds of times (e. g., in the nurse cells of the ovaries) (see, for instance, Choi and Nagl, 1977; Mermod *et al.*, 1977). As in rDNA amplification in oogenesis, such nuclear differentiation is necessary for ensuring intensive processes of transcription of various kinds of RNA.

For a number of invertebrates — worms and arthropoda — another process has been described. namely, the loss of a portion, sometimes a considerable one, of the DNA in the somatic cell nuclei. In some instances this is expressed as elimination of certain chromosomes, and in others, as the diminution of parts of chromosomes. Thus, for instance, in *Wachtliella persicaria* eight nuclei are formed as a result of the first three divisions; one of them migrates into the egg site containing the polar plasm and retains a full set (40) of chromosomes. In the remaining seven nuclei most of the chromosomes (30 to 32) are eliminated (Geyer-Duszynska, 1966). In various types of crustacean cyclops (Beermann, 1968; see Kiknadze, 1972)

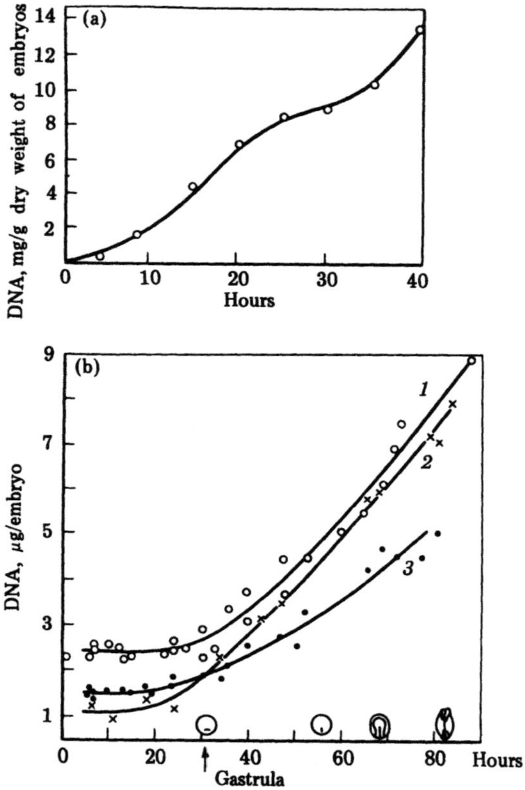

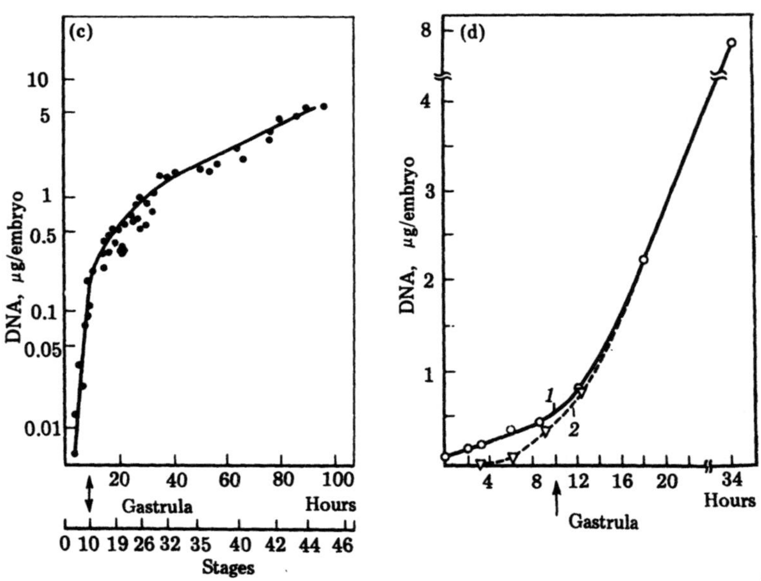

Fig. 4. Changes in DNA content of the embryos in the course of development of various animals: *a*, sea urchin; *b*, newts (*1*, *T. alpestris*, *2*, *T. cristatus*, *3*, *T. palmatus*); *c*, frog *Xenopus laevis*; *d*, loach *Misgurnus fossilis* H. (*1*, experimental curve, *2*, theoretically calculated curve based on the haploid DNA content). The DNA was determined in a fraction isolated after Ogur and Rosen by UV-absorption (*a*, *b*), by the diphenylamine (*c*) and the microbiological (*d*) methods (*a*, Mazia, 1949; *b*, Chen, 1960; *c*, Dawid, 1965; *d*, Timofeeva, 1967).

and ascarids (see Rossi *et al.*, 1975) the somatic cells of the embryo, unlike the oviduct cells, lose a considerable portion of chromatin. In *Cyclops furcifer* this occurs at the 6th to 7th division: at first, heterochromatinization of the terminal and centromeric regions of the chromosomes takes place, and then these heterochromatic regions experience diminution with a loss of up to 75% of the nuclear DNA. According to Beermann's hypothesis (1968), the molecular diminution mechanism consists in "looping out" part of the sequences of the chromosomal DNA because of the breaking by specific endonucleases of one of the DNA strands at definite sites, the drawing of these sites together, and their linking with the formation of loops which then separate from the main sequence in the form of circles by breaking and repairing the second DNA strand.

In *Ascaris megalocephala*, after the first division, one of the nuclei has a normal quantity of DNA, while in the second a diminution of chromatin takes place. Then, in the series of subsequent divisions of the normal nuclei, only one nucleus retains the initial quantity of DNA, while in the others a portion of chromatin is lost. As a result of five divisions, 31 blastomeres are formed, in which the DNA content is smaller than in the initial nucleus. These blastomeres then form the somatic cells of the embryo, while the normal nucleus forms the germ-line cells (see Rossi *et al.*, 1975). In *Ascaris suum* the quantity of DNA in the somatic cells is 33% less than in the germ-line cells as a result of diminution (Goldstein and Straus, 1978). In *Ascaris lumbricoides* this diminution is 27% (Tobler *et al.*, 1972). Finally, in *Parascaris equorum* the DNA losses reach 80% of its content in the egg (Moritz, 1970). The same authors (Moritz and Roth, 1976) found that only repetitive sequences are eliminated. These results are in contradiction with other measurements on ascarids, which may also be due to differences in the species used.

Tobler *et al.* (1972) showed that in the nuclear DNA of the somatic cells of the ascarid the repetitive sequences constitute 10%, whereas in the DNA of the spermatids (generative cells), as many as 23%. This suggested that the diminution is associated with the loss of DNA, half of which consists of repetitions and the other half of unique sequences. To support these data, the authors conducted a hybridization analysis of DNA from spermatids and somatic cells of ascarid larvae with labeled RNA copies (cRNA) obtained by *in vitro* transcription of these DNA's. During hybridization, at low DNA concentrations, only the repetitive DNA sequences can be studied. When spermatidal DNA was hybridized separately with larval and spermatidal cRNA, it was found that the binding of the larval cRNA was half that of the spermatidal. When, however, after complete binding of the larval cRNA the sepermatidal cRNA was added, additional binding of RNA took place. The results obtained proved that the larval cRNA and, hence, the larval DNA, lack part of the repetitive sequences which are present in the DNA of the generative cells (Tobler *et al.*, 1972).

The problem of chromatin diminution and chromosome elimination is still open: the biological significance of these phenomena is unclear and is widely discussed (see Sager and Kitchin, 1975). It is obvious that the genes of the DNA portion that is preserved only in the germ-line cells do not participate directly in the control of embryogenesis, larval development, and adult organism functions. It would seem natural to assume that these genes are involved in oogenesis and spermatogenesis. Indeed, the development of species in which chromosome elimination or diminution occurs generally belongs to the mosaic type. This means that they must have a quite pronounced differentiation of the egg cytoplasm (ooplasmic segregation) because the embryonic differentiation is largely defined already in oogenesis and therefore must require the participation of a considerable portion of the genome. It readily follows that elimination or diminution occurs precisely in those parts of the genome which fulfill their function only in the course of gametogenesis and would have been a dead weight in the somatic cells.

A special place in the question of the change of the DNA quantity in the embryonic cells is occupied by the problem of DNA amplification in somatic cells. The simple idea that selective replication of individual genes could be a mechanism of differentiation has been voiced repeatedly. This problem, as a part of the problem of differential activation of genes, is considered in detail in Volume II. Here we can only say that the experimental material in favor of the amplification theory is quite unreliable and contradictory and is usually found in single reports, not supported by further evidence either from the authors themselves, or from other investigators. Besides, in a number of reliable experiments it has been demonstrated that in some highly specialized cells synthesizing one or several types of protein, such as the red blood cells, the gland cells of the oviduct, etc., where the amplification of globin and ovalbumin genes would be expected to occur in the first place, this phenomenon is absent altogether. The number of genes also remains unchanged when they undergo a sharp activation in carcinogenesis (Sala-Trepat *et al.*, 1979).

Nevertheless, in recent years an ever-increasing body of data has been accumulating in the literature indicating that development is attended by frequent changes in DNA, either consistent or random. The most impressive and reliably documented examples of such changes are those occurring in the structure of immunoglobulin genes during lymphocyte differentiation. This phenomenon was discovered by Tonegawa and has been confirmed at many laboratories (see Volume II). The immunoglobulin molecule consists of heavy and light chains, each of which contains a variable region specific only to one type of molecule and produced by only one lymphocyte clone (offspring of one cell).

Besides, each chain contains a constant region which is identical for all the immunoglobulins of the given class. On the chromosome, the genes of the variable regions of the molecule (there are hundreds of them) are localized at a considerable distance from the corresponding constant region gene. During differentiation, an appreciable portion of DNA is deleted, as a result of which a gene of the variable region finds itself closer to the constant region. The actual process is more complicated and the variability is achieved by joining together one of the hundreds of variable genes, one of the 5 to 10 so-called J segments (junctions), and the constant region in numerous combinations. Thus, the number of variants of immunoglobulins is a product of the number of variants of light and heavy chains, since in each progenitor cell of the clone only one such opportunity is realized (the total number of possible types of antibodies is now estimated at 10^6). Another change in DNA, a more consistent one, but also apparently due to a specific deletion, occurs on transition from producing one class of antibodies to another. Thus, during lymphocyte differentiation one observes consistent changes in DNA structure, although the choice of some variable region or another in a given progenitor cell is random. Up to now this is the only known example of a change in DNA composition during development, but this discovery indicates the fundamental possibility of its occurrence in other, still unknown cases. As with rDNA amplification, it seems unlikely that a special mechanism responsible for processes like amplification or consistent deletion should be used in just one type of cell (oocytes or lymphocytes).

Indeed, in the past decade papers have been published now and then which would seem to indicate changes in the genome (Fitzmaurice and Bauer, 1972) and amplification of some genes (Strom and Dorfman, 1976) in the course of development. Besides, there are numerous data on quantitative changes in the content of satellite DNA and moderate repetitions in the course of development (see, for instance, Lozovskaya *et al.*, 1976). We do not mention here the changes in the number of genes, which, though occurring quite frequently, should be regarded as a phylogenetic rather than ontogenetic mechanism. By this we mean such phenomena as compensation for the deficiency of repetitive ribosomal or histone genes by magnification in *Drosophila*, migration of "unstable" genetic elements, increase in the number of some genes in the cell culture, etc. (see Khesin, 1980).

It can be concluded that now one can no longer speak of fundamental stability of DNA in the course of development. Part of such changes are known, while others are yet to be revealed. On the other hand, numerous data on the constant amount and composition of DNA ("stability") and on the preserved nuclear totipotency in many, if not all, differentiations, show that DNA remains substantially unchanged in the course of development. For example, an analysis of germ-line and somatic DNA of an adult nematode using restriction endonucleases indicated the identity of their DNA's to within several percent (Emmons *et al.*, 1979).

The changes that do occur probably refer to an insignificant portion of the genome and to occasional differentiations. True enough, even small changes in nucleotide sequence may play the decisive role in the regulation of the gene function.

Chapter Two

RNA

The synthesis and functions of RNA in oogenesis, embryogenesis, and subsequent stages of development have both common and specific features. The similarity consists in the fact that throughout the individual development the same main species of RNA are synthesized and perform the same functions in the oocyte, the embryo cell, and in the cells of an adult organism. Moreover, some mRNA's are synthesized throughout the course of development. But primary importance should be attached not so much to these common features of the synthesis and functioning of the main RNA species as to the differences and peculiarities which determine the specificity of RNA transcription in oogenesis and embryonic development.

In the cells of an adult organism the synthesis of all the RNA types is so coordinated as to maintain a definite level of protein synthesis. All the species of rRNA (28S, 18S, and 5S) are synthesized in equimolar ratios, in which they enter the ribosomes; a constant ratio of tRNA to the number of ribosomes is attained and maintained, and most of the newly formed mRNA become involved rapidly enough in the translation process. Conversely, no such coordination is observed in oogenesis and early embryogenesis. Only a small part of the protein-synthesizing apparatus formed, and also of the mRNA, is used in oogenesis for synthesizing protein in the oocyte itself. The remaining, greater, part is stored for prolonged periods, sometimes months or years, to be used in the first hours of embryonic development. Another difference lies in the fact that the whole RNA synthesis in oogenesis occurs in a single nucleus (or in a small number of nuclei of the nurse cells), and the RNA synthesized accumulates in a single cell, the egg. A set of active genes, which changes only slightly, apparently functions in oogenesis for a long time. In the embryonic development period, on the contrary, different sets of genes are active in different embryonic cells, replacing each other in the course of differentiation, i. e. relatively quickly. The peculiarities of RNA transcription at different stages of development evidently require specific control mechanisms. For these reasons one should consider RNA in oogenesis and in embryonic development sequentially and separately.

RNA IN OOGENESIS

The importance of RNA synthesis occurring in oogenesis becomes obvious in considering those mutations which, operating during the egg formation, lead to disturbances in development and to lethality.

Most of the induced mutations manifest themselves in disturbing the oogenesis processes: the formation of follicular cells, the yolk deposition, the formation of egg membranes, etc. (Bakken, 1973). In the axolotl, the o-gene mutation has been discovered which leads, in homozygous females, to the arrest of development not of the mutants, but of their offspring. During oogenesis the product of this gene accumulates in the germinal vesicle ensuring the normal course of gastrulation. Injection of nuclear sap from wild-type oocytes into the eggs of o-gene mutant females corrects defects in their development and ensures a normal course of gastrulation (Briggs and Cassens, 1966). The right- and left-coiling of the shells of some snails is determined similarly during oogenesis. These examples show that during oogenesis the function of the genes controlling both the oogenesis itself and the subsequent development is fulfilled. The function of these genes is manifested primarily in RNA synthesis.

RNA Content in the Egg

In the course of oogenesis the oocytes, which initially have a diameter of 10 to 20 μm, grow to the size of mature oocytes or eggs with diameter ranging from 50-100 μm (mammals, echinoderms, and some insects) to several centimeters (birds and sharks). This means that during oogenesis the oocyte volume increases by 10^2 to 10^{12} times (in most organisms, by 10^3 to 10^6). This increase is mainly due to yolk accumulation. But the number of many RNA-containing structures, such as ribosomes, mitochondria, etc., also increases simultaneously.

The absolute RNA content in mature eggs of various animal species correlates with the size of the eggs and ranges from 6×10^{-5} μg in the ascarid to 2×10^3 μg in birds.

Most of the egg RNA is contained in the ribosomes. In amphibians, at least 95% of it are represented by molecules of 28S and 18S rRNA (Brown and Littna, 1964, 1966). The relative content of rRNA's in the eggs of fishes (Timofeeva and Kafiani, 1964; Aitkhozhin *et al.*, 1964), sea urchins (Nemer, 1963; Slater and Spiegelman, 1966; see Giudice, 1973), worms (Schwartz, 1970), insects (Harris and Forrest, 1967; Lockshin, 1966), crustaceans (McClean and Warner, 1971), birds (Lerner *et al.*, 1963), and mammals (Mintz, 1964a, b) is as large (Table 5). The absolute rRNA content in the egg greatly exceeds that in the cells of an adult organism. This is true not only for large eggs, which are many times as large as somatic cells but also for small eggs. Nevertheless, RNA concentrations in small and medium-sized eggs and in somatic cells differ but slightly. Thus, the eggs of mammals contain 1 to 2×10^{-3} μg rRNA, i. e. 2.5 to 5×10^8 ribosomes, while liver cells have 10^6 to 10^7 ribosomes. The mammalian eggs are larger than somatic eggs by 10^2 to 10^3 times, therefore the RNA concentration is about the same in both.

4*

Table 5

The Content of Main RNA Species in Unfertilized Animal Eggs

Species	Total RNA content in egg, μg	28S+18S rRNA, μg	5S rRNA, μg	tRNA, μg	mRNA, μg	Reference
Xenopus laevis	4.0	3.8	0.06	0.04	4.7×10^{-2} (1.9%)	Brown and Littna, 1964, 1966; Davidson *et al.*, 1968
Misgurnus fossilis	2.4-2.6	2.3-2.6	0.042-0.052	0.021-0.035	—	Aitkhozhin *et al.*, 1964; Timofeeva and Kafiani, 1964; Solovyova *et al.*, 1973a; Timofeeva *et al.*, 1982
Drosophila melanogaster	0.19	—	—	—	—	Hough-Evans *et al.*, 1980
Musca domestica	1.2	—	—	—	—	Hough-Evans *et al.*, 1980
Urechis caupo	13.5×10^{-3}	12×10^{-3}	1.3×10^{-4}	1.3×10^{-3}	—	Schwartz, 1970
Strongylocentrotus nudus	7.8×10^{-3}	—	—	—	—	Timofeeva *et al.*, 1969
	2.0×10^{-3}	—	—	—	—	Davidson, 1969
	—	—	—	—	(4.5%)	Slater and Spiegelman, 1968
Lytechinus pictus	—	1.5×10^{-3}	—	—	4×10^{-6}	Humphreys, 1971

Xenopus eggs contain about 4 μg rRNA (Brown and Littna, 1964), which corresponds to 10^{12} ribosomes, i. e. is 10^5 to 10^6 times more than in somatic cells. As for the volume, these eggs are 2×10^5 times as large as liver cells, i.e. here, too, the rRNA concentrations differ only slightly. In large bird eggs, which are mainly filled with yolk, the rRNA concentration is one or two orders of magnitude lower.

In addition to the high-molecular-weight rRNA's, the egg also contains low-molecular-weight RNA's, which include ribosomal 5S rRNA and transfer 4S RNA.

In an unfertilized amphibian egg all the 5S rRNA is contained in the ribosomes (Brown and Littna, 1966). The molar ratio of 28S and 18S rRNA to 5S rRNA in the ribosomes is 1 and, in accordance with the molecular weights (1.5-1.7×10^6, 0.65×10^6, and 0.036×10^6 daltons), the weight content of 28S, 18S and 5S rRNA must be, on the average, 67, 31, and 1.6%, respectively. In *Xenopus* eggs the content of 5S rRNA is 0.060 μg, or 1.5 to 2% of the total RNA's. Thus, there is actually one molecule of 5S RNA per ribosomal particle (Brown and Littna, 1966). In the loach the content of 5S rRNA in the egg is 4.6-5.2×10^{-2} μg, or 1.7-2% of the total RNA (Solovyova *et al.*, 1973a; Timofeeva *et al.*, 1982). The content of 5S rRNA in the egg of the sea worm *Urechis* is 1.3×10^{-4} μg per 1.2×10^{-2} μg of 18S+28S rRNA (Schwartz, 1970). Calculations show that only 65% of *Urechis* ribosomes contain 5S rRNA. Since in the absence of 5S rRNA the ribosome is inactive, the 5S rRNA deficiency in the eggs of *Urechis* indicates that not all the egg ribosomes are functionally adequate.

The relative content of tRNA in the eggs of most animals is small. In the frog *Xenopus* (Brown and Littna, 1966) and the loach (Solovyova *et al.*, 1973a) it is 1%, in the sea worm *Urechis* (Schwartz, 1970) it is 10% of the total egg RNA. Although no such determinations of tRNA content in eggs have been made for other animals, some data, in particular the sedimentograms of total RNA preparations from sea urchin eggs (Slater and Spiegelman, 1966, 1968; Gross *et al.*, 1965) and birds (Lerner *et al.*, 1963), suggest that the total content of 4S+5S RNA is also not high and does not exceed 5% (Fig. 5). Since the molecular weight of tRNA is two orders of magnitude lower than that of rRNA, it may be assumed that there are about 1 to 2 tRNA molecules per one ribosome. This is at least 10 times less than in somatic cells.

According to Brown and Littna (1966), in *Xenopus* the bulk of tRNA in the egg is in the free state in the cytoplasm, and only 10% of it is bound with ribosomes. Such a tRNA : ribosome ratio ensures the normal functioning of only 10% of egg ribosomes. This does not seem to be the case with all species. Some data indicate that the eggs of some invertebrates contain much more tRNA. In the oocytes of *Urechis*, tRNA constitutes 10% of the total RNA, and there are 10 molecules of tRNA per one ribosome (Schwartz, 1970).

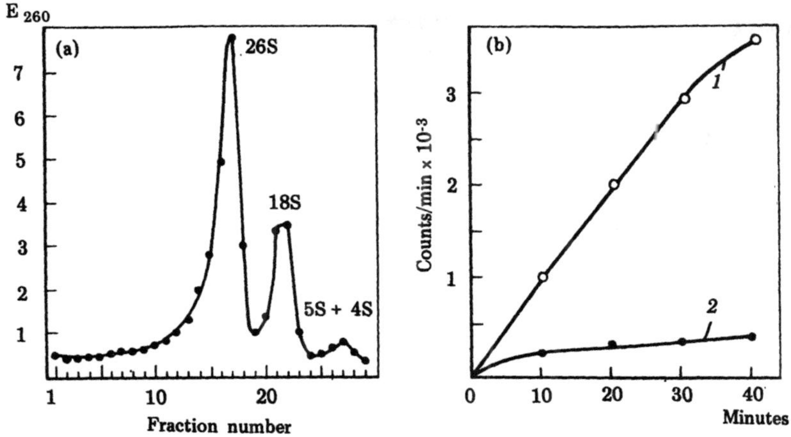

Fig. 5. Total and messenger RNA of the eggs of the sea urchin *Lytechinus pictus*. The RNA of unfertilized eggs of *Lytechinus pictus* was analyzed in a sucrose density gradient (*a*), and its template activity in a cell-free protein-synthesizing system determined by ^{14}C-serine incorporation (*b*); *1*, 50 μg of egg RNA was added to the system; *2*, control (without the egg RNA). It can be seen that the bulk of the egg RNA consists of rRNA (peaks of 26S, 18S, and 5S) and tRNA (4S). The RNA template activity is 4.5% of that of an equal amount of MS2 virus RNA (Slater and Spiegelman, 1966).

Of no less importance is the question of tRNA species composition of the egg and its effect on the synthesis of the various proteins. This question has been investigated in some detail only for sea urchins and amphibians.

The soluble fraction obtained by centrifuging the homogenate of unfertilized sea urchin eggs at 100,000g stimulated normally the incorporation of amino acids into the proteins in the presence of embryonic ribosomes (Timourian, 1967). This indicates that an unfertilized egg contains the whole set of tRNA's and of the appropriate aminoacylating enzymes.

Still earlier, Nemer and Bard (1963) had demonstrated the protein synthesis in the homogenate of unfertilized sea urchin eggs. Addition of exogenous tRNA from yeast had but a slight stimulating effect on protein synthesis. Thus, in an unfertilized egg various species of tRNA are present in quantities sufficient for sustaining the protein synthesis and are not among the synthesis-limiting factors.

That amphibian eggs possess the necessary set of tRNA's follows from the fact that they experience a sufficiently intensive protein synthesis (Smith *et al.*, 1966). Still more illustrative in this respect is the intensive synthesis of most diverse proteins on mRNA templates injected into amphibian oocytes and eggs (Gurdon, 1974).

The most important component of the RNA of unfertilized eggs is the messenger RNA. Its content is very small and it is heterogeneous in size. Therefore mRNA cannot be identified by specific optical density peaks among the other RNA's when they are separated by sucrose gradient sedimentation. A number of facts, however, point to the presence of mRNA in an unfertilized egg. Thus, RNA isolated from unfertilized sea urchin eggs may serve as template for protein synthesis in *in vitro* systems (Maggio *et al.*, 1964; Gross *et al.*, 1973a).

It was further shown that in enucleated loach embryos obtained by fusion of gametes whose nuclei had previously been inactivated by high X-ray doses, the protein synthesis proceeds throughout the prolonged period these embryos remain alive (Neyfakh *et al.*, 1968). In sea urchins, incorporation of labeled amino acids into an acid-insoluble material has been demonstrated in anucleate halves of the zygote (Brachet *et al.*, 1963 ; Denny and Tyler, 1964; Baltus *et al.*, 1965). The puromycin sensitivity of this incorporation suggested that it indeed reflects the protein synthesis occurring on functioning polysomal complexes (Burny *et al.*, 1965).

In frogs *Rana pipiens* mature oocytes with the germinal vesicle removed were capable of increased protein synthesis, provided their maturation was hormone induced (Ecker and Smith, 1968).

A slight protein synthesis is also detected in unfertilized sea urchin eggs (Timourian, 1967; MacKintosh and Bell, 1969a, b; Timourian and Watchmaker, 1970; Smith *et al.*, 1966).

Data on the presence of mRNA's in mature oocytes and in eggs were obtained in investigations of template activity of RNA's isolated from oocytes and eggs. The template activity was determined by its ability to stimulate the incorporation of labeled amino acids into acid-insoluble material in an *in vitro* system. In most cases a system of ribosomes and soluble components of the protein-synthesizing system of *E. coli* is used (S fraction obtained after Nirenberg and Matthei by centrifuging, at 30,000*g*, a DNase-treated bacterial homogenate; to eliminate the endogenous templates, the fraction was preincubated for 45 min in the presence of appropriate salts and cofactors). To find the quantitative expression of the activity, determinations of protein synthesis are carried out concomitantly under the same conditions, but in the presence of an RNA which is regarded as a 100% template. This is the RNA of the MS2 virus. Incorporation of amino acids into acid-insoluble material in the presence of RNA with an unknown template content is expressed as percent of such incorporation in the presence of MS2 RNA. Since under these conditions the template can be read out several times, Slater and Spiegelman (1968) proposed the term "relative template titer" to replace "template activity." On a bacterial protein-synthesizing system, 2.5 to 5% of the total RNA of the *Xenopus* egg shows template activity (Davidson *et al.*, 1966; Cape and Decroly, 1969). The number of transcripts from nonrepetitive nucleotide sequences (unique-type RNA) in the *Xenopus* egg has been determined as 7×10^{11} nucleotides, i.e. as low as 4×10^{-4} μg, or 0.01% of the total egg RNA.

The egg RNA partly consists of mitochondrial RNA. This RNA is mainly contained in the mitochondrial ribosomes. It differs from the eukaryotic and prokaryotic rRNA, its two components having a molecular weight of 5.3×10^5 and 3.0×10^5 daltons. Its quantity does not exceed 1-2% of the total egg RNA.

In sea urchin eggs the ratio of the RNA species is approximately the same as in amphibian and fish eggs: of the 2 pg of the RNA, 95% are rRNA and tRNA, and 4% is the DNA-like RNA (mRNA). Template activity is shown by 4.5% of the RNA (Slater and Spiegelman, 1968) (Fig. 5). A different, and presumably more accurate, calculation for fertilized eggs of another sea urchin species, *Lytechinus pictus*, shows (Humphreys, 1971) that out of the total quantity of rRNA (1.5×10^{-9} g) the polysomes contain about 2×10^{-10} g rRNA, while the quantity of mRNA (4×10^{-12} g) makes up 2% of this value. Some data on the content of various RNA species in eggs are listed in Table 5.

RNA Accumulation in Oogenesis

The rate of RNA accumulation in the oocyte is determined by the final amount of RNA required for the mature egg and by the oogenesis duration. There is no direct relationship between the egg size and the oogenesis duration, although species with large eggs rich in RNA generally have prolonged oogenesis. The oogenesis rate is determined to a great extent by the biology of reproduction, the life span, the hibernation duration, etc.

The shortest oogenesis times are probably in the order of several days: in *Drosophila* the whole developmental cycle takes 10 to 15 days. Some mammals have relatively short oogenesis: in the mouse it begins in mid-embryogenesis, 10 days before birth, and ends as early as 2.5 to 3 months after birth. This, however, is not true for many other mammals, where oogenesis may last for years.

The amphibians, in which oogenesis has been studied most comprehensively, can be divided into two groups: the hibernating forms with seasonal reproduction and tropical forms with all-year-round reproduction. In the hibernating amphibians, the oocyte development begins immediately after the metamorphosis and ends within 2 or 3 years. In the African frog *Xenopus laevis* it probably lasts at least 6 to 8 months, and in the tropical South-American frog *Engystomops pustulosus*, 2.5 to 3 months (Davidson and Hough, 1969).

A similar diversity is found in fishes. Some tropical aquarium fishes start breeding already at the age of 2.5 to 3 months, but in northern fishes the first spawning period does not begin until the age of one or two years. In migratory sturgeon fishes the first spawning occurs only at the age of 8 to 12 years, which is probably due to their prolonged growth in the sea.

In birds and reptiles, oogenesis usually lasts months or years. Different types of RNA have different patterns of accumulation and their synthesis does not continue throughout the whole oogenesis. In the sea urchin a considerable part of the mRNA is synthesized only toward the end of oogenesis (Piatigorsky and Tylor, 1967; Giudice *et al.*, 1972a; Sconzo *et al.*, 1972). In the worm *Urechis caupo,* rRNA synthesis proceeds throughout oogenesis, but in its last third its intensity sharply increases (by a factor of 4 to 5). At a later date it was shown that the synthesis and accumulation of RNA at the early stages of the *Urechis* oogenesis proceed somewhat faster than the increase in oocyte size (Miller and Epel, 1973; Schwartz, 1970; Davis and Wilt, 1972). In *Xenopus*, predominantly low-molecular-weight RNA (5S rRNA and tRNA) are synthesized first, followed, in the large-growth period, by the intensification of the 28S, 18S rRNA, and mRNA synthesis. The relative content of the various types of RNA in the course of *Xenopus* oogenesis is shown in Table 6 and in Figs. 6, 7, and 8. Formerly, it was assumed that in mature oocytes the synthesis of all the types of RNA ceases, and only during maturation a short-term burst of synthesis of DNA-like RNA occurs (Brown and Littna, 1964). But later studies of the sea urchin (Levner, 1974) and *Xenopus* (Colman, 1974; La Marka *et al.*, 1975) have shown that RNA synthesis still continues in full-sized oocytes. It is believed that this synthesis offsets the RNA decay, which also continues in oogenesis.

This problem was investigated more comprehensively at a later time (Dworkin and Infant, 1978). An intensive mRNA synthesis takes place in the nucleus of the unfertilized egg, but at least the bulk of this RNA

Table 6

Variation in Relative Content of Various RNA Species in *Xenopus* Oogenesis (Rosbash and Ford, 1974)

Stage	RNA in oocyte, μg	RNA, % 28S	18S	5S	4S
Early previtellogenesis	0.04	5	3	45	47
Late previtellogenesis	0.07	32	17	31	20
Early vitellogenesis ($d = 200\text{-}300$ μm)	1.2	56	24	11	9
Mid-vitellogenesis ($d = 500\text{-}600$ μm)	3.5	63	29	5	3
Mature oocyte ($d = 1000\text{-}1200$ μm)	4.3	65	29	3	2.9

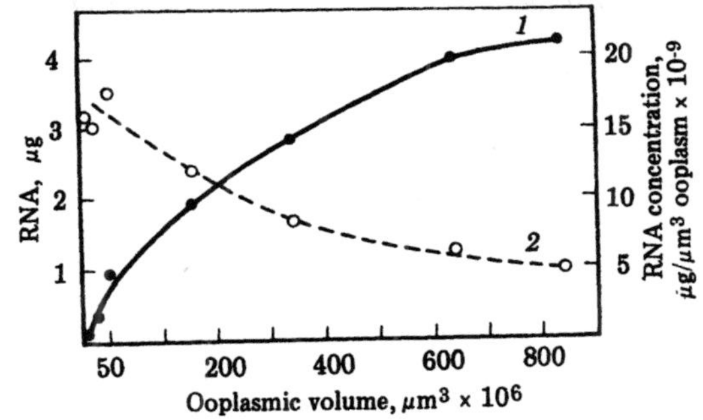

Fig. 6. Accumulation of RNA in *Xenopus* oogenesis. Increase in absolute quantity (*1*) and decrease in concentration (*2*) of the cytoplasm RNA vs oocyte size (Scheer, 1973).

decays with a half-life of 30 min and cannot provide the quantity of mRNA necessary for activation of protein synthesis after fertilization. Calculations show, however, that if just 10% of the newly synthesized RNA has a longer half-life (30 to 35 h), this synthesis rate is sufficient for continuous renewal of mRNA in the unfertilized egg (Dworkin and Infante, 1978).

However, a certain part of the mRNA's synthesized at the very outset of *Xenopus* oogenesis can persist until the stage of mature oocytes, i.e. for two to three years (Ford *et al.*, 1977). The bulk of this long-lived early synthesized mRNA consists of molecules with short (15 to 30 nucleotides) poly$(A)^+$ sequences, whereas the later-synthesized and shorter-lived mRNA's have longer poly$(A)^+$ (40 to 80 nucleotides) (Cabada *et al.*, 1977).

The rate of all the syntheses is so coordinated that a protein-synthesizing apparatus is created in the course of oogenesis which ensures not only protein synthesis during oogenesis, but also a considerable period of embryonic development.

Sources of RNA Accumulation

The various types of oogenesis can be divided into two groups: those in which RNA is synthesized by the oocyte's own nucleus, and those in which RNA is synthesized by accessory (nurse) cells. Some species have both ways of RNA accumulation.

Many insects (dipterans, bugs) and worms have nurse cells differing from follicle cells in that they are formed together with the oocyte from the same oogonium and remain joined to it by special channels. Thus, in *Drosophila*, four divisions result in 16 cells, of which one becomes an oocyte, and the remaining 15, nurse cells. In other insects, a complex of seven nurse cells and an oocyte arises similarly after three oogonial divisions.

The autoradiographic investigations carried out so far on a few insect species have shown that RNA synthesis occurs exclusively, or almost exclusively, in the nurse cell nuclei and is practically nil in the oocyte's own nucleus (Bonhag, 1955; Bier, 1963; Koch *et al.*, 1967). Tracing of a label has shown that from the nurse cell cytoplasm it passes through the channels into the oocyte cytoplasm (Choi and Nagl, 1977). It was shown that in the ovaries of the silk moth tRNA, rRNA, and poly(A)-containing mRNA (which constitutes 0.2 to 0.3% of the whole newly synthesized RNA) are synthesized in the nurse cells. The actinomycin D chase made it possible to observe the transport of mRNA and tRNA to the oocyte, which ends as early as 2 to 4 h after their synthesis. The transport of rRNA lasts longer (Paglia *et al.*, 1976a). In other insects, however, RNA synthesis takes place in the oocyte's own nuclei (Kunz, 1967). A short period of RNA synthesis in the oocyte nucleus has been observed in *Drosophila* as well (Mahowald and Tiefert, 1970), although the main RNA synthesis proceeds in the nurse cells (Mermod *et al.*, 1977).

Proceeding from electron-microscopic data (Zamboni and Gondos, 1968), Davidson (1969) assumes that in rabbits the oocytes are connected by similar channels, and that the content of some oocytes finds its way into others as a result of atresia.

Follicular cells are formed from the ovary stroma, i. e., unlike nurse cells, they do not belong among the germ-line cells. Maybe this difference has no profound biological significance and is due merely to the fact that syncytial connections are formed more readily between the daughter cells. But it may also be that these differences are fundamental, i. e. are determined by some genetic features of the germ-line cells. Thus, it is quite probable that in animals with chromosome elimination or chromatin diminution in the somatic cells the genetic material eliminated in these processes serves precisely for oogenesis (Geyer-Duszynska, 1966). For the follicular cells, which apparently perform only accessory functions, the presence of these "oogenesis genes" and the identification with the germ-line cells are evidently not mandatory. Conversely, the preservation of the complete genotype in the nurse cells, which synthesize RNA for the oocyte, is not only justified, but is absolutely necessary. This assumption would seem to have been disproved by the example of the diving beetle *Ditiscus* described by Wilson (1928) and discussed, from the standpoint of interest to us, by Davidson (1969).

The elimination (or diminution) of chromatin in this species occurs not in early embryonic development, as usual, but only prior to oogenesis in the course of four oogonial divisions. This gives rise to an oocyte with a complete chromosomal set and 15 nurse cells, in which a considerable part of chromatin is absent. It would be useful to know how the cells synthesizing RNA for the oocyte can do it without chromatin, which would seem to be particularly meant for oogenesis. In Davidson's opinion, these nuclei fulfill their genetic function before divisions. (A similar situation is observed in spermatogenesis, where the specific nuclear function is completed before meiosis.) We believe, however, that in this instance a different explanation is also possible: the far-reaching specialization of the nurse cell nuclei manifests itself in the inactivation of the chromatin uninvolved in oogenesis. This assumption is also supported by the fact that in *Ditiscus* "diminution" occurs only in the course of oogonial divisions and does not affect the somatic cells. Therefore it cannot be regarded as a process completely similar in the genetic sense to diminution and elimination in the course of cleavage. It would be extremely interesting to investigate this object many decades after Wilson with the use of modern methods.

A second important feature of nurse cells is their polyploidy (see Puckett and Snyder, 1974; Choi and Nagl, 1977). Indeed, rapid oogenesis cannot obviously be achieved just because RNA is synthesized not by one, but by 7 or 15 nuclei. An increase in the number of DNA templates for RNA synthesis is effected by polyploidization of nuclei, whose genetic apparatus is not involved in the further development of the embryo and is not inherited by the offspring. Apparently, polyploidy effectively replaces such a form of chromosome activization as lampbrush structures because they are never observed in the nurse cells. The specialization and active function of these nuclei is also expressed in their shape, which resembles that of the nuclei of irreversibly differentiated insect silk glands.

RNA synthesis also takes place in the follicular cells, but it is not regarded as RNA synthesis for the oocyte (see Webb, 1976). In these cases, the oocyte's own nucleus functions actively, and the cell membranes separating the eggs and the follicular cells usually remain undisturbed. But the possibility of the follicular cells serving as RNA sources for oogenesis cannot be ruled out completely and for all cases. The embryological literature contains dozens of reliable morphological descriptions of follicular cells of different species injecting their content into the oocyte or trapped by it completely. It is difficult to say at present to what degree this is a way of growth of the oocyte and whether functionally adequate RNA's also find their way into the oocyte.

The second, and evidently the main, way of RNA accumulation in oogenesis is its synthesis by the oocyte's own nucleus. We have already described in the preceding chapter the events occurring at the chromatin level in that period. Obviously, the lampbrush loops must be considered the site of synthesis for different types of RNA (Gall and Callan, 1962; Davidson *et al.*, 1964; Davidson, 1969). Only high-molecular-weight rRNA's (28S and 18S) are synthesized on amplified rDNA in the nucleoli, the intensity of the RNA synthesis in the nucleoli being much higher than on the lampbrush loops (Davidson and Mirsky, 1965). This has been shown not only for amphibians, but also for fishes, birds, and even some insects (Kunz, 1967) which utilize nurse cells more often. DNA-like RNA's largely accumulate in the germinal vesicle itself, remaining, in part, attached to the lampbrush loops.

Thus, while in the chromatin of somatic cells (liver) the RNA : DNA ratio is close to 1 : 15, the newt germinal vesicle contains 0.8 pg DNA and 70 pg RNA, i. e. the RNA : DNA ratio is about 100 : 1. The nucleotide composition of this RNA is DNA-like (Edstrom and Gall, 1963), and therefore the lampbrush loops could be regarded as the mRNA storage place (Hess, 1968; Davidson, 1969).

In this chapter we mainly consider the RNA synthesized in the oocyte germinal vesicle, since investigations in this area have been mainly conducted on *Xenopus*.

Synthesis of Transfer RNA's

The synthesis of tRNA in oogenesis has been investigated on *Xenopus* oocytes. The first investigations on tRNA synthesis were carried out by Brown and Littna in 1964-1966. More detailed and recent data on the same object were obtained by Denis *et al.* in later years (Mairy-von-Frenckell, 1970; Denis, 1974; Mairy and Denis, 1970, 1971, 1972; Wegnez and Denis, 1972; Wegnez *et al.*, 1972).

Denis divided the oocytes according to their sizes: the smallest, of average diameter 50, 90, and 150 μm, were obtained by mild pronase digestion and EDTA treatment of ovaries from young frogs sized 1.5, 2.5, and 4.0 cm, respectively. These oocytes were going through the small-growth period, and their lampbrush structures had not yet achieved full development. Larger oocytes were obtained from adult females; they were insulated from the follicular cells with pronase and sorted out by sieving into four grades with average diameters of 375, 600, 850, and 1100 μm. The last grade included oocytes which had completed their growth and were ready for ovulation. The beginning of large growth, the yolk and pigment deposition corresponded to the fourth grade, i. e. to the smallest oocytes (375 μm) obtained from adult females. This stage, with the lampbrush structures achieving their highest development, can be referred to as the fourth stage of oogenesis (after Duryee). It can be seen that while on transition from one stage to another the oocyte size increases by a factor of 3 to 4, on passing from the third (oocytes from young 4-cm-long frogs) to the fourth ("small" oocytes of an adult frog) grade the oocyte size increases almost 16-fold. The time lag between these stages is also the largest and lasts many months.

Denis separated the RNA's isolated from oocytes on a Sephadex G-100 column. High-molecular-weight RNA (28S, 18S, and mRNA) moves in this column with the excluded volume and emerges as a single sharp peak (Fig. 7), whereas 5S and 4S RNA are separated from one another and form a second (5S) and a third (4S) peak. The nature of the RNA of these peaks was identified and confirmed by their nucleotide composition and by the hybridization test.

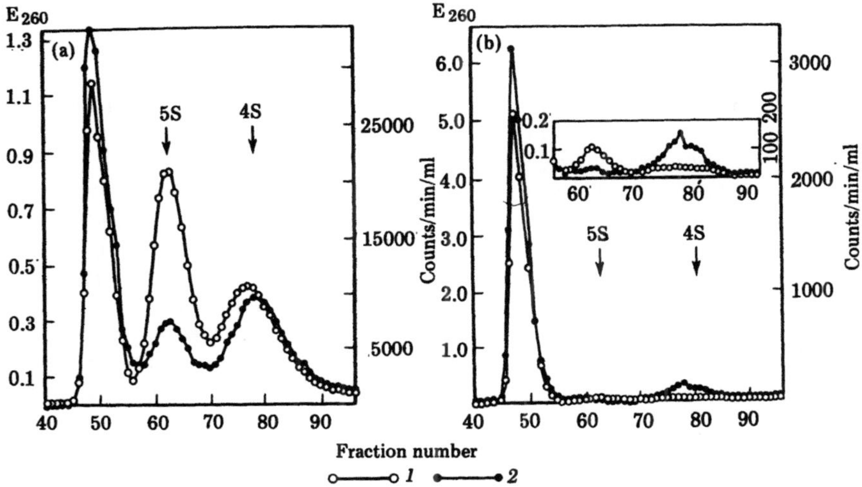

Fig. 7. Relative content and synthesis of 4S (tRNA) and 5S RNA in *Xenopus* oogenesis. Oocytes in early (*a*, 90 μm in diameter) and late (*b*, 600 μm) oogenesis were incubated for 4 h with ^{3}H-guanosine. The RNA was isolated and fractionated on a Sephadex G-100 column. The quantity (*1*) and radioactivity (*2*) of the RNA were determined. The relative quantity of 4S and 5S RNA's and their synthesis rate at the early stage are much higher than at the late one (Mairy and Denis, 1970).

Four hours before removing oocytes the frogs were injected with ^{3}H-guanosine, and its incorporation could be estimated by the RNA radioactivity. It is not easy, however, to correctly assess the change in the synthesis rate of different types of RNA in the course of oogenesis by their specific activity, as is usually done. The difficulty lies in the fact that, as a given type of RNA accumulates in the course of oogenesis, the radioactivity measured has to be related to an ever-increasing amount of RNA. It would doubtlessly be more correct to calculate the RNA radioactivity per DNA quantity, or per one genome, or simply per one oocyte.

The absolute amount of tRNA in the oocyte increases throughout oogenesis but in a nonuniform manner. The increase in the amount of tRNA begins even in the tiniest oocytes (first grade) soon after the metamorphosis. We do not know what quantity of tRNA is contained in the initial cells, oogonia, but it cannot be much larger than in ordinary cells, i. e. it cannot be more than 5 to 10% of the total RNA. However, already in first-grade oocytes (with a diameter ranging from 20 to 80 μm, 50 μm on the average) the tRNA constitutes about 40%, i.e. more than the rRNA contained in the oocyte at this stage. This extremely high tRNA concentration is also maintained at the subsequent stages of early oogenesis, although the oocyte size increases 27-fold within this time, which shows that the quantity of tRNA also increases considerably.

The high rate of tRNA synthesis at these stages of oogenesis is also evident from the radioactivity of the corresponding peaks (Mairy and Denis, 1971) (Fig. 7). In the large-growth stages (grades 4-6) the share of the tRNA in the oocyte gradually decreases to 2% (a mature egg contains about 1% of tRNA), but this does not mean that its synthesis and accumulation cease altogether. Direct measurements show that during large growth the absolute quantity of tRNA increases three-fold. If judged by the radioactivity of the third peak (Mairy and Denis, 1971) the tRNA synthesis continues up to the end of oogenesis, although its rate slightly decreases at the end of large growth. The rate of tRNA accumulation lags behind the oocyte growth rate. This is reflected both in the ratio of the number of tRNA molecules to that of the ribosomes and in the ratio of the growth rates for the tRNA quantity and for the oocyte mass.

At the early stages of oogenesis tRNA constitutes a considerable part of the total RNA synthesized: it accounts for about 36% of the total label detected. Part of the tRNA synthesized is preserved for many months of subsequent oogenesis. The ratio of the number of tRNA molecules to that of the ribosomes in the course of oogenesis has a specific pattern. At first it increases from 10 (the ratio typical of somatic cells) to 100 and more. But then, as the ribosomes accumulate, this ratio decreases to very low values (1 or 2). However, no such decrease in the relative content of tRNA is observed in *Drosophila*, for instance (Puckett and Snyder, 1974).

Not all the tRNA synthesized in oogenesis persists throughout the entire period of growth and maturation of the oocyte. A considerable portion of it decays, and the tRNA synthesis occurring in oogenesis must both offset this decay and ensure the accumulation of tRNA.

Preferential synthesis and accumulation of tRNA in early oocytes have also been found in nine fish species of different families (Mazabraud *et al.*, 1975). The tRNA content in the previtellogenic oocytes of these species ranges from 20 to 70%.

As shown by Denis and Mairy (1972), 90% of the total tRNA of small amphibian oocytes have the form of ribonucleoprotein (RNP) particles, which sediment at 42S. These particles also contain 5S rRNA but in smaller quantities. The 42S particles contain two types of protein (Wegnez and Denis, 1973). According to the latest data 42S particles of fishes and amphibians consist of four 15S subunits, each containing one molecule of 5S RNA, three molecules of tRNA, one molecule of 4×10^4-dalton protein and two molecules of 5×10^4-dalton protein (Picard *et al.*, 1980; Denis *et al.*, 1980). During large growth the 42S particles disappear and the tRNA accumulated is apparently released.

The set of tRNA's in mature *Xenopus* oocytes was found to be close to that in the tissues of adult animals. However, the tRNA's synthesized in small oocytes differ from those of late oogenesis. The authors believe that either the previously synthesized tRNA's are modified in the course of oogenesis, or new tRNA species appear (Denis *et al.*, 1975). It was shown in a special investigation that tRNA synthesis does not involve gene amplification, as was formerly thought to occur (Wegnez *et al.*, 1972).

Synthesis of 5S rRNA

The synthesis of 5S rRNA in oogenesis was investigated by Denis *et al.* using the same technique of oocyte separation by size they employed for the tRNA synthesis study (Mairy and Denis, 1971, 1972; Wegnez and Denis, 1973). Like 28S and 18S rRNA, 5S rRNA is a ribosome component and it constitutes only 1.6% of the total rRNA. At the end of oogenesis, when the ribosomes contain 96 to 98% of the total RNA of the *Xenopus* egg, the quantity of 5S RNA does not exceed 1.5 to 2%. But investigation into the synthesis and incorporation of this component in the ribosomes have revealed some new mechanisms of RNA storage and utilization in oogenesis.

At the early stages of oogenesis, which last as long as several weeks, the weight content of 5S rRNA in the first-through-third-grade oocytes is 30 to 40%, i. e. many times higher than in the somatic cells. The weight content of the 28S and 18S rRNA at these stages is also 30%. This means that in early oocytes there are 60 to 100 molecules of 5S RNA per one molecule of 28S or 18S RNA (Fig. 7). At the subsequent stages of oogenesis this relationship rapidly decreases owing to the synthesis of 28S and 18S rRNA, but even after the beginning of vitellogenesis the molar ratio 28S : 5S in the fourth-grade oocytes (average diameter 375 μm) is still close to 1 : 8. By this stage the oocyte, whose mass is still only 4% of that of a mature oocyte, contains at least half of the total 5S rRNA of the egg (Mairy and Denis, 1971, 1972). Its synthesis in early oogenesis proceeds at the same rate as that of tRNA, then slows down but persists to the end of oogenesis. Denis and coworkers showed that the molecules of 5S rRNA synthesized in oogenesis differ somewhat in nucleotide sequence from 5S rRNA which are synthesized in the cells of the embryo and of the adult organism (Denis *et al.*, 1972) (Fig. 8). These differences (six replacements) help separate 5S rRNA of the egg and the embryo on a MAK column (Wegnez *et al.*, 1972; Ford and Southern, 1973). The functional significance of these differences is unknown (Denis *et al.*, 1972).

As has already been mentioned above, the intensive synthesis of 5S rRNA in oogenesis is ensured by the special composition of the genome. The haploid set of chromosomes of *Xenopus laevis* contains about 2.4×10^4 genes coding for 5S RNA, of which 90% are oocyte-

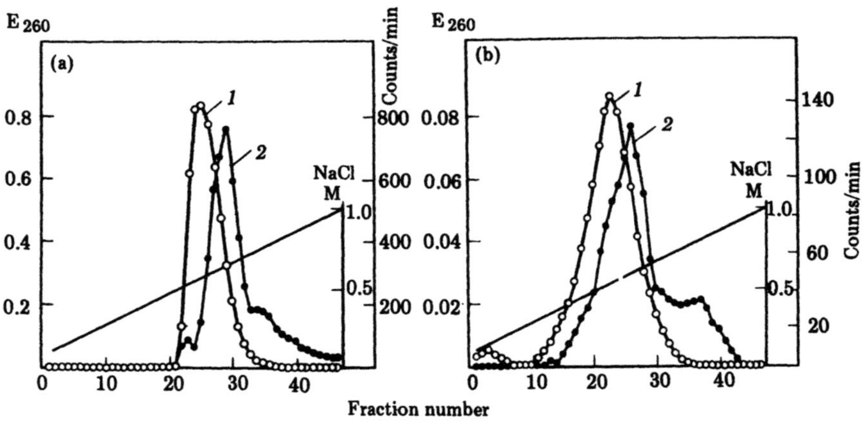

Fig. 8. Comparison of 5S rRNA's synthesized in oogenesis and in embryogenesis: *a*, labeled RNA from embryos was mixed with unlabeled RNA from early oocytes; *b*, up to the neurula stage, the embryos were grown with ^{32}P-phosphate injected into the egg. The 5S RNA was isolated and fractionated on a MAK column; *1*, RNA stored in oogenesis (optical density); *2*, RNA synthesized in embryos (radioactivity). The difference in the 5S RNA's synthesized in oocytes and in embryos can be seen (Denis, 1974).

type genes and less than 10% are somatic-type genes functioning at all the other stages of development (see Denis, 1974; Fedoroff, 1979).

The 5S rRNA synthesized in early oogenesis is largely preserved up to the late stages. Ten months after the synthesis ends, at least 50% of the label persists in the 5S rRNA (Denis *et al.*, 1972).

As well as tRNA, 5S rRNA synthesized during the small growth becomes incorporated in 42S RNP particles (Thomas, 1970; Denis and Mairy, 1972; Wegnez and Denis, 1973). These RNP particles contain 30% of the total 5S RNA, and therefore the molar ratio of 4S : 5S RNA in the 42S particles is 3 : 1 (Ford, 1971; Denis and Mairy, 1972). It was assumed that the rest of 5S RNA (over 50%) remained free. It was shown recently, however, that in amphibians the entire 5S RNA not included into the 42S particles formed 7S particles consisting of one molecule of 5S RNA and one molecule of protein with a molecular weight of 4×10^4 (Picard and Wegnez, 1979; Denis *et al.*, 1980; Picard *et al.*, 1980). This is, apparently, the same protein which is contained in the 42S particles. This protein, however, definitely differs in molecular weight and amino acid composition from the one forming a complex with 5S RNA in ribosomes. Thus, in early oogenesis 5S RNA is stored in two forms: in the 42S particles together with the tRNA and in the 7S particles. During the formation of ribosomes the protein bound with the 5S RNA for the duration of storage is replaced by a different, specifically ribosomal, protein (Picard and Wegnez, 1979).

At the beginning of vitellogenesis, when an intensive synthesis of 28S and 18S RNA begins in the oocyte, it is accompanied by the breakdown of the 42S particles, the release of the 5S RNA, and the formation of ribosomes. They include 5S RNA's — both those stored up previously and those synthesized during large growth. In the course of oogenesis all, or almost all, 5S rRNA synthesized enter into the ribosomes (Mairy and Denis, 1972). A considerable quantity of 5S RNA (from 20 to 30% of the total RNA) was also found in previtellogenic small oocytes of different fish species (Mazabraud *et al.*, 1975). The tench (Denis and Wegnez, 1977), dogfish (Wegnez *et al.*, 1978), and loach (Mashkova *et al.*, 1981) also showed differences between the 5S RNA accumulating in the oocytes and the 5S RNA of the somatic cells. Like the genes of tRNA, the genes encoding 5S RNA are not amplified in oogenesis (Wegnez *et al.*, 1972).

Synthesis of 28S and 18S rRNA

To assess the synthesis of 28S and 18S rRNA by sedimentation is more difficult than the synthesis of the low-molecular-weight RNA's, because the former is accompanied by the synthesis of mRNA with similar sedimentation characteristics. A correct idea of the ratio between rRNA and DNA-like RNA can be obtained by determining the nucleotide composition of the molecules synthesized. In *Xenopus,* the rRNA contains 61-62% of GC, whereas the DNA contains 44% GC pairs. It is, therefore, easy to calculate the fractions of rRNA and D-RNA in the high-molecular-weight RNA under investigation. Thus, according to Mairy and Denis, the high-molecular-weight RNA in the small-growth oocytes contains 49 to 50% of GC, which means that only 40% of this RNA is ribosomal, the remaining 60% being DNA-like. At the beginning of vitellogenesis the situation changes and the share of rRNA in the high-molecular-weight RNA fraction increases to 80% and then almost to 95% (Mairy and Denis, 1971).

To determine the nucleotide composition of the synthesized, rather than stored, RNA the authors used, as the precursor, ^{32}P-phosphate, which is incorporated uniformly in all the nucleotides. In this case, after a 4-hour incubation, the radioactivity of the GC nucleotides varied at all the stages of oogenesis from 45 to 51%, with an average value of 48%. This corresponds to only 30% of rRNA and may mean that 28S and 18S rRNA's constitute a smaller but relatively stable portion of the high-molecular-weight RNA synthesized. This can be seen from the sedimentation curves of the RNA preparations (Mairy and Denis, 1971). As regards the optical density, i. e. the mass, almost all the RNA's are 28S and 18S (at this late stage the high-molecular-weight RNA constitutes 94%). But the radioactivity profile in the sucrose gradient forms hardly noticeable peaks in the 28S and 18S region, the labeled high-molecular-weight RNA being distributed over a wide range of sedimentation constants with a maximum near 37 to 40S. These large RNA's may be precursors of rRNA. In these experiments, the incubation in the presence of labeled phosphate continued for 4 h. Usually, the high-molecular-weight precursors of rRNA manage to go through processing within this time, i. e. to turn into mature forms, 28S and 18S rRNA. In the fly ovaries, for instance, the entire processing of rRNA from the initial precursor 38S to the mature forms 28S and 18S lasts about 40 min (Hall and Cummings, 1975). The sedimentograms should have then revealed a considerable part of the label under the optical peaks of 28S and 18S rRNA.

However, the data of other authors suggest preferential synthesis during vitellogenesis of the large rRNA's. Thus, Davidson *et al.* believe that rRNA synthesis accounts for 90 to 98% of the whole transcription at the lampbrush stage (Davidson *et al.*, 1964, 1966). In later studies in which labeled GTP was injected into medium-sized and large oocytes (stages 4 and 6) and incubated for 12 h, it was shown that the major part of the label was incorporated in 28S and 18S rRNA (La Marca *et al.*, 1973), with only a small part of it incorporated in the range of 37-40S. It is possible that rRNA processing in amphibian oocytes takes more time than in other species: in Denis' experiments only a small part of the label is found in mature rRNA (28S and 18S) after 4 h.

According to the data for *Xenopus*, early oocytes (100 µm in diameter) account for 5% of the total rRNA synthesis, and by the beginning of vitellogenesis they accumulate only 0.3% of the total egg rRNA. During vitellogenesis (oocyte diameter 400 µm) the rRNA synthesis increases to 40% of the total RNA synthesis (Van Gansen *et al.*, 1976). Scheer *et al.* (1976) carried out more accurate measurements on the newt with an allowance for the UTP pool and the rate of ^{3}H-uridine uptake by the oocyte. They showed that the synthesis of 28S and 18S rRNA, if taken as 100% in mid-vitellogenesis, constitutes only 0.01% during previtellogenesis, and again decreases to 13% in mature oocytes. These differences in rRNA synthesis rate correlate with the number of transcribed rRNA filaments visible in the electron micrographs of the nucleoli. Scheer *et al.* conclude that the differences in rRNA synthesis rate during oogenesis are indeed achieved by the regulation of transcription and even, to be more precise, by varying the number of initiation sites.

During the large growth, the accumulation of rRNA occurs rapidly owing to its stability. By the onset of vitellogenesis (grade 4) the oocytes accumulate only about 9% of the total rRNA of the future egg. By that time the oocytes already contain over 30% of the tRNA and about 50% of the 5S RNA of the egg. Hence, during the period of large growth the oocyte accumulates the remaining 90% of the rRNA, 70% of the tRNA, and 50% of the 5S RNA (Denis and Mairy, 1972; Denis, 1974).

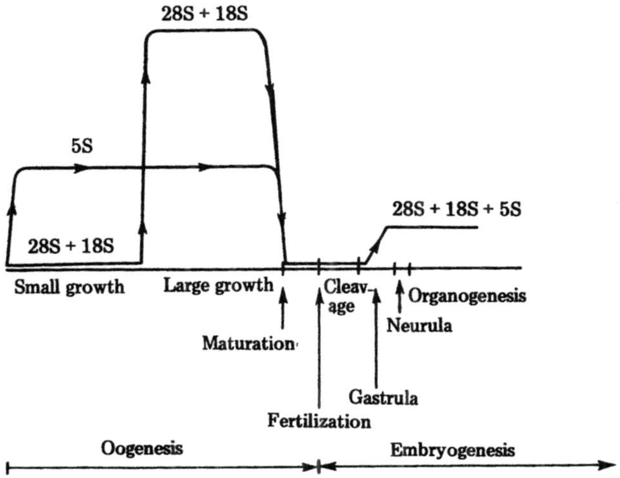

Fig. 9. Diagram of the change in synthesis intensity of ribosomal RNA's in *Xenopus* oogenesis and embryogenesis. The lack of coordination between the syntheses of 5S rRNA and 28S+18S rRNA in oogenesis and their coordination in embryogenesis are seen (Denis, 1974).

This means that while during small growth the 28S and 18S RNA accumulation rate was several times lower than that of the low-molecular-weight RNA, during large growth it became higher (Fig. 9). Grade 4 and 5 oocytes are especially active in the 18S and 28S RNA synthesis. As we know from the data of Callan (1966), MacGregor (1972) and Thomas (1972), it is at these stages that the additional nucleoli in the germinal vesicle take the form best adapted to active function. The rRNA synthesis rate does not slow down at stage 6 either (La Marca *et al.*, 1973). It may also go on for a long period (up to 10 days) in isolated oocytes incubated in a simple saline medium (Colman, 1974).

Some parameters of the synthesis of rRNA and of its passage from the nucleus into the cytoplasm during *Xenopus* oogenesis were determined by Scheer (1973). According to his calculations the transcription of the pre-rRNA chain (elongation) has a rate of 34 nucleotides per sec. This is slower than the elongation of rRNA in HeLa cells at 37°C (90 nucleotides per sec), but faster than in the cells of unicellular sea algae (12.5 nucleotides per sec) (see Scheer, 1973). At such a transcription rate and with 67×10^6 functioning RNA polymerase molecules (Miller and Beatty, 1969a, b) the quantity of rRNA synthesized in the oocyte nucleus in one second is sufficient for 3.13×10^5 ribosomes. In growing somatic cells of eukaryotes ribosomes are formed at a rate of 10 to 100 per second (Loening, 1970). These differences are in good agreement with the 2000-fold amplification found for rDNA in *Xenopus* oocytes.

As the oocyte grows, so does its nucleus, the nucleus surface area and the number of nuclear pores increasing accordingly. During large growth the number of nuclear pores increases from 10 to 37 million. This enables one to estimate the rate of flow of rRNA through the nuclear pores. Proceeding from the rRNA synthesis rate and the number of pores at different stages of oocyte development, Scheer (1973) could estimate the rRNA flow rate through the nuclear pores. It increases from 0.7 to 2.6 molecules per minute per pore from the beginning to the middle of the large-growth period, decreases by the end of oocyte growth, and drops to zero at the mature egg stage (Scheer, 1973).

At the beginning of large growth (grade 4 to 5 oocytes), at the time when the oocyte size increases only by a factor of 10 to 12, the quantity of rRNA is increasing at its former rate, which results in its highest concentration for the entire oogenesis period. The large rRNA's apparently do not form long-lived intermediate RNP complexes, but are immediately incorporated into the ribosomes formed at this time. Preferential rRNA synthesis has been found in the sea urchin (see Giudice, 1973) and the echiuroid worm *Urechis* oocytes (Davis and Wilt, 1972). Mermod and Crippa (1978) have also described the increase in the rRNA quantity in *Drosophila* oocytes.

Synthesis of Messenger RNA

The mRNA synthesis in amphibian oogenesis is correctly related to lampbrush structures since an active synthesis on thousands of DNA loops cannot be anything else (Izawa *et al.*, 1963a; Edstrom and Gall, 1963). Lampbrush loops gradually elongate and achieve maximum dimensions already at the beginning of vitellogenesis (stage 4 after Duryee and, probably, grade 4 after Denis). Autoradiographic data show that all these stages are attended by an intensive RNA synthesis (Gall and Callan, 1962).

As regards mRNA synthesis at earlier stages, we have Denis' data on the nucleotide composition of newly synthesized high-molecular-weight RNA. These data show that D-RNA's make up a considerable portion of the molecules transcribed. At later stages of oogenesis, judging from the nucleotide composition of the RNA synthesized, at least 60% of DNA-like RNA's are formed. In large oocytes, even at the end of oogenesis, when the intensity of all the syntheses falls drastically, the profile of the labeled RNA across the gradient, and its nucleotide composition, point to the synthesis of a considerable portion of the D-RNA. Thus, it can be concluded that a certain level of D-RNA synthesis is present at all the stages of amphibian oogenesis but it is the highest in mid-oogenesis, in the period of fullest development of the lampbrushes.

In sea urchin oogenesis, an mRNA synthesis has also been detected at different stages, but, surprisingly, it has been found to be considerable during the last months, or even weeks, of oogenesis. When ^{3}H-uridine is introduced into a sea urchin three months before spawning time, at least 10 to 20% of the labeled nonribosomal RNA is found in the mature eggs (Piatigorsky and Tyler, 1967). Further, if the label is introduced one week before spawning time, a considerable portion of it is found in the D-RNA (Gross *et al.*, 1965). This synthesis continues in the mature egg as well (Levner, 1974; Dworkin and Infante, 1978).

Mature sea urchin eggs undergo synthesis of short-lived mRNA (5-hour half-life) which makes up most of the mRNA translated on the ribosomes of unfertilized eggs. The synthesis rate of mRNA per haploid set of chromosomes in eggs is the same as at blastula. The bulk of these RNA's are histone mRNA's (Brandhorst, 1980).

A hybridization analysis of RNA composition in sea urchin oogenesis showed that the nucleus of early previtellogenic oocytes contains templates with a complexity of 1.6×10^8 nucleotides. In the cytoplasm of these oocytes the complexity of the mRNA is 10 times less, which corresponds to about 10^4 active genes. The average content of each RNA species is 600 copies per oocyte (Hough-Evans *et al.*, 1979). During the subsequent vitellogenesis, which lasts about three weeks, the number of mRNA species in the cytoplasm more than doubles (complexity 3.7×10^7 nucleotides), while the number of copies of each species increases to 1600 per oocyte (Hough-Evans *et al.*, 1977). The data on the mRNA composition are considered in more detail further on.

Brown and Littna (1966) detected D-RNA synthesis at the very end of *Xenopus* oogenesis during egg maturation. They studied the RNA synthesis after the females were administered ^{32}P-phosphate. During egg maturation, a considerable short-term activation of the synthesis takes place, producing from 1 to 6 ng of new RNA. In the sucrose gradient, labeled RNA does not form any noticeable peaks in the 28S and 18S region, although the percentage of G+C nucleotides in it is higher than in the total DNA. In an RNA with a sedimentation constant below 28S it is 44%, and in the region of heavier RNA's, 55%. It can be assumed that at least half of the RNA's synthesized during egg maturation are not rRNA's and have a DNA-like nucleotide composition. A similar RNA synthesis has been detected during the maturation of a mouse oocyte after it was removed from the follicle (Bloom and Mukherjee, 1972). RNA synthesis during the maturation of *Xenopus* eggs was confirmed recently by La Marca *et al.* (1973) (Fig. 10). They believe that RNA synthesis must take place at all the oogenesis stages, including the latest, so as to compensate for the natural decay of the previously accumulated RNA's. According to their data the half-life of rRNA in the oocyte is 9 to 30 days and, hence, the maintenance of a constant RNA level requires its continuous synthesis (La Marca *et al.*, 1973; Leonard and La Marca, 1975).

The quantity of D-RNA accumulated in the egg during oogenesis cannot be determined accurately enough from the nucleotide composi-

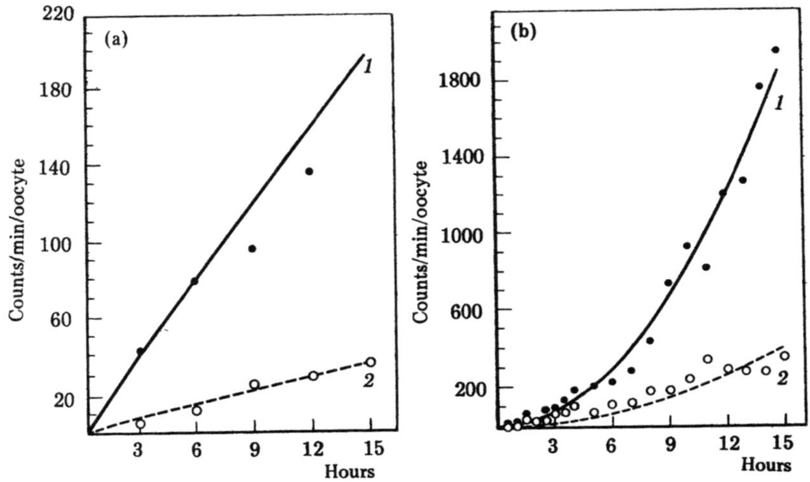

Fig. 10. Similarity of RNA synthesis rates in *Xenopus* oocytes at stages 4 (lampbrushes, diameter 500 to 600 μm) and 6 (growth is completed, 1000 to 1200 μm); *a*, radioactivity of the GTP pool in the oocyte; *b*, incorporation of ^{3}H-GTP into RNA; *1*, stage 4; *2*, stage 6. The ^{3}H-GTP pool in the small oocytes is much smaller, and so is the labeling of the RNA. Hence, the RNA synthesis rate at the two stages is approximately equal (La Marca *et al.*, 1973).

tion because of the large excess of rRNA. Therefore the mRNA content in *Xenopus* eggs was determined by the template activity. Davidson *et al.* (1966), in the stage 4 *Xenopus* oocyte, found that about 2%, or 42 ng, of the RNA's showed template activity. At stage 6 this value was slightly larger (47 ng). Later it was found by the same method that about 7% and 6% of the total RNA show template activity in the *Xenopus* oocyte and mature egg, respectively (Cape and Decroly, 1969). In sea urchin eggs, 4.5% of the RNA show template activity (Slater and Spiegelman, 1966). These figures, however, must be treated with caution, because the quantity of mRNA detected on the polysomes of sea urchin eggs after fertilization and the protein synthesis activation was found to be 10 times lower (Humphreys, 1971).

It is possible, of course, that not all the mRNA's are translated during the first hours of development.

The mRNA store in the egg has also been assessed using the RNA-DNA hybridization method. An important characteristic of mRNA is its diversity or, in other words, the number of genes on which it was synthesized during oogenesis. A crude estimate of the number of functioning genes is produced by counting the number of loops in the lampbrushes. The haploid set of the newt has about 5×10^3 loops (see Gall and Callan, 1962). During hybridization under saturation conditions it was found that in *Xenopus* the RNA stored is complementary to about 3% of the DNA. According to Davidson and coworkers' calculations this corresponds to 8×10^5 genes. This shows good agreement with the 5% of the DNA detected in the loops of lampbrushes.

Of more interest is, however, the estimate of mRNA obtained by hybridization with single-copy sequences, because the majority of all the structural genes probably have one or a few copies in the genome.

According to Davidson and Hough's data (1971), the oocyte RNA is complementary to 1.2% of single-copy DNA sequences (Fig. 11). A simple calculation shows that this quantity of DNA is equivalent to 2×10^7 nucleotide pairs, or to 4×10^4 genes, with an average size necessary for encoding beta-chains of the globin molecule. These figures, though closer to the number of lampbrush loops, also seem to be overstated.

In these experiments, a 6- to 10-fold excess of RNA over DNA was sufficient to ensure saturation in the hybridization of unique genes with oocyte RNA. This means that each type of mRNA is represented in the oocyte by a relatively large number of copies. Assuming that the oocyte contains no more than 1% of D-RNA and that nonrepetitive sequences constitute only 1/10 to 1/100 of them, it is easy to calculate that the oocyte contains a total of 7×10^{11} nucleotides of mRNA. If these RNA's are synthesized on the DNA, whose total complexity is 2×10^7 pairs of nucleotides (1.2% of the total single-copy DNA), then each type of mRNA transcripts from single-copy DNA in the oocyte should number as many as 10^4 copies. These data and Davidson's calculations can hardly be considered very reliable because they are based on a number of assumptions. There is also some uncertainty.

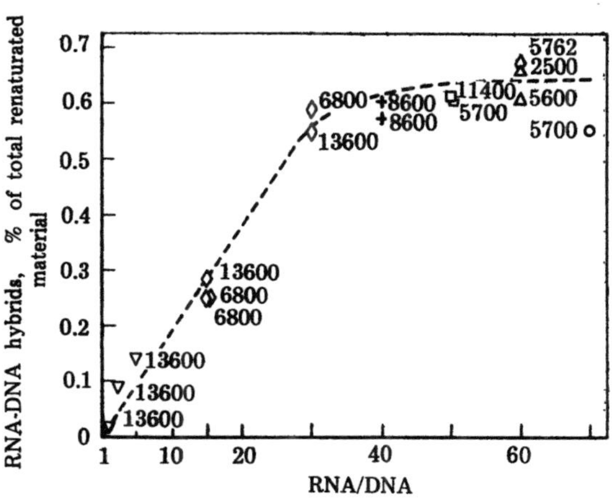

Fig. 11. Content of RNA transcribed from single-copy DNA sequences in a mature *Xenopus* oocyte. The single-copy fraction of ^{14}C-DNA was hybridized with the oocyte ^{3}H-RNA in different quantitative ratios at the C_0t (DNA) values indicated in the figure. The maximum hybridization level was about 0.62%. Thus, the fraction of single-copy DNA from which the RNA was transcribed in oogenesis is 1.2%. When expressed as the number of nucleotide pairs (2×10^7), it reflects the diversity of single-copy transcripts (Davidson and Hough, 1971).

If we assume that all these RNA's are synthesized during the lampbrush period and that it lasts 4 months, the synthesis of each of the RNA molecules must take about 10^3 sec ($10^7/10^4$ sec). This is definitely slower than is usually assumed. This suggests that either the number of copies is larger by an order of magnitude or that in actuality not all the RNA synthesized remains in the oocyte but its greater part decays soon after the synthesis, as is the case in somatic cells. Even if we consider the synthesis of long-lived mRNA's alone, it should be remembered that proteins are also synthesized in the oocyte, and part of the mRNA may degrade as a result of this process.

To estimate the fraction of the mRNA used, Davidson performed a competitive hybridization of the RNA synthesized at the lampbrush stage and the RNA present in the mature egg with DNA. These experiments, involving only repetitive sequences, showed that the competition level reached 65%, i.e. the oocyte RNA at stage 4 (lampbrushes) and 6 (mature oocyte) shared 65% of the sequences. Davidson concludes that in the course of oogenesis mRNA synthesized on 35% of the genes active at the lampbrush stage is utilized between the stages investigated. This conclusion seems to be correct if we assume that no new mRNA's are formed between stages 4 and 6. In addition, it should be remembered that some mRNA's that are utilized during oogenesis are synthesized before stage 4. Finally, the duration of ^{32}P labeling of the oocyte at the lampbrush stage is also important. If it is short (no more than tens of minutes), all the RNA's synthesized at this stage of oogenesis will be included in the total RNA. If, however, the labeling lasts for hours (as in Davidson's experiments), only the long-lived RNA's will be labeled perceptibly, whereas the short-lived ones may break down in the nucleus during this time or move into the cytoplasm, take part in translation, and decay after that.

An important feature of oocyte mRNA is the degree of its polyadenylation, which helps estimate the quantity of mRNA's as molecules bearing poly(A) fragments, although not all the mRNA's are polyadenylated.

In *Xenopus* oocytes the amount of poly(A) fragments bound with the RNA was determined by hybridization with poly(U). It has been shown that the quantity of poly(A)-containing RNA's reaches its maximum already at the beginning of vitellogenesis and then remains constant, although the total quantity of RNA in the oocyte continues to increase (Fig. 12). In a mature oocyte, the fraction of poly(A) RNA constitutes 0.7 to 1% of the total RNA (Rosbash and Ford, 1974). These figures are slightly less than those for *Xenopus* mRNA obtained by other methods, but they refer exclusively to poly(A)-containing mRNA's.

Attempts at estimating the quantity and metabolism of *Xenopus* mRNA by isolating poly(A)-containing RNA have been continued in recent years. Two classes of polyadenylated RNA containing, respectively, 40 to 80 and 15 to 30 adenylate residues each have been found in

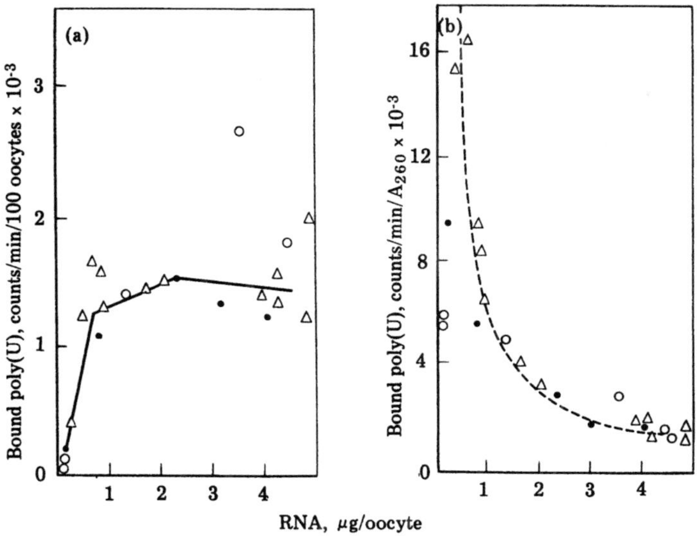

Fig. 12. Accumulation of poly(A)-containing RNA in *Xenopus* oogenesis. The number of poly(A) sequences in the RNA of oocytes of different size was determined by hybridization of labeled poly(U) and calculated per oocyte (*a*) or per RNA molecule (*b*). Accumulation of poly(A)-containing RNA occurs prior to vitellogenesis; after which this RNA is diluted due to the synthesis of RNA containing no poly(A) (the dashed curve in *b* is the theoretical dilution curve) (Rosbash and Ford, 1974).

the oocyte (Cabada *et al.*, 1977). Both classes encode the same set of proteins.

In a recent detailed investigation (Dolecki and Smith, 1979), previtellogenic oocytes with maximally developed lampbrushes (stage 3) and oocytes which have completed their growth (stage 6) revealed identical quantities of poly(A)$^+$ RNA equal to 86 ng, i. e. twice the previous estimates of the total mRNA content in the oocyte.

The synthesis rate of poly(A)$^+$ RNA at stage 6 is three times as high as that at stage 3. Most of the newly synthesized poly(A)$^+$ RNA is, however, unstable and degrades within a few hours (half-life of 5 h for stage 3 and 1.6 h for stage 6). Even the stable poly(A)$^+$ RNA is no longer accumulated in the oocyte at these stages but slowly degrades, apparently renewing all the time the previously synthesized poly(A)$^+$ RNA. Rough half-life estimates for long-lived poly(A)$^+$ RNA are 63 days for stage 3 and 38 days for stage 6 (Dolecki and Smith, 1979). The picture is further complicated by the fact that *Xenopus* has a large reserve of stage 2 oocytes and some of them resume, spontaneously and asynchronously, their further development.

During the maturation of large *Xenopus* oocytes induced by progesterone, the quantity of poly(A)'s first increases, mainly by elongation

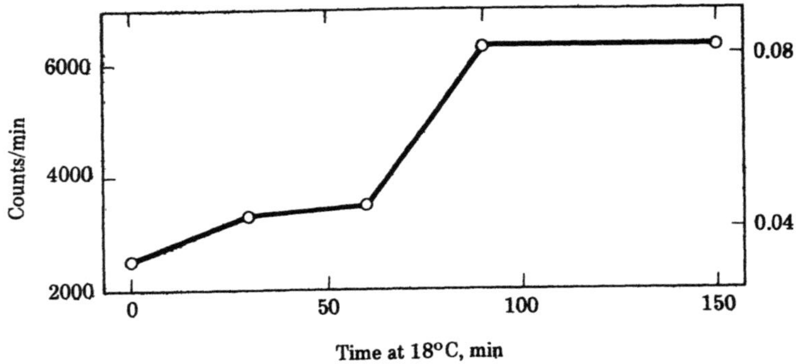

Fig. 13. Polyadenylation of RNA in sea urchin eggs after fertilization. RNA was isolated from unfertilized eggs and from *Lytechinus pictus* early embryos at various stages and then hybridized with ^{3}H-poly(U) to determine the quantity of poly(A). The quantity of poly(A)-containing RNA increases considerably prior to the first cleavage (Slater *et al.*, 1973).

of the existing poly(A)$^+$ mRNA's by 10 to 20 nucleotides. But after the rupture of the germinal vesicle the quantity of the poly(A) is almost halved within 10 to 15 h owing to the degradation of part of the poly (A)$^+$ mRNA's (Darnbrough and Ford, 1979). These changes are not controlled by the nucleus, since they occur in anucleate oocytes as well.

It has been shown for sea urchin eggs that a considerable polyadenylation of mRNA's which have been synthesized during oogenesis occurs in the egg cytoplasm after fertilization (Wilt, 1973; Slater *et al.*, 1973, 1974; Slater and Slater, 1974) (Fig. 13). The quantity of poly(A)-containing RNA's then increases 2- to 2.5-fold. This may mean that not all the stored mRNA's are adenylated during sea urchin oogenesis and, hence, their quantity, as determined by this method in the oocyte, is at least halved. The actual underestimation will be still greater if we consider that many histone mRNA's, which make up a considerable part of the stored mRNA's, contain no poly(A) fragments. Nor do some other mRNA's (Nemer *et al.*, 1975; Nemer, 1975).

In the oogenesis of *Urechis caupo* the number of poly(A) sequences increases by tens of times, but the ratio of the poly(A) to the total quantity of RNA decreases, i.e. the fraction of the poly(A)$^+$ RNA increases more slowly than does the rRNA, which makes up the bulk of the RNA (Davis and Davis, 1976, 1978). After fertilization the quantity of poly(A) increases five-fold within one hour. This means that it is nonadenylated mRNA that is mainly accumulated in oogenesis.

In mammals, a comparison of the quantity of poly(A)-containing RNA in rabbit eggs before and 10 h after fertilization (the first cleavage) showed identical values (0.25%) (Schultz, 1975), although a slightly larger portion of poly(A)-containing RNA's becomes bound with ribosomes within this time.

It was shown recently that in the mouse egg poly(A)$^+$ mRNA constitutes 5 to 8% of the total RNA, which greatly exceeds the corresponding values for other species (Bachvarova and DeLeon, 1980). Assuming that in the polysomes mRNA comprises 4% of the total RNA, these quantities of mRNA are sufficient for programming all the ribosomes.

In early development, the content of all the RNA species, including poly(A)$^+$ mRNA, decreases. This decrease may reach 30 to 40% on the first day of development (Bachvarova and DeLeon, 1980; Levey *et al.*, 1978). Consequently, no additional postfertilization polyadenylation occurs in mammals.

In the oocytes of the silk moth *Saturniidae*, the content of poly(A)-containing mRNA's is 0.2 to 0.3% of the total RNA (Paglia *et al.*, 1976a).

The Britten and Davidson's group investigated, in *Drosophila* oogenesis, the complexity of RNA synthesized mainly in the nurse cells and accumulated in the egg (Jacobs-Lorena *et al.*, 1980; Hough-Evans *et al.*, 1980). These experiments showed that the complexity of transcribed mRNA does not change substantially throughout oogenesis, i.e. it seems to be transcribed from the same single-copy genes. This complexity (1.2×10^7 nucleotides) is equivalent to the activity of about 6×10^3 genes, which is close to the total number of genes in *Drosophila* (see below). The average number of copies of each mRNA species in the mature egg was found to be 4.5×10^4 (actually, different mRNA species may widely differ in the number of copies in the egg). Similar data obtained on the house fly *Musca domestica* demonstrated that the RNA complexity in the egg is 2.4×10^7 nucleotides, i.e. about twice as many as in *Drosophila* (the size of the unique part of the genome in the fly is almost four times as large as in *Drosophila*). Somewhat different figures were, however, obtained by other investigators (Arthur *et al.*, 1979). Mature *Drosophila* oocytes revealed three classes of poly(A)$^+$ mRNA: about 10^2 species of sequences are represented by 2×10^7 copies each, about 3×10^3 species, by 8×10^5 copies, and 1.1×10^4 species, by 1.2×10^5 copies each. The total number of active genes and the average number of copies for each RNA species here proved to be, respectively, 2.5 and 6 times as many as those reported by Davidson, Britten *et al.*

Formation of Ribonucleoprotein Complexes

The formation of RNP complexes in oogenesis is of special interest because the ability of various RNA's to be stored for many months is usually associated with them. Various types of RNP have been found in the nuclei of somatic cells (Samarina *et al.*, 1965), in the cytoplasm of embryonic cells (Spirin, 1966), and in amphibian oocytes.

When investigating a homogenate of young oocytes or whole ovaries of *Xenopus* and *Tinca tinca* in a sucrose gradient, RNP particles were detected which sediment at 7S and 42S (Ford, 1971; Denis and Mairy, 1972; Mazabraud *et al.*, 1975) and contain specific proteins (Delaunay *et al.*, 1975). Throughout the first half of oogenesis (small-growth stages) these particles stored 5S rRNA's and tRNA's in complex with proteins (Fig. 14). The authors also mention particles with a constant 58S, which, however, does not figure any longer in their subsequent communications.

In the new papers of this series (Picard and Wegnez, 1979; Picard *et al.*, 1980; Denis *et al.*, 1980) these data were adjusted and supplem-

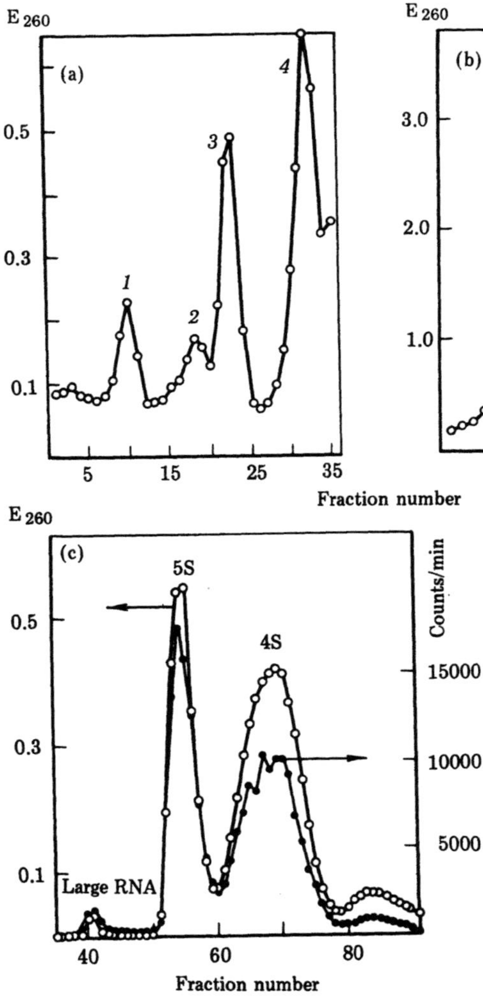

Fig. 14. Ribonucleoprotein particles in *Xenopus* oocytes. Oocytes of diameter 90 μm (*a*) and 375 μm (*b*) were homogenized, centrifuged at 10 000*g*, and the supernatant was fractionated in a sucrose density gradient (sedimentation coefficients of the fractions: *1*, 87S, *2*, 58S, *3*, 42S, and *4*, 4-5S). The 42S RNP particles (peak *3* in *a*) were isolated from 90 μm oocytes previously labeled for 8 h with ^{3}H-guanosine, RNA was obtained from them and fractionated on a Sephadex G-100 column (*c*). During oogenesis, the relative content of ribosomes (*1*) increases, while the 42S particles (*3*) disappear at the beginning of large growth. RNA is represented in these particles almost entirely by fast-synthesized 4S and 5S RNA's (*c*) (Denis and Mairy, 1972).

ented. The 7S and 42S particles of previtellogenic oocytes of amphibians and teleosts contain all the 5S rRNA's which are not present in the ribosomes, and almost all the tRNA. The 7S RNP particles of *Xenopus* contain one molecule of 5S RNA and one molecule of *b* protein with a molecular weight of 4×10^4 daltons (Picard and Wegnez, 1979). The 7S RNA particles of the fish *Tinca tinca* also contain one molecule of 5S RNA and one molecule of protein but with a lower molecular weight of 3.2×10^4 daltons (*c* protein) (Denis *et al.*, 1980). It is believed that in fishes, as in amphibians, 5S RNA molecules form, immediately after their synthesis, complexes with *b* protein which is later replaced by *c* protein specific to fishes.

In amphibians, as in fishes, the 42S particles have a molecular weight of about 1×10^6 daltons and consist of four subunits with a sedimentation constant of 15S. The gradual dissociation of 42S tetramer into 34S trimers, 25S dimers, and 15S monomers takes place upon increase in KCl concentration from 25m*M* to 125m*M* (Picard *et al.*, 1980; Denis *et al.*, 1980). Each 15S subunit contains one molecule of *b* protein (4×10^4 daltons) and two molecules of *a* protein (5×10^4 daltons); also, part of the 42S particles of fishes contain about 20 species of minor proteins (1.5×10^4 to 1×10^5 daltons) (Denis *et al.*, 1980). In amphibians (but not in fishes) 7S particles are a constituent of the 42S particles (Picard *et al.*, 1980).

At the beginning of vitellogenesis (large growth) the 42S and 7S particles disintegrate: 5S RNA becomes incorporated into the ribosomes, and the tRNA's are released. This process was apparently observed under an electron microscope (Thomas, 1969, 1970).

No reliable data on the existence, in oocytes, of RNP particles containing mRNA are available as yet. But in the nuclei of amphibian oocytes, RNP particles have been discovered which sediment at 50-55S (Rogers, 1968). Hess, followed by Davidson, believes that the lampbrush loops themselves may be a storage place (Hess, 1968; Davidson, 1969). These concepts are supported by the unusually large quantity of RNA and protein in the loops of lampbrush chromosomes. While in the chromatin of the normal liver cell the RNA : RNA : protein ratio is, according to Izawa, 1 : 0.06 : 2.8, for chromatin of the newt germinal vesicle this ratio is 1 : 9 : 500 (Izawa *et al.*, 1963b). It is interesting to note that in the oocyte the RNA and protein content per unit DNA increases more than 100-fold, but the RNA : protein ratio remains unchanged (1 : 47 in the liver and 1 : 50 in the oocytes). The RNA : DNA ratio is found to be still higher if only the lampbrush loops, and not the chromatin as a whole, are considered: they contain 780 times more RNA than DNA (Davidson, 1969). The chromatin of *Triturus* oocytes contains 0.007 μg RNA (Izawa *et al.*, 1963b) with a DNA-like nucleotide composition, while the entire egg contains 4 μg RNA, of which D-RNA constitutes no more than 1%. In other words, a considerable part (about 20%) of the D-RNA is contained in the chromatin loops of the germinal vesicle and is bound with the protein there. Sommerville described specific RNP particles in lampbrush chromosomes, which contain protein and about 3% D-RNA. It is probably in these particles that mRNA is stored in the vitellogenesis period (Sommerville, 1973). It has been shown that the oocyte germinal vesicle of amphibians contains 3×10^3 times more RNA than the liver cell nucleus (Finamore *et al.*, 1960).

At the end of stage 4 of oogenesis, when the chromosome loops pull in, the particles containing D-RNA probably remain in the germinal vesicle and emerge when it ruptures during maturation.

It has been shown that in silk moth oocytes, poly(A)-containing mRNA is hardly detectable in polysomes, i.e. it is not translated (Paglia *et al.*, 1976a, b) but exists in RNP complexes with a density of 1.42 to 1.48 g/ cm^3, which corresponds to the RNA/ protein ratio of 1:3 to 1 : 4. On electron micrographs these particles have a diameter of 180 to 250 Å.

RNP complexes are probably not only a form of storage but also a way of function control for a given type of RNA. At any rate, these complexes are stable enough to remain unchanged for many months, but later they may decay in a matter of hours or minutes, as is probably the case in the sea urchin during fertilization. As manifested by 42S particles, their existence in oogenesis itself may depend on other processes such as ribosome formation. It may be assumed that the decay of RNP particles is somewhat selective with respect to mRNA, i.e. some mRNA's are released at certain stages of early development, while others, at other stages.

The synthesis of RNA in oogenesis serves two purposes: ensuring the protein synthesis in oogenesis itself, and the building up of a protein-synthesizing system (PSS) to operate during embryogenesis. But a considerable part of the proteins synthesized in the oocyte (proteins of RNP complexes, ribosomal proteins, etc.) also serves, directly or indirectly, the purpose of building up the PSS. PSS is not built up uniformly in the course of oogenesis, i. e. by proportional synthesis of all the RNA types. In amphibians and fishes, where this process has been studied more thoroughly, a complex schedule of the synthesis of the various types of RNA has been established. Subordinated to RNA synthesis are complicated transformations of the nuclear apparatus, such as rDNA amplification, changes in nucleolar DNA, the lampbrush formation, which also proceed according to a definite time program.

At the early stages of oogenesis, when the rDNA's have already been amplified, but the nucleolar DNA's have not yet decondensed into circles and the lampbrush loops have not yet been formed, there is a preferential synthesis of low-molecular-weight 4S and 5S RNA's which cannot yet be used and are stored in the form of 42S particles. The fact that tens of identical RNA molecules are contained in a single particle indicates that 42S particles are not functioning units, but only serve for storage purposes. Although early oogenesis is definitely attended by the synthesis of rRNA and probably mRNA, the bulk (75%) of the total RNA consists of low-molecular-weight species. Within this previtellogenic stage of oogenesis, over 1/3 of the total egg tRNA and almost half of all the 5S rRNA are synthesized. At the next stage of oogenesis, during the large-growth period, the circular rDNA in the nucleoli assumes a decondensed structure associated with the maximum activity, and an intensive synthesis of rRNA begins on them. The rate of this synthesis exceeds many times over the rate of the synthesis on lampbrush loops. At this time the bulk of the mature oocyte rRNA is synthesized, as compared to less than 1/10 of its total amount produced before this stage. The synthesis of tRNA and 5S RNA also continues, apparently at a rate not lower than before. Within the remaining time, twice as much tRNA and as much 5S rRNA as before must accumulate.

The 5S rRNA, both newly synthesized and released from 42S particles, is incorporated into the newly formed ribosomes, which evidently appear just after the synthesis of 28S and 18S rRNA. The intensive synthesis of 5S RNA in oogenesis is ensured by a specific set of repetitive genes which differ from the genes functioning in the embryonic and somatic cells. A considerable amount of ribosomal proteins must also take part in the assembly of ribosomes. These proteins appear to be synthesized at the same time (Hallberg, 1974).

The lampbrush loops also achieve their maximum development by mid-oogenesis, and an intensive synthesis of mRNA takes place on them; this mRNA apparently forms the RNP, which remains attached to the lampbrush loops. Part of the mRNA is undoubtedly utilized in the course of oogenesis itself. According to calculations based on data on repetitive sequences, 1/3 of the templates formed at the lampbrush stage functions in oogenesis, and the other 2/3 remain in the egg and are used in the course of embryonic development.

As the growth of the oocyte is drawing to an end, the intensity of the RNA synthesis in it decreases, although it does not terminate completely. A large oocyte can remain in this state for a rather long time (in hibernating frogs, throughout the winter) until the pituitary hormones stimulate the beginning of maturation. In amphibians, the gonadotropic hormones act on the follicular cells, and the latter then synthesize a progesterone-like hormone which directly affects the egg. In a mature oocyte, progesterone primarily causes the activation of the genome and a temporary activation of the RNA synthesis. The nature of this RNA has not been determined conclusively, but it evidently contains D-RNA as well.

A small part of the total RNA produced during oogenesis (about 2%) is mitochondrial RNA. However, after the RNA synthesis has ceased in the oocyte nucleus, the synthesis of mitochondrial

RNA continues resulting in an appreciable contribution to mature oocyte mRNA. Therefore part of the RNA synthesized in the oocyte during maturation must be ascribed to the mitochondrial synthesis. In the mitochondria of the *Xenopus* oocyte 25 to 45 pg RNA is synthesized per hour, to which the rRNA fraction contributes 6 to 13 pg (Webb *et al.*, 1975).

Toward the end of maturation the short-term synthesis of RNA ceases, the germinal vesicle ruptures, and its contents mix with the cytoplasm. It is possible that the stored mRNA's find their way into the cytoplasm only at this stage of oogenesis. We do not know precisely whether they are distributed uniformly throughout the egg and whether identical quantities of each type of mRNA enter the future blastomeres. There are, however, data favoring a uniform distribution. In amphibians the globin mRNA of the rabbit, injected into an egg, is then translated uniformly in all the parts of the embryo, i.e. is distributed as uniformly throughout the egg (Woodland *et al.*, 1974; Gurdon, 1975). It has been shown for sea urchin eggs, too, that the composition of the mRNA in micromeres and in the macro- plus mesomere fraction is the same up to the 16-cell stage, i.e. the mRNA is distributed uniformly throughout the egg (Mizuno *et al.*, 1974; Rossi *et al.*, 1975).

Such a detailed investigation of oogenesis has been carried out mostly for amphibians, and it is hardly probable that the same pattern should exist in all the animal species. It may be different for those animals where the RNA synthesis occurs in the polyploid nurse cells. It is clearly different in mammals and in those species where the egg is very small and has hardly any RNA store. But the scattered data available for echinoderms, worms, and molluscs indicate that they also have periods of more or less intensive RNA synthesis in oogenesis resulting in the formation of the PSS in their eggs.

RNA IN EMBRYONIC DEVELOPMENT

Onset of Nuclear RNA Synthesis

RNA synthesis can be determined biochemically and autoradiographically. Autoradiography at the earliest stages of development has an important advantage in that it helps distinguish between nuclear and cytoplasmic synthesis. At the early stages, when the total RNA synthesis in the embryo is quite weak and the number of nuclei is also very low, even a weak nuclear synthesis of RNA would be expected to manifest itself on autoradiographs, because the incorporated radioactivity is concentrated in a few nuclei. Conversely, if it is found that the number of silver grains over the nuclei does not exceed the baseline level, this will surely indicate that the incorporation of the radioactive label into the embryo does not reflect the synthesis of high-molecular-weight RNA in the nuclei. In such cases, however, one may fail to detect a very weak synthesis or a synthesis of low-molecular-weight RNA's, which might have been washed out during standard histological treatment of preparations. The biochemical determination of a very weak RNA synthesis at the early stages of development requires the use of a number of techniques to prove convincingly the nuclear (or mitochondrial) origin of the RNA formed, the absence of bacterial contamination, etc.

The RNA synthesis rate is usually expressed in specific activity units, counts/ min, per unit quantity of the RNA. Since the synthesis of a new RNA, usually D-RNA, is investigated, and since in calculations it is referred to the quantity of ribosomal RNA previously stored in the egg, this value has no biochemical meaning. But as the quantity of RNA in the embryos of many animal species remains practically constant in the course of early development, this value is equivalent to the incorporation of a precursor into the RNA per embryo.

The embryonic RNA labeling, when referred to the quantity of DNA (preferably nuclear) or to the number of cells at each stage, gives an idea of the RNA synthesis rate per nucleus. It would have seemed that the latter value indicates more directly the changes in synthesis intensity during development. Indeed, if one is interested in the genomic activity, the value of the RNA synthesis per nucleus reflects this process most accurately. But if one wishes to know to what extent the nuclei provide the embryo with new templates, it would be more correct to relate the RNA labeling to the whole embryo or, what is the same, to the unit mass of the cytoplasm (per 1 mg protein or per 1 mg total RNA).

Brown and Littna (1964) administered labeled orthophosphate to a frog before the beginning of ovulation and investigated the RNA at the various stages of development. Their data on the RNA synthesis during cleavage are rather inconsistent but are considered to indicate that a moderate synthesis does occur at the cleavage stages. The figures presented by the authors do not show this "early" synthesis, and a definite RNA labeling was detected by them only beginning with stage 9 (late blastula) (8th hour of development at 22°C). In the subsequent hours the RNA labeling increases sharply up to early gastrula (stages 10-11), after which it falls off slowly. A similar increase in the RNA synthesis from the late blastula to gastrula stages was observed by other authors as well (Decroly *et al.*, 1964).

These stages are followed by a gradual increase of the RNA quantity in the amphibian embryo, which doubles only by the feeding tadpole stage (Bristow and Deuchar, 1964).

The RNA synthesis in *Xenopus* embryos was investigated autoradiographically and biochemically, and its activation was detected beginning with stages 8-8.5 (Bachvarova *et al.*, 1966). At stage 7 (early blastula) the RNA labeling could not be shown. But other authors did detect biochemically a low-level RNA labeling at this stage constituting about 1/10 of that of stage 9 (Woodland and Gurdon, 1969). At the same time, since the number of cells at stage 7 is much less than at stage 9, silver grains should have been detected over the nuclei. Since the RNA synthesis is not detected on autoradiographs in the nuclei of early blastula (stage 7), it is reasonable to assume that the nuclear synthesis of RNA up to stage 8-8.5 is insignificant. Similar data were obtained on axolotl embryos (Strelkov and Ignatyeva, 1975).

According to Bachvarova *et al.*, the activation of RNA synthesis does not occur simultaneously in the different parts of the embryo. It begins first of all (at stage 8.5) in the equatorial zone and in the primary endoderm cells, and later (at stage 9) in the primary ectoderm. The RNA synthesis is at its maximum at stage 10 in the primary prechordal mesoderm and in the endoderm, and at its minimum in the ectoderm (Bachvarova *et al.*, 1966).

A comparison of the intensity of the RNA synthesis in the dorsal and ventral parts of the gastrula ectoderm failed to reveal any differences between them as normalized to the quantity of DNA (Flickinger, 1972).

Thus, the above data show that in Anura amphibians a very weak nuclear synthesis may exist at the cleavage stages, but this cannot be considered proved. The beginning of the RNA synthesis should be related to stage 8.5, i.e. to the mid-late blastula. This stage coincides rather closely with the beginning of the morphogenetic function of nuclei, as determined by their inactivation by irradiation (Neyfakh, 1960, 1961).

In the loach embryo the RNA synthesis was investigated biochemically (Aitkhozhin *et al.*, 1964; Kafiani and Timofeeva, 1964, 1965; Timofeeva *et al.*, 1967; Kafiani *et al.*, 1969) and autoradiographically (Kostomarova and Korobtsova, 1970). Isolated blastoderms, incubated in a saline medium, were studied using labeled nucleosides as precursors. The nonspecific precursor $^{14}CO_2$ can freely penetrate the intact roes, but it cannot be used for autoradiography, and biochemical investigations using this precursor require more thorough removal of labeled high-molecular-weight impurities (proteins, DNA, polysaccharides) from the RNA.

Experiments have shown that a slight RNA synthesis is detected already at the very earliest stages of development (Fig. 15). Within the first 6 h (mid-blastula) it increases very insignificantly, although the number of nuclei reaches 1000 by this time. The per-cell activity of label incorporation into the RNA in the course of cleavage decreases, reaching a minimum at the 6-hour stage. This decrease is due to the fact that at these stages the small values of the labeling rate are divided by the rapidly increasing number of cells. These results suggest the absence of nuclear synthesis of RNA in the course of cleavages. Nevertheless, in biochemical investigations the early incorporation of precursors into the RNA is inhibited by actinomycin D while it is only slightly affected by ethidium bromide, which points to nuclear localization of the RNA synthesis (Timofeeva *et al.*, 1977). The activation ot the synthesis of

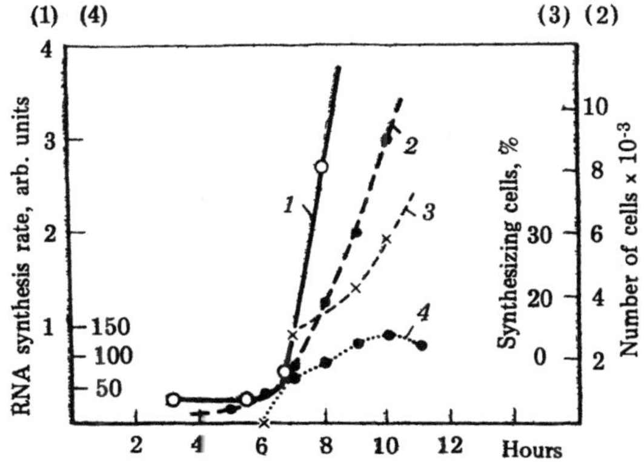

Fig. 15. Components of the total RNA synthesis during early development of the loach. Total synthesis rate of RNA (*1*) is determined by the number of cells (*2*), the fraction of synthesizing cells (*3*), and the synthesis intensity in each cell (*4*) (Neyfakh, 1971).

high-molecular-weight RNA's begins by the 6th hour. Autoradiographic studies have shown that during incubation of isolated blastoderms with ^3H-uridine the silver grains over the nuclei appear in a number exceeding the baseline level only by the 6th to 6.5th hour (Kostomarova and Korobtsova, 1970). These investigations did not, however, include any special precautions to prevent the loss of such low-molecular-weight RNA's as tRNA's and others, and their early synthesis may not have been taken into account when studied by autoradiography.

The stage at which nuclear synthesis in the loach is activated (6th hour of development) closely coincides with the switching on of the morphogenetic function of the nuclei, as determined by the developmental ability of "anucleate" embryos (Neyfakh, 1959, 1960).

As in amphibians, the RNA synthesis in the nuclei of the loach blastoderm does not begin simultaneously in all the cells. It first begins in the basal-layer cells and then slowly spreads out from the yolk to the animal pole of the blastoderm (Kostomarova and Rott, 1970). It might have been thought that preferential RNA synthesis in the basal cells is due to the fact that the label does not enter the isolated blastoderm uniformly from all sides but only from the side of the basal level. However, the cutting of the blastoderm in two showed the retention of the preferential synthesis in the basal part. This was confirmed for dissociated cells of the loach blastoderm. The activation of the RNA synthesis begins in them at the same time as in the intact blastoderm. It has been shown by autoradiography that RNA is not synthesized in all the dissociated cells and that the fraction of the synthesizing cells increases with development (Kostomarova and Korobtsova, 1970). The overall pattern of RNA synthesis changes is presented in Fig. 15 (Neyfakh, 1971).

Among the teleosts, the beginning of the RNA synthesis has been investigated in the embryos of the trout, a fish with more yolk than in the loach. The RNA synthesis, as detected biochemically and autoradiographically, starts at the blastula stage (17 τ_0), before the switching on of the morphogenetic function of the nuclei, as determined by their radiation inactivation (Dontsova *et al.*, 1970; Melnikova *et al.*, 1972). At this stage the main synthesis product is ribosomal RNA.

The RNA synthesis has probably been studied most thoroughly in the development of the sea urchin. Glišin and Glišin (1964) found that the first cleavages are accompanied only by the exchange in the CCA-ends of tRNA, while the true RNA synthesis begins only from the 32-blastomere stage. Similar data have been obtained in other early investigations (Comb *et al.*, 1965). However, Nemer (1963, 1967), and Slater and Spiegelman (1970) detected an RNA synthesis at the 4-blastomere stage, Kedes and Gross (1969) at the 2-blastomere stage, and, finally, Wilt (1964), Rinaldi and Monroy (1969), and others (Siekevitz *et al.*, 1966), before the first division. An early RNA synthesis is also revealed in isolated nuclei (Løvtrup-Rein, 1972).

Chamberlain (1970), however, discovered that an RNA synthesis also occurs in anucleate egg fragments. In its hybridization properties, early stage RNA proved to be complementary to mitochondrial, rather than nuclear, DNA (Craig, 1970). At later stages, when a nuclear RNA synthesis already definitely takes place, the mitochondrial RNA synthesis still constitutes a considerable part of the total synthesis (Hartman and Comb, 1969). Later, Wilt (1970) repeated his own earlier investigation using improved techniques and allowing for the specific radioactivity of the immediate precursor (GTP). He showed that the rate of its incorporation into RNA remains very low until the 8-blastomere stage and begins to increase from the 16-blastomere stage (Fig. 16). Our data show that a considerable enhancement of the RNA synthesis in the sea urchin embryo occurs at the beginning of the 5th division (32 blastomeres) (Timofeeva *et al.*, 1969). This is just a little earlier than the switching on of the morphogenetic function of the nuclei, as determined with the aid of actinomycin D or irradiation.

It can evidently be concluded that in the sea urchin a weak nuclear synthesis indeed begins very early, but reaches a considerable level (per whole embryo and per nucleus) beginning with the stage of 16 to 32 blastomeres. These stages may differ somewhat in different sea urchin species. Besides the nuclear synthesis there is a concomitant cytoplasmic (mitochondrial) RNA synthesis (Hartmann and Comb, 1969; Selvig *et al.*, 1970, 1972; Hartmann *et al.*, 1971; Craig and Piatigorsky, 1971; Chamberlain and Metz, 1972). Wilt and coworkers recently used a new technique for determining the rate of the nuclear RNA synthesis. Sea urchin blastulae were incubated in a high-concentration mixture of nucleosides labeled with heavy stable isotopes ^{13}C and ^{15}N. This made it possible to separate, in the density gradient, the newly synthesized heavier RNA from those present before. The determination of the precursor pool and of the exact quantity of the RNA synthesized enabled the authors to calculate the synthesis rate: 9.3×10^{-15} moles of nucleotides per embryo per minute (Grainger and Wilt, 1976).

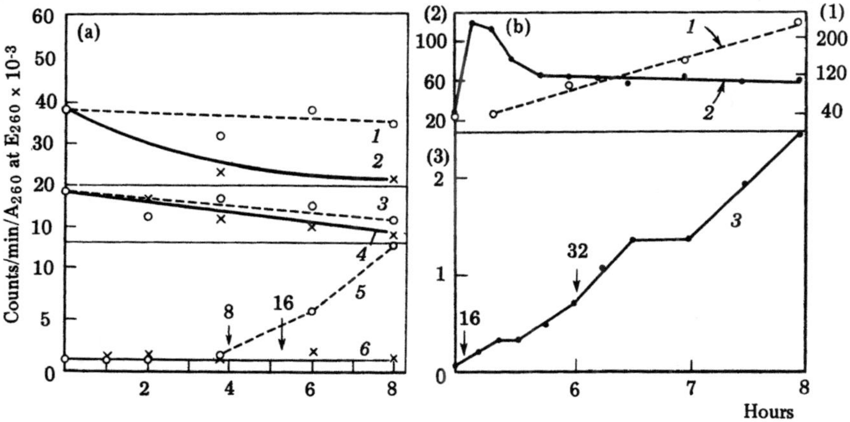

Fig. 16. RNA synthesis during early development of the sea urchin. The eggs (*a*) and the embryos 5 h after fertilization (*b*) of *Strongylocentrotus purpuratus* were incubated with ^{3}H-guanosine, and the radioactivity of the acid-soluble fraction, of GTP and RNA was determined in them at different time intervals. *a*, Radioactivity of GTP (*1, 2*), of the acid-soluble fraction (*3, 4*), of RNA (*5, 6*) of fertilized (*1, 3, 5*) and unfertilized (*2, 4, 6*) eggs; *b*, radioactivity of GTP (*1*), of the acid-soluble fraction (*2*), and of RNA (*3*) (Wilt, 1970).

In insects, the RNA synthesis apparently begins rather late, after the end of the division of nuclei in the central cytoplasm and their transfer into the upper layer of the cytoplasm, which then forms the blastoderm (Lockshin, 1966). This was recently shown reliably on *Drosophila* eggs (Zalokar, 1976). However, according to other recent results, a low-level RNA synthesis in *Drosophila* begins already in the dividing nuclei at the egg center and by the moment of their transfer to the egg surface and the blastoderm formation it increases several-fold per nucleus (Lamb and Laird, 1976; Anderson and Lengyel, 1979). During the subsequent embryonic development the RNA synthesis in the embryo increases 6-fold (from 2nd to 15th hour of development at 25°C), but grows only slightly in per-nucleus terms (Anderson and Lengyel, 1979). In bugs *Oncopeltus* (Harris and Forrest, 1967) the RNA synthesis is found to begin prior to gastrulation. Its intensity sharply increases, then falls as sharply.

Among the eggs with the mosaic type of development, those of the mollusc *Ilyanassa* have been studied most thoroughly. The RNA synthesis in this species apparently begins already during cleavage (Davidson *et al.*, 1965; Collier, 1975), although Collier detected it at a much later stage in his earlier work (1966).

In another mollusc, the pond snail *Limnaea stagnalis*, according to autoradiographic data (Van den Biggelaar, 1971) there is no RNA synthesis up to the 8-blastomere stage. At the 8- to 12-blastomere stage the synthesis begins in all the nuclei though with a different intensity. A little later, at the 12- to 16-blastomere stage, the synthesis increases by two or three times, particularly in the derivatives of those blastomeres where it has been weaker than in the others. Later (16- to 24-blastomere stage) the synthesis becomes somewhat more intensive and practically similar in all the nuclei. An early RNA synthesis in the pond snail *Limnaea stagnalis* has also been shown biochemically (Brahmachary *et al.*, 1971; Brahmachary, 1974). Its beginning coincides with the switching-on of the morphogenetic function of the nuclei (Neyfakh, 1976) or is slightly ahead of it. In other molluscs, the RNA synthesis is observed already at the earliest stages (Firtel and Monroy, 1970; McLean and Whiteley, 1974), but even when it is repressed by actinomycin D the early development proceeds normally.

In ascarid eggs, RNA is synthesized earliest of all. A reliable synthesis has been detected in them already in the pronuclei, prior to their fusion. It is to be noted that at these stages the synthesis of rRNA prevails being succeeded by the predominant D-RNA synthesis only at the subsequent stages (Kaulenas and Fairbairn, 1966, 1968; Kaulenas *et al.*, 1969). This synthesis (up to meiosis) should be more correctly related to the end of oogenesis, rather than to the beginning of embryogenesis; this, however, has only a formal significance. In other worms (*Urechis*) the RNA synthesis begins at later stages (Das *et al.*, 1965).

According to several authors, in ascidians no RNA synthesis is observed up to the hatching of the larva (Lambert, 1971) and their embryonic development is resistant to actinomycin D (Librera, 1964). This placed the ascidians in an exclusive position since it implied that the whole course of their embryonic development is predetermined genetically by RNA accumulated in oogenesis. Later, however, some sensitivity of ascidians to inhibitors of RNA synthesis was shown (Whittaker, 1977), and more sensitive methods revealed the presence of RNA synthesis (Puccia *et al.*, 1976; Meedel and Whittaker, 1978) already in the first hours of embryogenesis.

In mammals, the RNA synthesis also begins very early: it is detected even before the 2-blastomere stage (Woodland and Graham, 1969; Knowland and Graham, 1972; see the reviews by Graham, 1973; Pikó, 1975). Up to the 8-blastomere stage the quantity of RNA per nucleus increases at least ten-fold, and then several-fold again up to the blastocyst stage (30 cells) (Mintz, 1964a, b; Pikó, 1970, 1975). As regards the whole embryo (its mass at these stages does not increase) this means that the RNA synthesis rate increases by more than two orders of magnitude. But these investigators usually neglect the egg permeability to uridine and the activity of uridine kinase which is responsible for its phosphorylation to UTP; so these figures cannot be regarded as final (Daentl and Epstein, 1971). Nevertheless, even early RNA plays a substantial role in the earliest development, as seen from the fact that the usual doses of actinomycin D administered at the 2-blastomere stage arrest the development right away or at the 4-blastomere stage.

In summing up this section it can probably be said that the RNA synthesis in mosaic-type eggs begins much earlier than in regulation-type eggs (the exception being mammalian eggs), and in large eggs the RNA synthesis begins later than in small ones. In many cases the RNA synthesis in nuclei is slight at first, so its determination requires special efforts. At a definite stage it intensifies sharply, reaches a certain maximum, and then slowly decreases with respect to one nucleus while continuing to increase with respect to the entire embryo. Quantitative data on the RNA synthesis during early development are still insufficient, they concern only a few species and are often contradictory. Therefore it is too early for any generalizations to be made.

Pattern of Synthesis for Different RNA Types

Although there is a large arsenal of methods (fractionation in sucrose gradient or on Sephadex G-100, determination of the nucleotide composition after phosphorus labeling, etc.) to determine the RNA's synthesized (mRNA, rRNA, ortRNA), their utilization at the beginning of the development does not always yield reliable results. This is due to the low activity of the synthesis itself and to different, as compared with somatic cells, relationships between the synthesis level for various types of RNA.

To detect the synthesis of rRNA and tRNA, methionine with a labeled methyl group is used as precursor. It is known that tRNA and pre-rRNA contain methylated nucleotides, in contrast to most of the D-RNA's. Therefore incorporation of a methyl label into RNA is a good indicator of rRNA and tRNA synthesis.

An example of the difficulties involved in such investigations is the history of the studies into the pattern of the synthesis of different RNA types in the sea urchin. At one time it was believed to be a firmly established fact that the synthesis of heterogeneous RNA (D-RNA and mRNA) begins at very early stages, whereas the beginning of the rRNA synthesis is delayed until the gastrula and prism stages (Fig. 17) Nemer, 1963; Giudice and Mutolo, 1967, 1969; Sconzo *et al.*, 1970; Sconzo and Giudice, 1971; for more details see Giudice's review, 1973). But Emerson and Humphreys (1970, 1971) have shown that at the early stages, the synthesis of rRNA proceeding at a constant rate may remain unnoticed against the background of the very intensive and, with respect to one nucleus, gradually slowing down synthesis of D-RNA. The experimental data obtained by them seem to show that the rRNA synthesis begins at the very early stages, but can hardly be detected against the background of the intensive synthesis of D-RNA. An early rRNA synthesis has also been observed in early isolated nuclei (Løvtrup-Rein, 1972). At later stages the D-RNA synthesis (per nucleus) decreases, while the rRNA synthesis remains at the former level. The

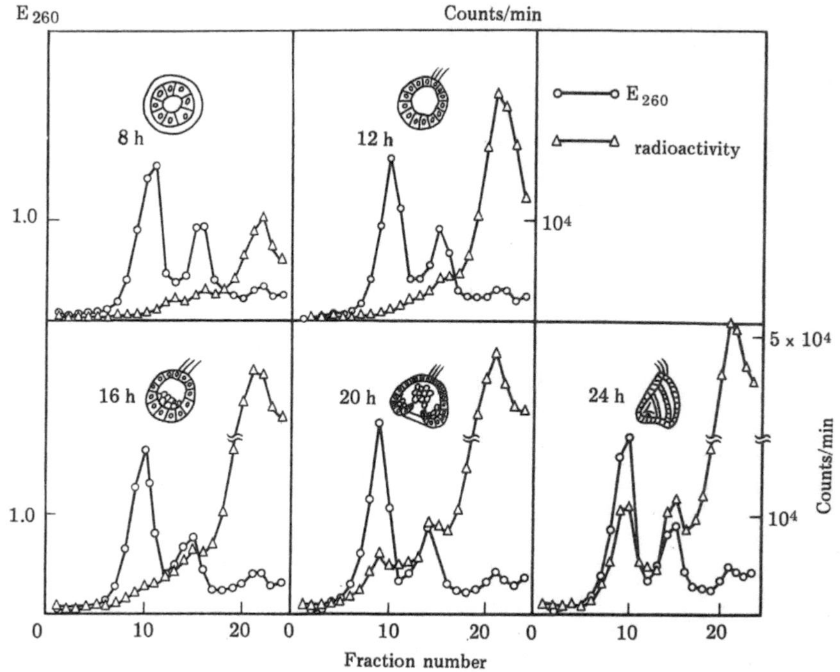

Fig. 17. Synthesis of rRNA in sea urchin embryos. The embryos were incubated with ^{32}P-phosphate, and their RNA was analyzed in a sucrose gradient. The peaks (left to right) are 26S, 18S, and low-molecular-weight RNA's. Each graph schematically represents the following stages of development: the early, swimming, and mesenchyme blastulae, mid-gastrula, and prism. A noticeable radioactivity appears in peak 26S after the 12-hour stage (swimming blastula) (Sconzo *et al.*, 1970).

increase in the number of nuclei in the course of development leads to an increase in the total rRNA synthesis per embryo, and most of the researchers believe this to be a late beginning of the rRNA synthesis at the gastrula stage.

But the studies did not stop at that point. Sconzo and Giudice (1971) determined the synthesis of rRNA's by their methylation on addition of ^{3}H-methionine at two stages: swimming blastula and gastrula. With regard to the number of cells (and nuclei) these stages differ only by a factor of two. At the same time, the radioactivity of the 26S and 18S peaks was found to be 14 times higher at the gastrula stage than at blastula stage. Therefore we can hardly agree with Giudice's calculations of the relative intensity of the rRNA synthesis based on the quantity of DNA (Sconzo *et al.*, 1970; Giudice, 1973). These calculations take into account the total DNA of the embryo, while at the early stages a considerable part of it is mitochondrial and does not reflect the number of nuclei in the embryo. Therefore, the question as to the beginning of rRNA synthesis in sea urchin embryos cannot be considered resolved, although it is clear that the synthesis of significant amounts of rRNA begins only at the stage of gastrulation.

A synthesis of tRNA and 5S RNA is detected in sea urchin embryos during cleavage (O'Melia and Villee, 1972). However, Glišin *et al.* (1972) and Nemer (1967) related the beginning of tRNA synthesis in the sea urchin to later stages.

Synthesis of three low-molecular-weight RNA species, one of which is localized in the cytoplasm and the other two in the nucleus, begins at the late stages of cleavage (after 64 blastomeres). This synthesis achieves a very high rate at the morula-blastula stages (1-2 molecules per gene per minute) and then lowers considerably (to 0.2 molecule per gene per minute) by gastrula (Nijhawan and Marzluff, 1979).

A classical object for investigating the pattern of the synthesis of different RNA types is the frog (usually *Xenopus*). It has been shown by Brown and Littna (1964, 1966) that the synthesis of D-RNA comes first, followed by that of tRNA at the late blastula-early gastrula stage and, much later (at the end of gastrulation), by that of rRNA (Figs. 18 and 19). The synthesis of pre-rRNA (in amphibians it sediments at 40S) and its processing in embryogenesis have been investigated on this object with the use of ^{3}CH-methionine. At stage 20 (late neurula) the half-life of pre-rRNA is 15 to 20 min (Landesman and Gross, 1969). The same method was used to show the absence of rRNA synthesis in *Xenopus* up to the gastrula stage and the existence of rRNA synthesis at postgastrulation stages (Landesman, 1972).

The pattern of the synthesis of *Xenopus* RNA's has also been investigated in detail by Knowland (1970), who showed that the tRNA synthesis begins at stage 9 (late blastula), i.e. later than the synthesis of D-RNA (activated at stage 8-8.5, though a very weak synthesis was detected already at stage 7). The synthesis of rRNA begins at the gas-

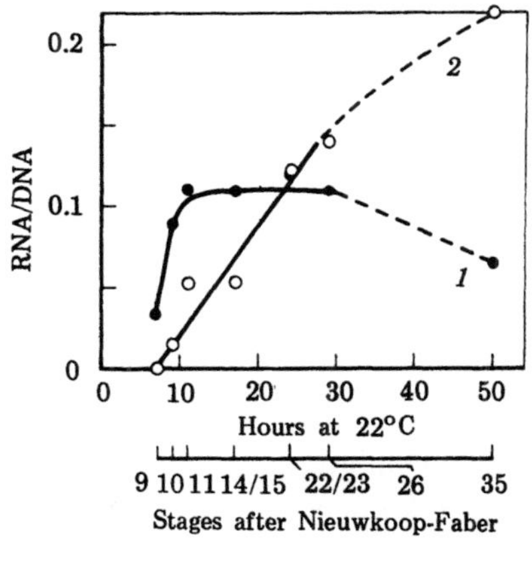

Fig. 18. Beginning of D-RNA and rRNA synthesis during early development of *Xenopus*. Relative synthesis of D-RNA (*1*) and rRNA (*2*) after incubation with ^{32}P-phosphate was calculated from the content of G+C bases (in %) and normalized to DNA quantity to estimate the synthesis intensity. The D-RNA synthesis is sharply activated from stage 9 (late blastula), while rRNA synthesis reaches measurable values during gastrulation (Brown and Littna, 1966).

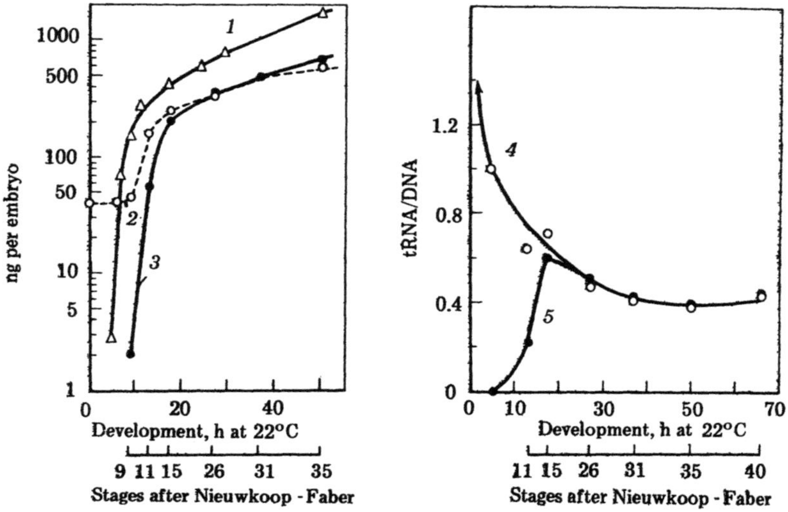

Fig. 19. Content and synthesis of tRNA in *Xenopus* embryos. Females were injected with ^{32}P-phosphate during ovulation, tRNA was isolated from the embryos at various stages, its total content determined (*2*), and the quantity of newly synthesized tRNA calculated from the radioactivity (*3*). *1*, Quantity of DNA; *4*, ratio of the total tRNA to DNA; *5*, ratio of the synthesized tRNA to DNA. The tRNA synthesis begins at early gastrula, and by early neurula the entire tRNA is represented by newly synthesized molecules; later on, the increase in the quantity of tRNA and DNA occurs simultaneously and a constant ratio is maintained between them (Brown and Littna, 1966).

trula stage (stage 12). The rRNA radioactivity at this stage constitutes only 8% of the total radioactivity. The share of the rRNA synthesis gradually increases reaching at the tail-bud stage the level of the rRNA synthesis in the tadpole.

Recently Shiokawa *et al.* (1979) detected a synthesis of poly(A)$^+$ RNA, tRNA, 5S RNA, and low-molecular-weight nuclear RNA's in isolated *Xenopus* cells dissociated at stage 8 (blastula). The poly(A)$^+$ RNA and tRNA's soon move into the cytoplasm, whereas the 5S RNA's remain in the nuclei until the beginning of synthesis and transport of the other rRNA's (28S and 18S) (for data on transport see below). The synthesis of the 40S pre-rRNA in the cells isolated at blastula (stage 8) does not begin until after a 5-hour cultivation.

The sequence of the RNA synthesis in teleosts proved to be different. It has been shown for the loach that D-RNA is synthesized at the early stages, while the rRNA synthesis begins at the mid-late gastrula (Aitkhozhin *et al.*, 1964; Belitsina *et al.*, 1964; Timofeeva and Kafiani, 1965) (Fig. 20). It would have seemed that the synthesis pattern in fishes should be similar to that established for amphibians. However, studies of RNA species synthesized in the trout blastoderm (Melnikova *et al.*,

1972) showed that here the RNA synthesis begins with the synthesis of rRNA (probably accompanied by D-RNA) (Fig. 21).

The pattern of the tRNA synthesis in teleosts also turned out to be unusual (Fig. 22). In the loach, the tRNA synthesis begins prior to the activation of the D-RNA synthesis, already at the early blastula stage (4.5 to 5 h of development at 21°C). At first, only the precursors of tRNA (4.5-5S) can be detected (Solovyova *et al.*, 1973b, 1976), because at

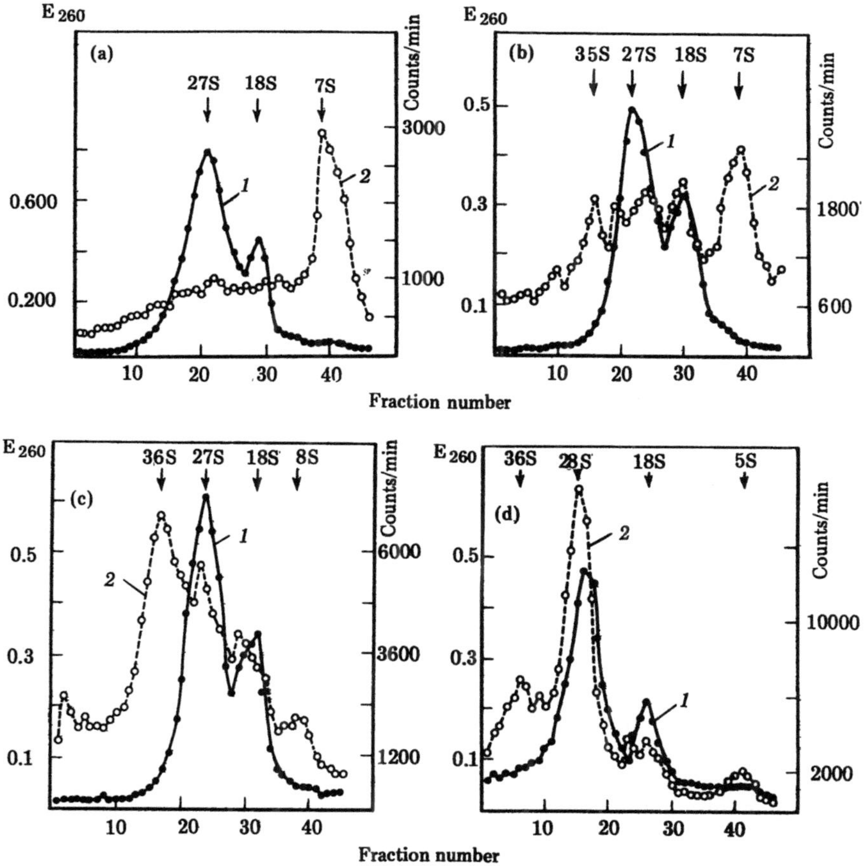

Fig. 20. Synthesis of various RNA's during early development of the loach. Loach embryos were incubated with $Na_2{}^{14}CO_3$ for 2 h at late blastula (8-10 h) (*a*), gastrula (11-13 h) (*b*), early organogenesis (18-20 h) (*c*), and late organogenesis (31-33 h) (*d*). Salt-insoluble fraction of RNA was isolated and analyzed by sucrose gradient centrifugation; *1*, quantity of RNA ($E_{260\,nm}$); *2*, RNA radioactivity. The sedimentation coefficients are shown at the top. Early gastrulation mainly shows the synthesis of heterogeneous RNA, at later stages, the synthesis of rRNA's and their precursors which predominate at the organogenesis stages (Rachkus *et al.*, 1971).

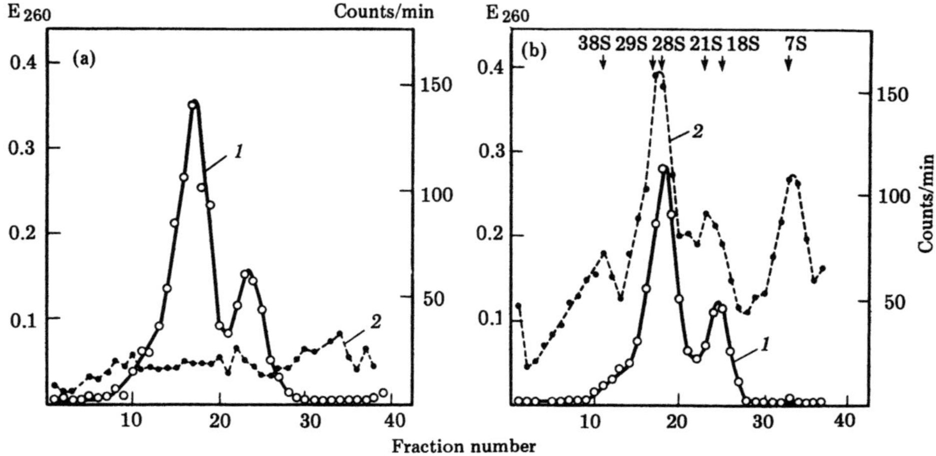

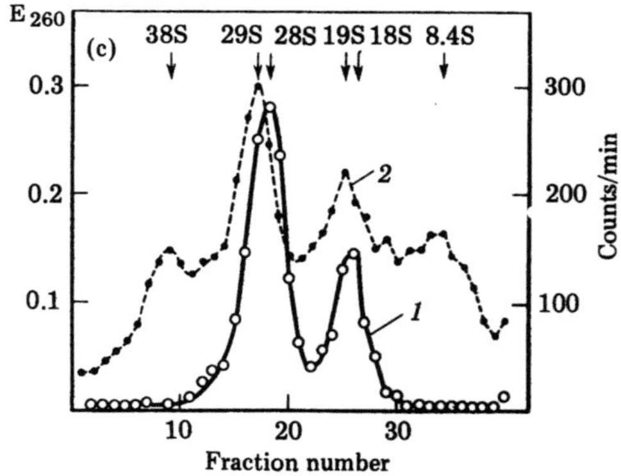

Fig. 21. Synthesis of high-molecular-weight RNA's during early development of the trout. Trout embryos were incubated with $Na_2{}^{14}CO_3$ for 20 h at 4°C at early (*a*) and late (*b*) blastula and at the beginning of gastrulation (*c*), then RNA was isolated and fractionated in a sucrose gradient; *1*, quantity of RNA ($E_{260\ nm}$); *2*, radioactivity. Synthesis of rRNA's and their precursors predominates already at blastula (Melnikova *et al.*, 1972).

these stages processing and the formation of "mature" tRNA's (4S) proceed slowly (Solovyova *et al.*, 1973b).

In birds, which would seem to have conditions for a large stockpile of ribosomes in the yolk, the rRNA synthesis begins very early, during cleavage (Lerner *et al.*, 1963; Raveh *et al.*, 1976).

In the bug *Oncopeltus*, the RNA synthesis begins at the blastoderm stage with all the types of RNA (D-RNA, rRNA, and tRNA) being synthesized simultaneously (Harris and Forrest, 1967).

In *Drosophila* embryos all the types of RNA are apparently also synthesized simultaneously. A detailed quantitative study of the absolute synthesis rates of various RNA species in *Drosophila* embryos showed that, calculated per nucleus per minute, it is 46×10^{-17} g for rRNA, 86×10^{-17} g for nuclear RNA, and 6×10^{-17} g for poly$(A)^+$ RNA transported into the cytoplasm (i.e. mRNA) (Anderson and Lengyel, 1979). Altogether, 160×10^{-9} g rRNA and 3.8×10^{-9} g mRNA is synthesized in the embryo within the first 15 h of embryogenesis, but because of the short life span of these RNA's (mRNA has a half-life of 3 to 6 h in the cytoplasm and about 20 min in the nucleus) the total quantity of RNA in the egg is 180 ng (Anderson and Lengyel, 1979).

In the bivalve mollusc *Mulinia* (Kidder, 1972), the D-RNA synthesis begins during cleavage, and that of tRNA, slightly later, at mid-late cleavage. The rRNA synthesis in *Mulinia*, as well as in amphibians, begins only during gastrulation. In embryos with mosaic-type development (molluscs, worms, and others) gastrulation occurs with a considerably smaller number of cells than in regulation-type embryos. In worms *Urechis* the synthesis of all the RNA types begins simultaneously but the relative synthesis rates are different. Thus, the rate of tRNA synthesis per nucleus remains constant throughout the development, i.e. it increases with respect to the entire embryo with increasing number of cells. On the other hand, the synthesis of rRNA in each nucleus rapidly increases until the gastrula stage and becomes 13 times as intensive as during cleavage (Schwartz, 1970).

In the ascarid, the rRNA synthesis (as well as that of D-RNA) begins as early as the pronucleus stage, i.e. it is apparently not interrupted at the end of oogenesis. This can be ascribed to the very small stockpile

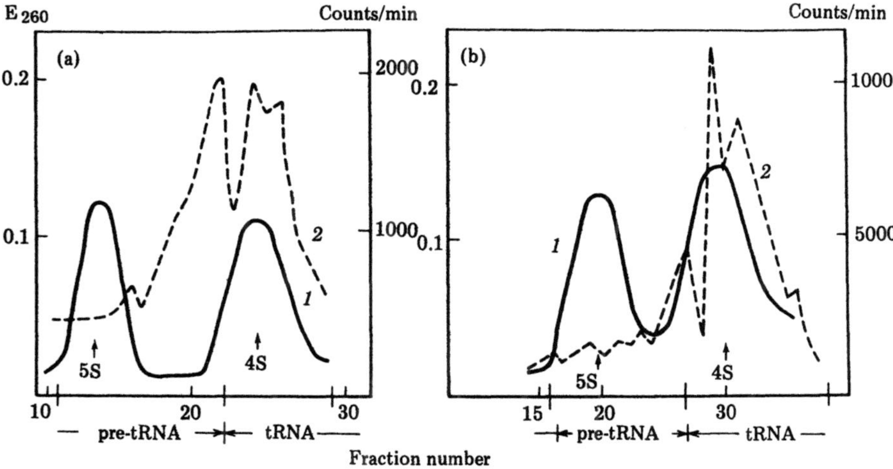

Fig. 22. Synthesis and processing of tRNA during early development of the loach. Loach blastoderms at the 5-hour (*a*) and 6.5-hour (*b*) stages were incubated for 1 h with ^{14}C-uridine, then salt-soluble (mainly low-molecular-weight) RNA's were isolated and fractionated on Sephadex G-100 column. *1*, Quantity of RNA ($E_{260\ nm}$); *2*, radioactivity. At the 5-hour stage the tRNA precursor is synthesized, its processing is retarded, and half of the labeled product is detected in the 4.5-4.3S region, whereas at the 6.5-hour stage the labeled RNA's are localized in the 4S region, i. e. the synthesis and processing are faster (Timofeeva and Solovyova, 1973).

of all the RNA types (ribosomes, in the first place) in the ascarid egg (0.06 pg per egg) (Kaulenas *et al.*, 1969).

In ascidians, some earlier works (Puccia *et al.*, 1976) have shown the predominance of rRNA synthesis already at gastrula. Later, however, it was established in a thorough investigation (Meedel and Whittaker, 1978) that already at early stages of development the bulk of the newly synthesized RNA is in the polysomes and is mostly polyadenylated, i. e. is actually mRNA. The rRNA synthesis was shown to dominate only at the postembryonic stages.

Finally, in mammals, where a slight RNA synthesis begins early (Mintz, 1964a, b; Ellem and Gwatkin, 1968), an rRNA synthesis is detected at the stage of 4 and, definitely, 8 blastomeres (Pikó, 1970, 1975; Knowland and Graham, 1972). The synthesis of all the other RNA types in mouse embryos occurs at the same stages (Ellem and Gwatkin, 1968).

In the rabbit, at the early stages (2 to 16 blastomeres) the synthesis of low-molecular-weight RNA's (tRNA's) predominates, increasing 12-fold within this time and again 13-fold by the 64- to 128-cell stage (Fig. 23). The synthesis of rRNA and of heterogeneous RNA is insignificant up to the 16-blastomere stage but by 64- to 128-cell stage it intensifies by hundreds of times (Manes, 1975).

In considering the data given in this section one can note that more often than not the synthesis begins with D-RNA. But it is probably different with fishes: in the loach it may be tRNA, in the trout, rRNA. The synthesis of tRNA begins simultaneously with that of D-RNA or a little later (in amphibians). In many species the rRNA synthesis apparently begins much later, but this must be proved conclusively in each case since the rRNA synthesis may not be noticed, being weaker as compared with the other types of RNA. In a number of instances, particularly in animals with mosaic-type development and an early synthesis onset, the rRNA synthesis starts from the very beginning of development.

The factors determining the pattern of the switching-on of different RNA's synthesis in different species are unknown. This pattern is probably determined by the ratio between the embryo requirements and the potentialities of the nuclear apparatus of the cells. The embryo requirements for RNA synthesis must be in accord with the state of the protein-synthesizing apparatus, which is built up in oogenesis and

Fig. 23. RNA synthesis during early development of the rabbit. Embryos at the stages of 2 blastomeres (*a*), 16 cells (*b*), and 64 to 128 cells (*c*) were incubated *in vitro* with ^{3}H-uridine, then RNA was isolated and subjected to electrophoresis. *1*, Optical density ($E_{260\,nm}$); *2*, radioactivity. The tRNA synthesis predominates at the early stages, the synthesis of rRNA and heterogeneous RNA is activated and increases hundreds of times by the 64- to 128-cell stage (Manes, 1975).

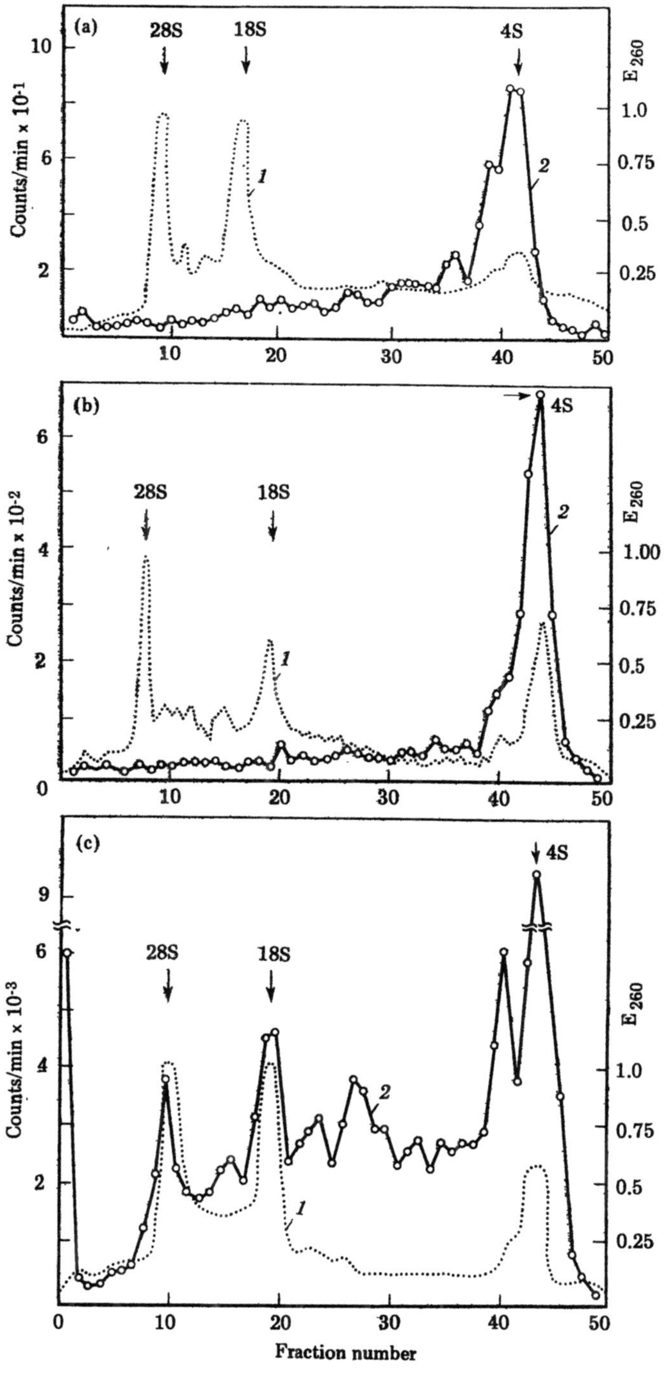
(a)
28S
18S
4S
Counts/min x 10-1
E 260
10
6
2
1.0
0.75
0.5
0.25
1
2
10
20
30
40
50
(b)
4S
28S
18S
Counts/min x 10-2
E 260
6
4
2
0
1.00
0.75
0.5
0.25
1
2
10
20
30
40
50
(c)
4S
28S
18S
Counts/min x 10-3
E 260
9
6
4
2
1.0
0.75
0.5
0.25
1
2
0
10
20
30
40
50
Fraction number

is inherited by the embryo from the maternal organism. Thus, amphibian embryos, which are homozygous for the deficiency of the nucleolar organizer (rDNA), do not synthesize rRNA, do not form new ribosomes yet reach the swimming tadpole stage (Brown and Gurdon, 1964). This means that their ribosome stockpile built up in oogenesis is sufficient and, naturally, that an intensive synthesis of new rRNA's begins late, at the end of gastrulation.

The tRNA : ribosome ratio in a mature amphibian and fish oocyte, is, on the contrary, such that it can ensure the utilization of only 1/10 of all the ribosomes in the protein synthesis. This is 10 to 15 times less than what is necessary for an intensive protein synthesis and the accumulation of the amount of tRNA that is normally present in embryos by the tail-bud stage. Therefore it seems natural to assume that the tRNA synthesis begins quite early in fishes (Solovyova *et al.*, 1973b) and almost immediately following the D-RNA synthesis in amphibians (Brown and Littna, 1966). It is possible, however, that the composition of the tRNA's synthesized plays the decisive role in the fact that few tRNA's are produced in oogenesis and their number increases at the beginning of differentiation. It is also possible that the protein synthesis in the different organ anlages requires different sets of tRNA's and their synthesis proceeds differently in different cells. Brown and Littna (1966) draw attention to the constancy of tRNA : DNA ratios discovered by them (see Fig. 19) and attach a regulatory significance to this quantity.

These considerations apply even more to the stockpiles and onset of mRNA synthesis. In oogenesis, those mRNA's are evidently accumulated for which no specific distribution between the anlages is required. The synthesis of new mRNA's in the embryo must probably be differential from the very outset, i. e. must be different in the different anlages. It is clear that the differential activation of genes in the embryonic cells can set in only when the fate of these cells has already been determined. The differential activation of genes, which is necessary for morphogenesis, is not, however, detected, at the earliest stages of development of sea urchin eggs, where transcription of histone mRNA, which is equally needed by all the types of cells, takes place immediately upon fertilization. This mRNA synthesis is, however, weak, and its activation occurs nonuniformly in different types of cell (see below) already from the 16-blastomere stage apparently ensuring subsequent morphogenesis.

At the early stages, the effect of the nuclear apparatus on the developmental processes is not strong, because the amount of the cytoplasm around each nucleus is very large, and the concentration of new RNA is small accordingly. A considerable part of the cleavage time is taken by mitosis processes, which are hardly compatible with transcription. But there are species, such as ascarids and mammals, in which even the first cleavage divisions last hours, rather than minutes, and the duration of the interphase after each mitosis is very long. In such a case, with small egg sizes an early activity of the nuclei and even the pronuclei is capable of building up significant amounts of synthesized RNA's.

Mitochondrial RNA's

The synthesis of mitochondrial RNA was detected and determined in the embryogenesis of the sea urchin, amphibians, and mammals. In sea urchin eggs the RNA synthesis begins at the earliest stages of development, and in the first biochemical investigations the principal problem was to distinguish between the nuclear and the mitochondrial synthesis. The mitochondrial synthesis of RNA was proved reliably by experiments on separation of sea urchin eggs into the nucleate and the anucleate halves (Baltus *et al.*, 1965; Chamberlain, 1970; Craig, 1970). A simultaneous nuclear and mitochondrial synthesis of RNA at the early stages of development was also shown by autoradiography (Selvig *et al.*, 1970, 1972). By hybridizing the newly synthesized RNA with DNA from nuclei and from mitochondria it was also shown that at the blastula and gastrula stages both genomes contribute to the transcription (Hartmann and Comb, 1969; Hartmann *et al.*, 1971).

It is, however, impossible to compare the RNA synthesis in nuclei and in mitochondria quantitatively, because the intranuclear and intramitochondrial precursor pool has not been determined. It is believed that in the sea urchin, a considerable share of transcription is effected in mitochondria during the first cleavage divisions, and in the nuclei, beginning with the 8- to 16-blastomere stage (Selvig *et al.*, 1970, 1972) (Fig. 24). The sedimentation pattern shows that the RNA's synthesized in the mitochondria of sea urchin embryos are heterogeneous but also contain the RNA's of mitochondrial ribosomes (Selvig *et al.*, 1972; Chamberlain and Metz, 1972).

In *Drosophila* embryos the mitochondrial synthesis of RNA is assumed to correspond to incorporation of 0.11×10^{-12} moles of UTP per embryo per hour and to constitute a considerable part of the total RNA synthesis during the first hours of development and less than 5% at the end of embryogenesis (Anderson and Lengyel, 1979).

The mitochondrial RNA's in amphibian oocytes include mainly the mitochondrial rRNA's and tRNA's differing from the corresponding cytoplasmic RNA's (Dawid, 1970, 1972a; Swanson and Dawid, 1970). The RNA synthesis in mitochondria during amphibian oogenesis constitutes only 2% of the RNA synthesis in nuclei, but at the end of egg maturation, when the nuclear RNA synthesis ceases, that of mitochondrial RNA continues at the previous rate of 21 to 45 pg/h per one oocyte, of which 6 to 13 pg/h are accounted for by rRNA (Webb *et al.*, 1975). It has been shown for dissociated cells of *Xenopus* embryos that the

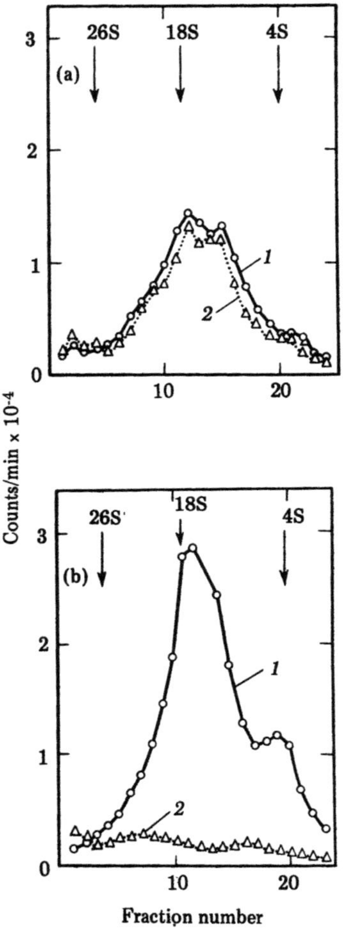

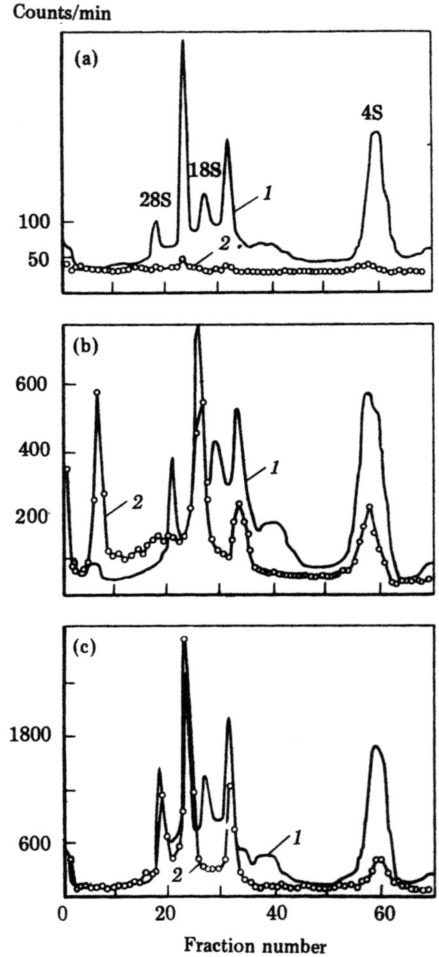

Fig. 24 (*left*). Synthesis of mitochondrial RNA in sea urchin embryos. Embryos of *Lytechinus pictus* at the 8-blastomere stage were incubated with ^{3}H-uridine in the presence of colchicine (10^{-3}M) (*a*) or ethidium bromide (5 μg/ml) (*b*), then RNA was isolated from mitochondria and fractionated in a sucrose gradient. *1*, Control; *2*, inhibitor effect. The RNA synthesis remains insensitive to colchicine (*a*), which arrests the cell division at the metaphase and represses transcription in the nuclei, but is itself repressed by ethidium bromide (*b*), the specific inhibitor of mitochondrial transcription (Selvig *et al.*, 1972).

Fig. 25 (*right*). Synthesis of mitochondrial RNA during development of *Xenopus*. Embryos were labeled with $^{14}CO_2$ at the 3- to 8-hour (cleavage) (*a*), 8- to 12-hour (blastula-gastrula) (*b*), and 25- to 27-hour (neurula) (*c*) stages, then RNA was isolated from mitochondria and analyzed by electrophoresis. *1*, Quantity of RNA ($E_{260\,nm}$); 2, radioactivity. The preparation contains mitochondrial rRNA's (large and small), 4S RNA, and an admixture of cytoplasmic rRNA's (28S and 18S). A considerable synthesis of mitochondrial rRNA takes place at the gastrula and neurula stages and is nearly zero during cleavage (Chase and Dawid, 1972).

synthesis of all the RNA types in mitochondria begins at the blastula stage and gradually increases (Young and Zimmerman, 1973). According to Chase and Dawid (1972) the activation of the synthesis of mitochondrial RNA occurs at the gastrula stage. The main contributor is then the rRNA of mitochondria (Fig. 25). The quantity of mitochondrial rRNA's doubles in the course of *Xenopus* embryogenesis (Chase and Dawid, 1972).

If it is confirmed that at the end of oogenesis the share of rRNA in the mitochondrial RNA synthesized does not exceed 25 to 30% (Webb *et al.*, 1975) while it predominates during embryogenesis, it will be possible to speak of the differential activity of mitochondrial genes.

During incubation of unfertilized and fertilized eggs of the trout for 12 h (at 10°C) with ^{3}H-uridine its incorporation is effected predominantly (60%) by the mitochondrial RNA's (16S, 12S, and 4S) and is repressed by ethidium bromide (Bourassa and Terner, 1979). After fertilization, incorporation of ^{3}H-uridine into 16S and 12S rRNA falls off and is completely repressed in mitochondrial 4S RNA.

In mammals, during the early development from the 2-blastomere stage to that of the blastocyst the mitochondria experience a number of morphological transformations (increase in size and then in the number of crysts). In the mouse the number of mitochondrial ribosomes increases considerably between the morula (3 days) and the blastocyst (4 days) stages (Pikó and Chase, 1973). This process is suppressed by inhibitors of replication and transcription of mitochondrial DNA, such as ethidium bromide. The synthesis of mitochondrial RNA, mainly rRNA's (16S and 12S) and tRNA's (4S), takes place at the same stages (Pikó and Chase, 1973; Pikó, 1975).

Other Types of RNA

Low-molecular-weight RNA's. Various types of low-molecular-weight RNA's consisting of 100 to 300 nucleotides and sedimenting at 4.5-8S have been discovered in the last few years. They differ from tRNA's (4S) and 5S rRNA's. This category includes 5.5-5.8S RNA's of HeLa cells, which attach to 28S in the nucleoli (Weinberg and Penman, 1968; Rubin, 1973). Similar RNA's, bound with 28S rRNA's and involved in the processing, are found in the nucleoli of other cells. It has been shown that these RNA's are encoded in the genome together with 28S and 18S rRNA's (Speirs and Birnstiel, 1974; Maden and Robertson, 1974) and their actual sizes correspond to 5.8S.

Besides, other strongly methylated low-molecular-weight RNA's, also localized in the nucleus, have been detected in the animal cells (Larsen *et al.*, 1969, 1970; Enger and Walters, 1970). These RNA's are also transcribed from repetitive sequences (Rachkus *et al.*, 1971; Marzluff *et al.*, 1975).

Low-molecular-weight RNA's have been found in unfertilized sea urchin eggs (Frederiksen and Hellung-Larsen, 1972). Sea urchin embryos contain five fractions of low-molecular-weight RNA's, which differ in size. Each of these fractions constitutes only a small part of the total quantity of embryonic RNA (Frederiksen and Hellung-Larsen, 1972, 1973, 1974). A recent thorough investigation of a sea urchin embryo revealed only three species of low-molecular-weight RNA (250 to 300 nucleotides), one of which is localized in the cytoplasm and the other two in the nuclei (Nijhawan and Marzluff, 1979). The nuclear low-molecular-weight RNA's are transcribed on genes which are repeated 50 to 60 times in the genome. The 5′-end of these RNA's has a cap of somewhat different structure than that of mRNA. In chick embryos, seven fractions of these RNA's are observed (Marzluff *et al.*, 1975) while loach eggs and embryos contain nine of them (Timofeeva *et al.*, 1977, 1981).

The synthesis of low-molecular-weight RNA's in the sea urchin is detected at the beginning of development, simultaneously with the synthesis of mRNA's and tRNA's. At the late stages, with the onset of the rRNA synthesis, the relative share of the synthesis of low-molecular-weight RNA's decreases (Frederiksen and Hellung-Larsen, 1973, 1974). The absolute synthesis rate of these RNA's is very high at the morula-blastula stages but lowers considerably by gastrula (Nijhawan and Marzluff, 1979). They are stable and their quantity in the embryo increases with the number of nuclei. In loach embryos, the synthesis of these RNA's has been found already during cleavage (4 to 5 h of development at 21°C) continuing into later stages (Solovyova and Timofeeva, 1973; Timofeeva *et al.*, 1977).

The function of these RNA's is being studied intensively. Because of their sizes they evidently cannot encode the protein synthesis since even histone mRNA's are larger (9S). It is assumed that some of these RNA's can participate in the processing of pre-mRNA. They have been detected as part of nuclear RNP particles. For one of the low-molecular-weight RNA species (U_1) a high complementarity at the 5′-end sequence (3 to 8 nucleotides) with the 5′-end of 43 introns of different genes has been established (Lerner *et al.*, 1980). On the other hand, the sequences between the 9th and 11th nucleotides (and also, to some extent, between the 12th and 20th nucleotides) proved to be complementary to the 3′-end sequences of the same introns. This enabled the authors to assume that these RNA's "recognize" the 3′- and 5′-ends of the introns and draw them together, thereby determining the specificity of splicing of the pre-mRNA's (see below). This is corroborated by detection of low-molecular-weight RNA's in the enzymes which accomplish the tRNA processing (Venkstern, 1980).

A special type of low-molecular-weight RNA's which regulate translation (tsRNA) has been detected in Haywood's laboratory (Bester *et al.*, 1975).

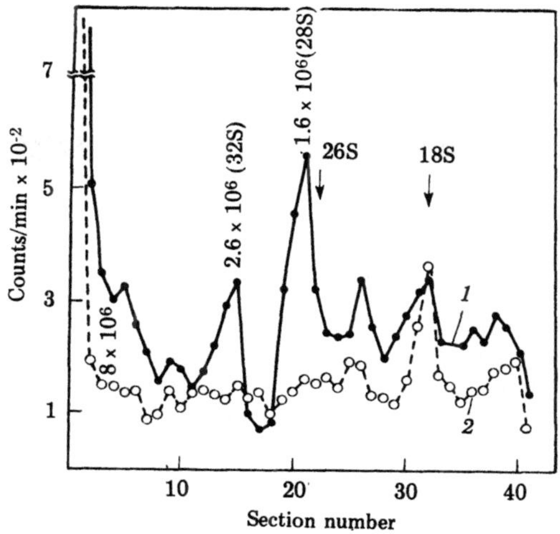

Fig. 26. Giant RNA's in the cytoplasm of sea urchin oocytes. The oocytes were incubated for 6 h with ^{3}H-uridine and divided microsurgically into nucleate (*1*) and anucleate (*2*) halves, then RNA was isolated and analyzed electrophoretically. The numerals over the peaks are the sedimentation coefficients and the molecular weights. An appreciable part of the labeled RNA in the cytoplasm has a molecular weight of 1.6×10^6 to 8×10^6, which is higher than that of 26S rRNA (Giudice *et al.*, 1974).

Giant RNA's in the cytoplasm. RNA's sedimenting with constants above 28S are found in nuclei and regarded as precursors of ribosomal and messenger RNA's (pre-rRNA's and pre-mRNA's). But in some cases RNA's of this size are also found in the cytoplasm. Thus, in the cytoplasm of the cells of *Chironomus* salivary glands 75S RNA has been detected, which is a polycistronic RNA transcribed on Balbiani rings (Daneholt, 1975). This RNA is transported from the nuclei without going through processing, although it is polyadenylated after the transcription (this RNA bears information for the main saliva proteins and is probably polycistronic) (Daneholt, 1975).

As will be mentioned below, during mitosis the newly synthesized RNA's, which have not yet undergone processing, are transported into the cytoplasm on disappearance of the nuclear membrane, and after the mitosis is completed return to the nuclei of the daughter cells (Neyfakh, 1970). Their size exceeds 40-60S (Abramova and Neyfakh, 1973).

Sea urchin embryos show giant cytoplasmic RNA's which sediment with constants above 28S (Giudice, 1973; Giudice *et al.*, 1972b, 1974; Sconzo *et al.*, 1974; Rinaldi *et al.*, 1974). These authors have obtained evidence for the fact that these RNA's are not an artefact, i.e. the result of partial disintegration of the nuclei (Fig. 26). Part of these RNA's contain poly(A) sequences and can be extracted from polysomal complexes. Nothing is known yet about the involvement of these RNA's in translation and which particular types of protein they encode.

SPECIES OF mRNA SYNTHESIZED IN THE EMBRYO

Methods of Identification of mRNA Species

Determination of the stages at which the synthesis of RNA in embryonic development begins, the intensity of this synthesis and the pattern of switching-on of synthesis for different types of RNA is just the first step in the investigation. The next objective in the study of embryonic transcription is to analyze the types of RNA synthesized and thus determine the set of active genes functioning at different stages of development. But in order to obtain a complete picture of the genetic control of development, one must know not only which mRNA's, when, and in what quantity are synthesized, but also how they are utilized at the subsequent stages of development, i. e. how long they are preserved and when their translation starts.

The bulk of the embryonic RNA's, as determined by the optical density, consists of rRNA's and tRNA's, with almost no mRNA's being detected. It is no wonder that the sedimentation analysis of RNA's from sea urchin eggs does not reveal any differences between the optical density profiles of RNA preparations from an unfertilized egg and from a late gastrula (Slater and Spiegelman, 1966). A sedimentation analysis of the radioactivity distribution in the RNA synthesized already indicates some differences in the nature of the synthesis at the early stages. As the sea urchin develops from the 4-blastomere to the blastula stage, the heterogeneity of the newly synthesized RNA increases, which evidently points to an increase in the number of RNA species, i. e. the number of active genes (Slater and Spiegelman, 1970). At the blastula stage, heavy RNA with a molecular weight of 1.5 to 2×10^6 appears and persists throughout later stages. This may possibly be a precursor of rRNA. At the gastrula stage, the RNA heterogeneity slightly decreases, and 28S and 18S rRNA's begin to make up a considerable part of the label.

This and other similar works illustrate the not-too-unexpected fact that the number of functioning genes increases with development. The average molecular weight of the RNA synthesized increases in the course of embryonic development. This is probably due to the fact that at the earliest stages of development most of the new templates determine the synthesis of histones — nuclear proteins with a low molecular weight and low sedimentation constants. The transcription of other, larger-sized, mRNA's begins later.

The most straightforward method for identifying the types of mRNA's is to determine those proteins which are encoded by these mRNA's. The appearance of new proteins in itself undoubtedly points to the presence of the corresponding mRNA's in the cell, but does not indicate whether these mRNA's have been stored in oogenesis or transcribed *de novo* on the embryo genes. A number of techniques, such as hybridization, inactivation of nuclei, etc., answer this question. The best results have been obtained for histone mRNA's transcribed and translated at the early stages of development of the sea urchin and other objects. Since protein synthesis always serves as a criterion of the mRNA species in these experiments, all these data are given below, in the sections dedicated to the synthesis and genetic control of proteins.

The data on the RNA species synthesized in the course of embryonic development have been obtained mostly by different versions of the molecular hybridization method.

The share of repetitive sequences in the genome can be established by the kinetics of DNA renaturation. In most of the species this share varies from 40 to 70%. The old methods of RNA-DNA hybridization produce information almost exclusively on the transcription of this fraction of the genome. Obviously, the transcription of single-copy genes cannot be evaluated in this way. Methods for isolating unique fractions of the genome, based on the differences in the renaturation rate of single-copy and repetitive sequences, have been designed and introduced in recent years (Davidson and Hough, 1971).

It has already been mentioned that these methods were successfully used to determine the number of active unique genes which act in oogenesis at the lampbrush stage. The share of RNA synthesized on these genes corresponds to about 1.2% of the unique fraction of the DNA but it must be precisely this part of the RNA that determines the synthesis of most proteins at the early stages of development. The importance of transcription of single-copy genes in the development of mammals is vividly illustrated by the works of Church *et al.* (Church and Brown, 1972; Brown and Church, 1972; Church and Schultz, 1974). The share of repetitive sequences transcribed at all the stages of mouse development (from the 8- to 16-blastomere stage to the adult animal) actually proved to be a constant value constituting 4 to 5% of the whole genome. At the same time the share of transcribed single-copy genes at these stages varies considerably in the course of development. At the early stages (from the 8- to 16-blastomere to the blastocyst stage) it is very small and does not exceed 1% of the total DNA. But then, during organogenesis, by the 7th to 14th day of embryonic development it sharply increases and reaches, at birth, 10% of the total DNA and 16% of the total number of single-copy sequences. The percentage curves for transcribed unique sequences differ for different organs. For the liver — a morphologically simple organ — the share of active unique sequences is small (1.2 to 1.7%) and changes but little from the moment of formation in the embryo to the adult organ. In the brain cells, however, this share in a 14-day embryo is 3%, reaching 5% at birth and 10% in the subsequent development.

These data show that the high relative share of active repetitive sequences as compared with the unique sequences is specific only to early embryonic development. It is hardly probable that the repetitive genes could be responsible for specific synthesis, even in the early embryos. Nevertheless, a substantial part of the data now available refer almost exclusively to repetitive sequences, which considerably restricts the possibilities of interpretation. But still, they shed some light on the genomic function at the early stages of development.

By analyzing the dependence of the hybridization percentage on the time of annealing of RNA with DNA, on the DNA : RNA ratio, and on the presence of unlabeled RNA's obtained from embryos at different stages, one can evaluate accurately enough several parameters, pertaining, however, only to repetitive sequences. The first of them is the share of active genes at the stage investigated. It corresponds to the quantity of DNA (in %) hybridized under conditions of saturation with RNA. A second parameter is the degree of RNA hybridizability, which is determined at considerable DNA excesses. This parameter indicates the degree of diversity of the RNA synthesized — in the first place the presence of heterogeneous nonribosomal RNA. It is obvious that large quantities of rRNA synthesized on the minor DNA fraction which constitutes only 0.02% of the genome will hybridize with it in negligible quantities. An increase in percentage of hybridizable RNA means that it contains molecules synthesized on a large number of different regions of DNA (genes). A third parameter is the homology of the two types of RNA, i. e. of the nucleotide sequences of DNA, from which they have been transcribed. This parameter can be estimated by the degree of competition, which manifests itself in the reduction of the percentage of hybridization of labeled RNA of the first species in the presence of increasing quantities of unlabeled RNA of the second species. The higher the competition, i. e. the more the unlabeled RNA blocks the hybridization of the labeled one, the greater is their homology, or the more sequences in the labeled and the unlabeled preparation are identical to each other. Comparing in this way the RNA preparations obtained at different stages of development (one of them labeled), one can see to what extent the same genes function at both stages (Fig. 27).

Additional information can be obtained by comparing the slopes of RNA competition hybridization curves and the RNA concentrations at which these curves reach the plateau. They indicate the relative number of identical RNA copies present in the two preparations compared. This enables one to determine, to some extent, whether the degree of competition is associated with the fact that a small number of identical genes, on which, however, an appreciable number of RNA copies are transcribed, are active at both stages, or with the fact that

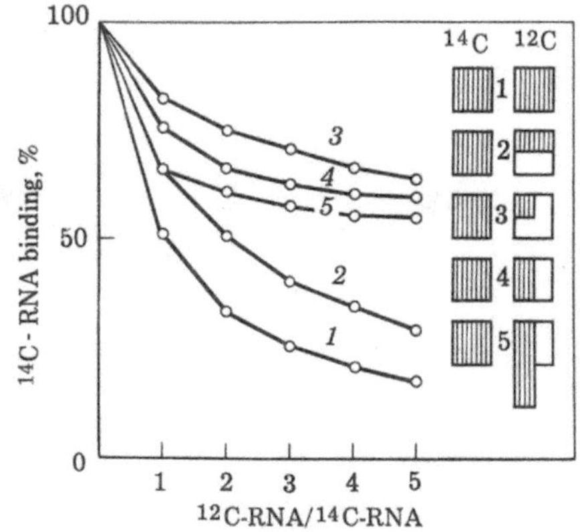

Fig. 27. Modeling of competitive hybridization with DNA. Mathematical modeling of the competition of labeled (^{14}C) and unlabeled (^{12}C) RNA's in relation to their quantitative ratios and to the content of identical sequences. *1*, Unlabeled RNA is identical to labeled RNA; *2*, unlabeled RNA contains the same fractions as labeled RNA, but in halved concentration; *3*, *4*, and *5*, unlabeled RNA contains only part of fractions contained by labeled RNA in lower (*3*), equal (*4*), or higher (*5*) concentrations. The curves differ in slope and in hybridization level at which they reach a plateau (Rachkus *et al.*, 1969a).

many identical genes, on each of which a small number of copies are synthesized, function at these stages.

Labeled RNA for molecular hybridization experiments can be obtained *in vivo* by short-term (tens of minutes) or long-term (hours, days) incubation of the embryo with a precursor, or, finally, *in vitro* by different chemical and enzymatic methods.

In the first instance, the radioactivity will reflect the composition of the RNA synthesized, in which short-lived fractions make up a considerable contribution. In the second instance, after a long-term label incorporation the absolute quantity of labeled short-lived RNA remains the same, while its relative share will drop sharply due to the accumulation of long-lived fractions during the prolonged incubation. Finally, with *in vitro* labeling of the isolated RNA, the hybridization will reflect the composition of the RNA synthesized at all the previous stages of development and preserved until the moment of isolation rather than the RNA synthesis level. It is obvious that with the second and third methods of labeling the RNA preparations must, to a large extent, include the long-lived mRNA's which have gone through the processing and function in the cytoplasm.

The presently available data have mainly been obtained on *Xenopus*, sea urchin, and loach. The data obtained agree in many respects, but differ in others.

Amphibians

First of all, we shall consider the situation arising in amphibian embryos at the blastula-gastrula stage with the onset of new RNA transcription, while the maternal mRNA's stored in oogenesis are still contained in embryonic cells (Fig. 28). Crippa *et al.* (1967)

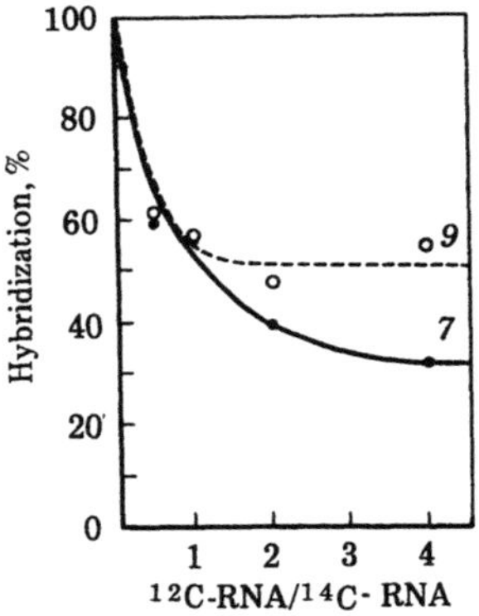

Fig. 28. Utilization, during early *Xenopus* development, of RNA transcribed at the lampbrush stage. At the beginning of large growth oocytes were incubated with ^{32}P-phosphate and the isolated ^{32}P-RNA was hybridized with DNA in the presence of various quantities of unlabeled RNA obtained from stage 7 (early blastula) and from stage 9 (late blastula). Like egg RNA, the stage-7 RNA contains 70% of the same sequences as the oocyte RNA at the lampbrush stage. At stage 9 this quantity decreases appreciably (Crippa *et al.*, 1967).

and Crippa and Gross (1969) carried out competitive hybridization of RNA, isolated at the blastula-gastrula stage, with labeled RNA of the egg obtained by methylation *in vitro* using ^{14}C-dimenthylsulfate. They showed that the egg stores RNA transcribed on genes which constitute 4.1% of the genome. This exceeds the values obtained by Davidson *et al.* (1966) at the lampbrush stage (2.7%), which may be attributed to an additional RNA synthesis in late oogenesis and during maturation. The above values refer to repetitive sequences, whereas the adult oocyte contains RNA transcribed on only 1.2% of single-copy DNA sequences (Davidson and Hough, 1971).

The RNA obtained at stages 7 to 10 (early blastula-early gastrula) competes with the egg RNA, thus showing that maternal species of RNA still persist at these stages. However, the number of these species increases gradually during development (Crippa and Gross, 1969). By early-mid-blastula (stage 7) the embryo retains all the RNA species stored in the oocyte, with 3/5 of them localized in the polysomes and apparently undergoing translation. By late blastula (stage 9) the embryo retains only 75% of the egg RNA species, with 2/3 of them detected in polysomes. At early gastrula (stage 10) only 50% of the RNA species stored in the egg remain. At this stage, all the maternal RNA's are already found in the polysomes. More recent data obtained on the South-American frog *Engystomops* showed that some RNA's transcribed in oogenesis persist until the tadpole stage (Hough *et al.*, 1973).

The mixture of the RNA obtained from polysomes at stages 7 and 9 competes with the egg RNA better than each of them does separately. This points to some differences in the species of RNA translated at these stages. It has been shown that the new embryonic mRNA, synthesized at stage 9, is also contained in polysomes and is translated together with the maternal RNA (Crippa and Gross, 1969). The polysomes contain the stored maternal RNA, transcribed on 2.8% of the maternal genome, and also the newly formed RNA synthesized on other DNA sequences which constitute 2.2% of the embryonic genome.

The same stages of development were investigated by Davidson *et al.* (1968). They injected ^{32}P-phosphate to a frog prior to ovulation, thus obtaining RNA labeled at the stages from ovulation to late blastula. The predominant part of the labeled RNA consisted of RNA's synthesized during an intensive increase in transcription at the late-blastula stage (stages 8 to 9). Experiments on the competition of the RNA thus labeled with the unlabeled oocyte RNA pointed to their complete difference, which means that other genes than those which operated during oogenesis are activated at blastula stages. Amphibians were shown to differ in this sense from the sea urchin, in which the genes active at the mesenchyme blastula stage are homologous in many respects to the genes functioning in oogenesis (Glišin *et al.,* 1966). The competition of the same RNA labeled at stages 0 to 9 with the RNA isolated at stage 7 (early-mid-blastula) was also quite insignificant. This is not surprising because the synthesis before stage 7 is negligible, and therefore the RNA at the early-blastula stage must not differ from the egg RNA to any considerable extent.

After that the competition of the RNA synthesized before stage 9 with the RNA isolated at the early-gastrula stage (stage 10) was investigated. Although these stages differ from one another by just a few hours of development, the competition proved to be small: only 28%, i. e. less than 1/3. This must mean that most of the RNA synthesized at the late-blastula stage (stage 9) is lost by the embryo already by stage 10. Less than 1/3 of its species remains in the cells or continues to be synthesized on the same genes. Thus, amphibians, prior to gastrulation, synthesize the RNA which presumably serves to ensure gastrulation. This RNA is essentially stage-specific, i. e. it is not produced either at the previous or at the subsequent stages and is used to a considerable extent at the same pregastrulation stages. The question as to whether its synthesis ceases at the subsequent stages remains open, because the 28% of the same RNA species remaining by stage 10 could have been created, wholly or to a great extent, by the continuation of activity of the genes operating at the blastula stage.

Investigations of Davidson *et al.* (1968) also provide some data on the number of RNA copies synthesized in oogenesis and at the blastula stages. This can be estimated by the DNA : RNA ratio at which hybridization saturation is reached. For the oocyte RNA, saturation is reached already at a DNA : RNA ratio of 1 : 2; the same ratio for the blastula RNA is 1 : 400. This means that each type of egg mRNA is represented by a very large number of copies (10^4, according to Davidson), while at the blastula stage only a small number of copies manage to be transcribed on each active gene.

Denis (1966a, b, 1968), also using *Xenopus* embryos, studied the mutual competition of RNA obtained at numerous stages of development, including the late-embryonic and larval stages. Using a relatively short-labeled RNA, Denis obtained data on how the different

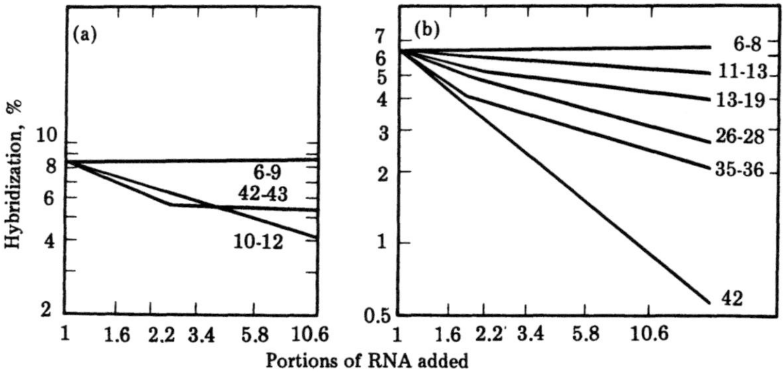

Fig. 29. Appearance of new RNA species during development of *Xenopus*. Embryos at stages 10 to 12 (gastrula) (*a*) or 42 (tadpole) (*b*) were incubated with $^{14}CO_2$, then RNA was isolated and hybridized with DNA in the presence of various quantities of unlabeled RNA (in portions of 100-108 μg) isolated at different stages. These stages are marked at corresponding curves. The RNA synthesized at gastrula does not compete with the RNA stored in the egg and at early stages (stages 6 to 9). With the beginning of RNA synthesis in the embryo the number of RNA species identical to those of the tadpole increases gradually (Denis, 1966b).

species of RNA succeed each other in the course of development. Data on a parallel series with a long-labeled RNA characterized the succession of the long-lived species of RNA and the duration of their utilization after the synthesis (Fig. 29).

Short-labeled RNA in a swimming tadpole (stage 42) does not compete with the egg or blastula RNA. The competition begins at the gastrula stage, gradually increasing with each subsequent stage of embryos and larvae, from which the unlabeled RNA was isolated (Fig. 29*b*). Hence, the development is attended by incorporation of ever-increasing number of new genes which are active throughout the subsequent development.

An appreciable competition with the RNA, short-labeled at the onset of organogenesis (stage 26 to 28, the tail bud), also begins during gastrulation and increases gradually to the tail-bud stage, persisting throughout all the subsequent stages.

Finally, the RNA short-labeled at the gastrula stage does not compete with the egg or blastula RNA but does so successfully with the RNA of its own stage and of all the following stages up to the late tadpole. The competition curves for the unlabeled RNA taken from the gastrula and the tadpole stages slightly differ in shape. The competitive repression of hybridization of the labeled gastrula RNA with unlabeled RNA of the same stage slowly increases with increasing concentration of the latter. In Denis' experiments this curve never reaches a plateau. This means that the number of RNA copies synthesized at the gastrula stage is relatively small. At the same time, an RNA preparation isolated from a swimming tadpole competes as successfully with the labeled gastrula RNA. In this instance, however, the slope of the curve is steeper, and even at small concentrations of the unlabeled RNA the curve reaches a plateau, which lies considerably higher than the curve for the competition of the gastrula RNA. This means that although at stage 42 the set of sequences (RNA fractions) common with those synthesized at the gastrula stage is smaller, the numbers of copies of each RNA type are larger.

Denis' data (1966a, b, 1968) show that development is attended by an increase in the number of active genes, and many of them have no strict stage-specificity: once incorporated at some stage or other, they continue being transcribed at the subsequent stages. But their share in the total number of active genes decreases at the later stages, because they are joined by new active genes in the course of development. It should be remembered that, first, here we deal, as before, only with repetitive sequences and, second, that the RNA preparations were isolated from whole embryos and larvae. Therefore it is hard to say whether the addition of new active genes reflects the events occurring in each cell or whether it indicates the appearance of new specialized cell types with their specific sets of functioning genes.

When comparing the fraction of the genome involved in transcription at different stages, one observes a considerable difference between the short-term and the long-term label. In the former case, hybridization hardly changes from stage to stage (1.8-2.4%), whereas with long-term labeling it increases from 1% during ovulation to 2.4% at the gastrula stage, to 5% at the tail-bud stage and to 8-9% at the tadpole stage. This means that the activation of the new genes leads to a synthesis mainly of long-lived RNA's whereas the fraction of the genes on which short-lived RNA's are transcribed remains the same.

An increase in the number of active genes in the course of development of the frog *Rana pipiens* is also shown by Greene and Flickinger (1970). As in the preceding investigations, these authors found that it is addition of new genes, rather than their replacement, that occurs in the course of development; this addition is expressed in the fact that the RNA's isolated from the older embryos and larvae compete successfully with the labeled RNA of the early stages. Greene and Flickinger labeled the isolated RNA with ^{3}H-dimethylsulfate, i.e. they investigated not the RNA synthesized within some period or other, but all the RNA present in the cells. It is obvious that the relationship between the RNA types synthesized at the preceding stages depended not only on the synthesis intensity, but just as much on the life span of these RNA's and on the time lapse since the moment of their synthesis. Under these conditions of experiment the short-lived RNA's constituted a small fraction, and only at the expense of the molecules synthesized immediately before the stage at which the RNA's were isolated. With this set-up the experiment enabled one to evaluate not so much the changes in the set of active genes, as the changes in the composition of the mRNA's ready for translation.

The development is attended by an increase in DNA-RNA hybridization and a change in the shape of the curve reflecting the dependence of the hybridization percentage on the quantity of RNA. At the gastrula, and even neurula stage, the curve forms a plateau already at small quantities of RNA, i. e. all the complementary sequences of DNA are covered by RNA, and the new portions of RNA can no longer bind with the DNA. This means not only that the number of genes on which these RNA's were synthesized is small, but also that each of these RNA's is represented by a large number of copies, only part of which can be hybridized. At the subsequent stages (tail-bud, tadpole), however, the hybridization curve not only rises, demonstrating a larger number of active genes, but does not form a plateau at the highest RNA concentrations. This means that the RNA contains a small number of copies of each species, and, as the amount of added RNA is increased, more and more new RNA molecules find DNA sequences complementary to themselves. The authors draw the same conclusion on the basis of experiments on the depletion of RNA by passing it successively through filters with new portions of DNA. The RNA obtained at the earlier stages requires a very large quantity of DNA to be bound in an appreciable amount, whereas the RNA of the later stages is extracted from solution in a substantial amount already by the first portions of DNA. It is characteristic, however, that the late RNA also contains a part not extractable even by a considerable amount of DNA. This part probably contains RNA transcribed from single-copy sequences.

The difference in the composition of the mRNA's formed in different parts of the amphibian embryo is revealed by the competitive hybridization method already from the late-blastula stage. RNA's isolated from the vegetal and animal parts of the blastula, from the dorsal and ventral regions of the embryo at the tail-bud stage, differ markedly (Flickinger *et al.*, 1966).

Utilization of techinques of isolation and hybridization of poly$(A)^+$ RNA with cDNA provided data for *Xenopus* eggs and embryos as to the diversity of sequences and the number of their copies at different stages of development (Perlman and Rosbash, 1978). In the ovary, in mature and immature *Xenopus* oocytes, the poly$(A)^+$ RNA is represented by 2×10^4 sequences, of which 1.1-1.2×10^3 are rather abundant (about 2×10^7 per oocyte) and 1.9×10^4 are present in several times fewer copies. In the tadpole and in the tissues of an adult animal (kidney culture) about the same number of types of RNA sequences were detected, and most of them were the same as in the egg. Only 3×10^3 RNA sequences present in the tadpole tissues are missing in the oocytes. These sequences belong to the type represented by a small number of copies per cell (Perlman and Rosbash, 1978).

On the other hand, the poly$(A)^+$ mRNA comprises only a part of all mRNA of the amphibian embryo. To demonstrate this, Miller (1978) utilized anucleolate *Xenopus* mutants which had no rRNA synthesis and in which all the labeled polysomal RNA was mRNA. This labeled RNA was separated on a column with oligo(dT)-cellulose into poly(A)$^+$ and poly(A)$^-$ RNA. The poly(A)$^+$ RNA was found to constitute about 20%, and the poly(A)$^-$ RNA, the remaining 80%.

New possibilities are being opened up by the utilization of a library of cloned cDNA sequences, which enables one to compare the composition of mRNA at different stages of development (Dworkin and Dawid, 1980). The first attempts at using these methods showed that the qualitative and quantitative composition of the population of abundantly represented mRNA species changes only slightly up to gastrula. Later, at the tail-bud stage, the mRNA composition changes substantially and the nature of the new population remains unaltered until the larval stages.

The same paper demonstrates that the composition of mRNA populations translated on polysomes and stored in the form of RNP's (informosomes) is different, and that only a small number of mRNA species move from RNP into polysomes in the course of development (Dworkin and Dawid, 1980).

The change in mRNA composition at the gastrula-tail-bud stages was also reflected in the increased size of nuclear RNA's during this period (Shepherd and Flickinger, 1979).

Sea Urchin

There is a fundamental difference between the results of hybridization experiments on amphibians and on sea urchins: while in amphibians, new species of mRNA nontranscribable in oogenesis form in the period of early development, in the sea urchin a considerable part of active genes are the same at all the stages from oogenesis to the differentiated larva (prism). This was shown already in the first experiments on competitive hybridization carried out by Whiteley *et al.* (1966) and Glišin *et al.* (1966).

In Whiteley's experiments, ^{32}P-labeled RNA was obtained at the prism stage and then the dependence of hybridization on the addition of unlabeled RNA taken at earlier stages was investigated. It was found that the RNA obtained from an unfertilized egg or at the blastula stage competes considerably (by 50%) with the prism RNA, i. e. at these stages half of the RNA is synthesized on the same DNA sequences. In the course of development the degree of competition increases, reaching a maximum when the labeled and unlabeled RNA obtained at the same larval stage compete with one another. A slightly lesser but quite definite competition with the prism RNA is exhibited by mRNA isolated from adult testis and intestine tissues (Whiteley *et al.*, 1966).

Similar conclusions are drawn by Glišin *et al.* (1966), who investigat-

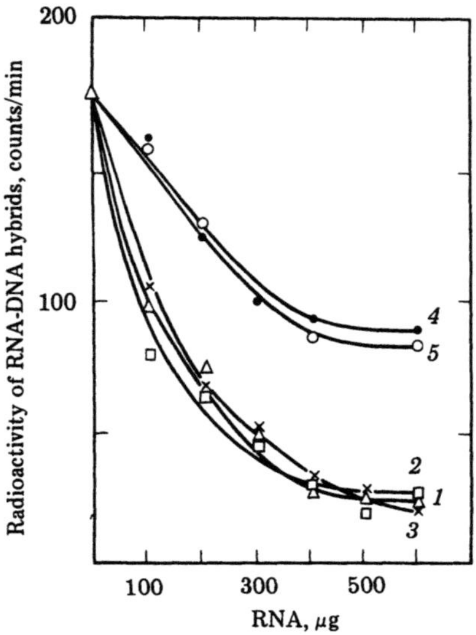

Fig. 30. Appearance of new RNA species during development of the sea urchin. Labeled RNA was obtained at late blastula and hybridized with DNA in the presence of different quantities of unlabeled RNA taken from the same stage (*1*), 2 h later (*2*), from an unfertilized egg (*3*), and from the gastrula (*4*) and prism (*5*) stages. The RNA composition in the unfertilized egg and at blastula is the same, but new RNA species appear in the embryo beginning with gastrula (Glišin *et al.*, 1966).

ed the competition between RNA labeled at the late-blastula stage and RNA obtained at earlier and later stages. As in Whiteley's experiments, RNA from an unfertilized egg, RNA taken at the early-cleavage stage, and RNA at the blastula stage proved to be similar when this hybridization method was used, i. e. they competed equally well with the labeled blastula RNA. Preparations obtained at the gastrula stage and from larvae at the prism stage competed less, though still considerably (by 40%), with the same RNA (Fig. 30).

From these early works on RNA hybridization in which repetitive sequences were taken into account preferentially or exclusively, it follows that, beginning with oogenesis and ending with the blastula stage, the active reading out is confined to the same sequences, and only at later stages are they supplemented by the same number of new ones.

In a later paper (Hynes and Gross, 1972) it was shown, however, that already at the morula stage a considerable part of the RNA synthesized (70%) differs from the RNA stored in oogenesis.

Volfson and Vorobyov (1971, 1972) and Volfson *et al.* (1971) made the next step: they obtained labeled RNA from whole embryos at the blastula and the gastrula stage or synthesized it *in vitro* on chromatin isolated at the same stages, or, finally, isolated it from the polysomes, i. e. they took not the total but apriori translatable mRNA. Of special interest in these investigations is the comparison between the degree of hybridization with DNA of RNA preparations synthesized on the blastula chromatin, on the gastrula chromatin, and on pure DNA. The quantity of DNA was excessive, i. e. all the RNA which could be hybrid-

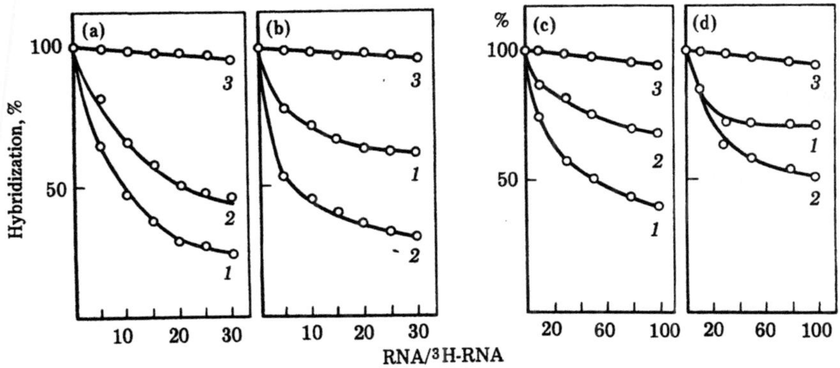

Fig. 31. Differences between RNA's synthesized in sea urchin embryos at blastula and gastrula. Embryos of *Strongylocentrotus droebachiensis* at late blastula (*a*) and late gastrula (*b*) were incubated with ^{3}H-uridine to obtain labeled RNA. Similar RNA preparations were produced by synthesizing RNA *in vitro* on chromatin isolated from the blastula (*c*) and gastrula (*d*). The labeled RNA was hybridized with DNA in the presence of unlabeled total (*a, b*) or nuclear (*c, d*) RNA from the blastula (*1*) or gastrula (*2*). Unlabeled foreign RNA from *Tetrahymena pyriformis* was used for control (*3*). The RNA's synthesized at blastula and gastrula differ from one another, although they have some sequences in common. These differences persist in RNA synthesized *in vitro* on chromatin (Volfson and Vorobyov, 1972).

ized under these conditions (with repetitive sequences) was extracted from solution, and the percentage of this bound RNA was determined. It was found that the RNA synthesized on pure DNA was hybridized least of all (less than by 20%). The RNA synthesized on pure DNA deprived of protein was transcribed uniformly along the entire length of the DNA, and therefore it should have been hybridized with the whole of DNA (i. e. the hybridization level should have been close to 100%). This does not occur, however, because under the conditions used in the study only repetitive sequences, whose share in the sea urchin is small (less than 20%), become involved in hybridization.

RNA which hybridizes to a much higher degree (45%) was synthesized on the blastula chromatin (*in vitro*) or in the blastula itself *in vivo*. This was due to the fact that at the early stages the fraction of actively transcribed repetitive genes is very large. RNA synthesized at the gastrula stage *in vivo* or on chromatin *in vitro* is already less involved in hybridization: at saturating quantities of DNA, no more than 25% RNA binds with it. The share of read-out repeats in the RNA population decreases, evidently owing to the increase in share of active unique genes. Mutual competitive hybridization of RNA's synthesized at the blastula and the gastrula stage shows that about half of the sequences active at these stages are homologous. But at least half of the transcribable repetitive sequences are nonhomologous, i. e. the transition from blastula to gastrula is attended by replacement or addition of a considerable part of active genes (Figs 31*a, b*).

Since RNA was synthesized on isolated chromatin in a cell-free system not by a homologous enzyme but by the *E. coli* RNA polymerase, the question naturally arises as to to what degree this RNA is identical in its composition with that synthesized in whole cells at the same stage. The presently available data are contradictory and fail to give a conclusive answer to this question, but the hybridization analysis methods used so far do not enable one to detect appreciable differences between the RNA's synthesized *in vivo* and *in vitro*. In Volfson and Vorobyov's investigations (1972), such RNA transcribed on chromatin at the blastula and the gastrula stage retained the same interstage differences which had been shown for the RNA's synthesized *in vivo* in intact embryos at these stages. Moreover, the production of RNA by synthesis on chromatin offers some advantages when the object of investigation is a set of active genes rather than the RNA composition in the cell. The point is that no matter how short the incubation period in an intact embryo is, the forming RNA's have enough time to undergo certain changes, such as processing or degradation of short-lived templates, and after isolation these RNA's are found to be mixed with the excess of the competing, unlabeled, long-lived mRNA and strongly diluted with rRNA. When transcription occurs *in vitro* all these secondary processes and impurities are minimized, and the products of active genes are investigated as if they were just formed. If, however, one is concerned with the composition of the RNA in the cell rather than with the gene activity as such, the labeled RNA is best obtained either by prolonged incubation with a precursor or by *in vitro* labeling after isolation, as has already been discussed with reference to experiments on amphibians.

In their experiments Volfson and Vorobyov (1972) also investigated the transport of RNA from the nucleus into the cytoplasm. To this end they compared the hybridization ability of RNA's obtained from different cell fractions: nuclei, cytoplasm, and such partly purified fractions as the chromosomal-nucleolar and polysomal fractions. These fractions help trace the order of changes in the RNA immediately after its synthesis (RNA from the chromosomal-nucleolar preparation), in the course of its change inside the nucleus (RNA from whole nuclei), after it leaves the nuclei (cytoplasmic RNA), and at the final stage, at the moment of translation (polysomal RNA). It turned out that the RNA hybridization ability throughout these successive stages increases from 15% in chromosomes to 35% in whole nuclei and to 55-60% in the cytoplasm and polysomes. These figures suggest important conclusions. The first is that the RNA undergoes considerable changes between transcription and translation. Secondly, in the course of these changes the share of repetitive sequences increases appreciably (as is convincingly indicated by the enhanced hybridization). This may be attributed to the fact that a considerable part of the RNA formed at the blastula stage consists of templates for histone synthesis (Gross *et al.*, 1973a, b). With their transport out of the nucleus and incorporation into the polysomes the relative content of histone mRNA's increases, and so does the percentage of their hybridization with the repetitive histone genes.

Of more interest, perhaps, are those experiments of Volfson (1972) where labeled blastula RNA from the chromosomal-nucleolar apparatus and RNA from the embryonic cytoplasm was investigated by its competition with unlabeled RNA of the same cell fractions obtained at the blastula and gastrula stages. It turned out that chromosomal-nucleolar RNA's from blastula and gastrula differ only slightly in the degree of competition (80 and 70%, respectively). At the same time, for the cytoplasmic RNA these differences in competition are much greater: 70% for the blastula and as low as 30% for the gastrula cytoplasmic RNA's. This means that the differences in RNA composition in the course of development appear to a great extent already during the processing and the transport of the RNA into the cytoplasm, or, what is probably more accurate, two fractions of the RNA are synthesized during development: one remaining in the nucleus and changing little in the course of development, and the other (mRNA proper) transported into the cytoplasm, whose composition changes considerably from one stage to another (Volfson, 1972).

The differences between the nuclear and cytoplasmic mRNA's in sea urchin embryos have been investigated in more detail in the laboratory of Britten and Davidson, where only transcripts from single-copy DNA sequences were considered (Galau *et al.*, 1974; Hough *et al.*, 1975). The results of these investigations contain important information, but are not conclusive. The authors studied the complexity of the RNA composition, i. e. the degree of its diversity, which can be measured by the number of the nucleotides making up the totality of molecular species of RNA in a given preparation. This diversity clearly reflects the number of genes on which these RNA's were transcribed.

At the gastrula stage of the sea urchin embryo the cytoplasmic RNA's were isolated from the polysomes, thus representing the true mRNA's. They were hybridized with nonrepetitive sequences of denatured DNA, and the DNA thus bound was then reassociated with the total DNA. The resulting data only indicated the share of single-copy DNA sequences responsible for the transcription of the embryonic mRNA's at the gastrula stage. It was 2.7%, which corresponds to 1.7×10^7 pairs of nucleotides or 1.4×10^4 medium-sized structural genes (1.2×10^3 pairs of nucleotides) (Galau *et al.*, 1974). A similar investigation of nuclear RNA's has yielded much greater values: nuclear RNA's are transcribed by 28% of the single-copy portion of the genome, i.e. they show a 10-fold increase in complexity as compared with polysomal mRNA's (Hough *et al.*, 1975).

According to Hough and coworkers' calculations, a considerable part of the mRNA in the cytoplasm consists of a few types of templates

8*.

represented by a large number of copies, while 8% of the total mRNA consists of very many species of template, each having no more than 500 copies per embryo. At the gastrula stage, the embryos of the sea urchin *Strongylocentrotus purpuratus* consist of 600 cells, and, according to these calculations, there is less than one copy of each of the numerous mRNA species per cell. The authors reasonably assume that different mRNA's are transcribed and reside in diversely differentiated cells of the embryo.

The conclusion that the complexity of the RNA's in the nuclei is much greater than in the cytoplasm is of fundamental importance. One of the possible explanations is a high level of mRNA selection in the nuclei, i.e. that only a preliminary selection of activated genes is effected at the transcription level, with a stricter regulation process taking place in the nuclei, at the level of the posttranscription selection of templates transported into the cytoplasm.

It should apparently be assumed now that an appreciable part of the RNA sequences belonging to nuclear RNA's (hnRNA's) are noncoding sequences (introns), which are transcribed from DNA and have not yet undergone splicing, during which they are cut out and degrade.

A comparison of the composition of the single-copy RNA in the cytoplasm of the oocyte and the embryo at different stages of development (Hough-Evans *et al.*, 1977, 1979) carried out by hybridization with the single-copy ^{3}H-DNA demonstrated that in the course of oogenesis the cytoplasm shows an increase in the number of transcripts from single-copy sequences and in the number of copies of each of them. Within a few weeks in which previtellogenic oocytes turn into mature ones, the complexity of RNA increases from 1.6×10^{7} to 3.7×10^{7} nucleotides (i.e. approximately from 10^{4} to 2×10^{4} active genes) and the average number of each mRNA species per oocyte, from 6×10^{2} to 16×10^{2}. The complexity of the nuclear RNA is an order of magnitude higher. In the course of embryonic development both the total number of species of cytoplasmic mRNA and the number of those found in mature oocytes drops off gradually. If we assume the oocyte RNA complexity to be 100%, then about half of all the maternal RNA are retained at blastula-gastrula species and about 40% at pluteus. The authors believe that in sea urchin embryos the maternal templates are kept not later than blastula, after which they are partly replaced by homologous newly transcribed mRNA's (Hough-Evans *et al.*, 1977).

In another work of this series Davidson and his coworkers (Galau *et al.*, 1976) separated DNA containing single-copy sequences complementary to the mRNA's of the gastrula from the rest of the single-copy DNA (not complementary to the gastrula RNA). These two types of DNA were hybridized with the mRNA of embryos and various adult tissues. In this way the authors determined the complexity of the transcribed part of the genome, both homologous and nonhomologous to the gastrula mRNA. While at the gastrula stage, this complexity, as measured by the number of nucleotides, was 1.7×10^7; at the blastula stage the number of nucleotides, homologous and nonhomologous to the gastrula mRNA, was 1.2×10^7 and 1.5×10^7, at the pluteus stage, 1.4×10^7 and about 6×10^5, and in the total oocyte RNA, 1.7×10^7 and 2×10^7, respectively. For adult tissues these figures were: 2.1×10^6 and 3.7×10^6 for the gut, 3.5×10^6 and 1.4×10^6 for coelomic cells, and 1.3×10^7 and 6.7×10^6 for the ovary. Thus, the overall complexity of the RNA sequences is higher in embryonic tissues and lower in differentiated ones, but is particularly high in the oocyte. Many genes which are active at the gastrula stage function at other stages and in adult tissues as well. DNA sequences of about 2×10^6 nucleotides (about 2×10^3 genes) are active in all the tissues investigated (Galau *et al.*, 1976).

Significant results were obtained by the Britten-Davidson group by using two DNA sequences isolated from the library of cloned cDNA's. These single-copy DNA's are complementary to a complex mRNA class, i.e. many thousands of species of these RNA's are transcribed in oocytes, and each of them is represented by a very few copies. It has been shown (Lev *et al.*, 1980) that each egg contains 1×10^3 and 1.4×10^3 molecules of each of these two RNA species, and that a considerable part of them are contained in the polysomes, i.e. are translated. In the course of early development the quantity of these two RNA species decreases. At blastula, the total number of these species of molecules is still large (520 and 550, respectively), but the cytoplasm and the polysomes retain less than 1/10 of them. This means that at blastula-gastrula they are practically no longer involved in protein synthesis. However, an appreciable quantity of one of these RNA species is again found in the gut of adult sea urchins. In the nuclear fraction both species are detected at least before pluteus. The authors believe that both RNA's are transcribed continuously but their transfer into the cytoplasm and translation occur only at the earliest stages of development and, for one of them, are resumed in the adult gut. These results are in good agreement with Davidson and Britten's contention as to the regulation of gene expression at the level of processing and transport into the cytoplasm, rather than at the level of translation (Davidson and Britten, 1979).

Local differences in the composition of mRNA in sea urchin embryos were found very early, at the 16-blastomere stage: before this stage the mRNA species, as compared by the competitive hybridization method, are identical in all the blastomeres, whereas beginning from the 16-cell stage the RNA transcribed in micromeres differs from the RNA of the macro- and mesomeres (Mizuno *et al.*, 1974).

These observations were further developed in another series of investigation, where the RNA populations from micro-, meso-, and macromeres transcribed from a single-copy DNA were compared at the 16-cell stage (Rodgers and Gross, 1978; Ernst *et al.*, 1980).

The following differences were shown: (1) the micromere nuclei lacked complex RNA, which is present in the macro- and mesomere nuclei; (2) the micromere cytoplasm lacks polysomal maternal RNA, which is detected in whole embryos; and (3) the complexity of the micromere RNA corresponds to only 75% of the total set of the embryonic RNA. The authors attribute this to the fact that the activation of RNA synthesis is delayed by one or two divisions in the micromeres; in other types of cell (macro- and mesomeres) it takes place at the 16-cell stage. Besides, in the micromeres the nuclei are surrounded by a very small amount of cytoplasm, which results in the underrepresentation of the stored maternal RNA's in the cytoplasm and polysomes. Thus, conditions are created in the micromeres for a change in translation plan, which determine the specialization of these cells during development.

Fishes

Investigation of DNA-RNA hybridization in fishes has been carried out on loach embryos by Timofeeva, Rachkus, Kupriyanova, and Kafiani (Rachkus *et al.*, 1969a, b, 1971). By the beginning of this investigation, the above-listed, widely diverging data on amphibians and sea urchins were already available. During embryonic development of amphibians, newly synthesized RNA's practically do not compete with the RNA's stored up in oogenesis, i.e. transcription proceeds on absolutely new sequences (genes). Investigations on the sea urchin showed no less convincingly that at the early stages (up to blastula, inclusive) the RNA synthesis proceeds on almost the same sequences as in oogenesis. Taxonomically, the fishes are closer to the amphibians, although the teleosts differ from them considerably as to the type of development. Nevertheless, hybridization experiments have demonstrated their similarity to the sea urchin as regards the nature of transcription.

RNA labeled at late blastula (the earliest stage when sufficient precursor incorporation into RNA can be achieved) competes successfully with the RNA of the unfertilized egg. This is also true for RNA synthesized at gastrula or in organogenesis (Rachkus *et al.*, 1969a, b) (Fig. 32). The similarity of fishes to echinoderms and their difference from amphibians are amazing. The explanation of this phenomenon will obviously be found only after other — related and distant — taxonomic groups are investigated from this point of view. It would be more interesting to conduct experiments on sturgeon family fishes: they are very close to amphibians in the egg structure and the type of development. It would also be useful to ascertain that urodeles differ but little from anurans (it would be surprising if it were found that the degree of similarity between the egg and

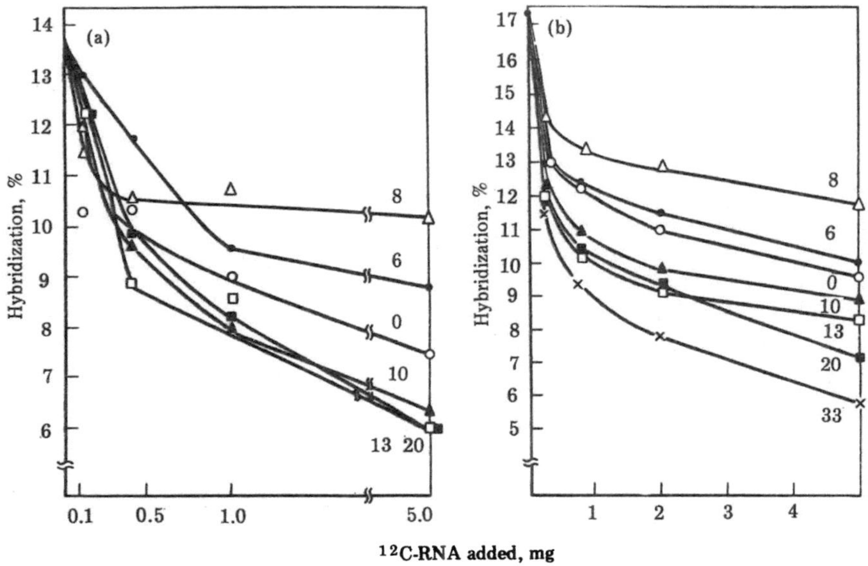

Fig. 32. Changes in RNA composition during embryonic development of the loach. ^{14}C-RNA was obtained at the 13-hour stage (gastrulation, *a*) or at the 33-hour stage (late organogenesis, *b*) and then hybridized with DNA in the presence of unlabeled RNA taken from an unfertilized egg, the 6- and 8-hour stages (blastula), 10- and 13-hour stages (gastrulation), and 20- and 33-hour stages (organogenesis). During early development (from 0 to 8 h) the embryo loses some of the RNA species stored in oogenesis, after which the diversity of these RNA species increases (Rachkus *et al.*, 1969a).

embryo RNA varies within related taxonomic groups). Next in the line for investigation are such more accessible species as sea worms and molluscs, and, among the vertebrates, birds. We can only advance more or less plausible hypotheses. One of them is that the general laws of molecular developmental mechanisms must be the same for all the animals, and all the more so for fishes and amphibians. A second is that many proteins, for instance, histones or some enzymes, are synthesized to an equal extent in oogenesis and at the later stages of development, hence their synthesis is determined by the activity of the same genes. In other words, a certain part of the RNA sequences must be common to all the stages of oogenesis and development.

The fact that these common sequences are not detected in the mRNA of amphibians indicates that they are transcribed from rare or unique genes and hence are not detectable by the hybridization methods used until recently. Advancement of hybridization methods which will enable one to operate with single-copy sequences will soon bring to light new facts which will bridge this gap. As has already been mentioned above, it is assumed that a considerable part of repeated sequences play some kind of auxiliary role, and the data on hybridization point to possible functional differences between different, even not very distant, groups of animals.

Rachkus *et al.* (1969b) (see Fig. 32) showed that the loach RNA labeled at the gastrula stage competes appreciably better with RNA from the unfertilized egg than with RNA at mid-late blastula (6- to 8-hour stages). To put it differently, in the course of the earliest development, when mRNA is practically not synthesized (the first 0 to 6 h) or when the synthesis is just beginning (the first 6 to 8 h), the number of RNA species contained in the embryo decreases. The use of the RNA stored in oogenesis in the course of early development has already been considered by us for amphibians. (In this respect the fishes and amphibians proved to be similar.) Later on, however, at the 8- to 10-hour stages, a quantitative shift is observed, when the competition of the RNA obtained at the 10-hour stage with that synthesized at the 13-hour stage or later increases again, i.e. when the number of common species of RNA increases. This increase reflects the active synthesis (occurring at the 8- to 10-hour stages) of RNA's, which apparently include many of the sequences they have at later stages.

Further on, one observes the picture described for amphibians: in the course of development the number of RNA species increases by addition of new genes, rather than by replacement of previously active ones. The increase in the number of active genes coincides as regards the stages of development with the progressing differentiation of the embryo and is presumably determined, to a considerable extent, by the increase in the number of cell types and not only by the increased diversity of active genes in each cell.

Additional data are produced by a separate analysis of the slope of the competition curves and the height of the plateau, i.e. of the level of competition (percentage of hybridization which can no longer be reduced by the addition of unlabeled RNA). In the course of development the plateau level decreases continuously, i.e. an ever-growing number of RNA species become the same as at the later stages, i.e. an ever-growing number of genes are activated. The slope of these curves also becomes progressively steeper. This means that fewer and fewer competing molecules are required to appreciably reduce the hybridization of labeled molecules. Hence, along with the increase in the number of RNA species, the number of copies of each species increases as well. This conclusion also agrees with the fact that the diversity of RNA's is often achieved owing to the diversity of cell types. The number of copies of many RNA species may then be quite small because many cell types are represented in the embryo by a very small number of cells, which constitute fractions of a few tenths of percent of all the cells. Some species of RNA from these cells can simply remain undetected.

A comparison of nuclear and cytoplasmic RNA in fishes showed that they have some differences and that the content of repetitive sequences in the cytoplasmic fraction is high. As we know now, similar data were subsequently obtained on sea urchin embryos. These results may be ascribed, to a certain extent, to the considerable transcription of the mRNA of histones encoded by repetitive genes.

Mammals

Mammalian embryos are an extremely difficult object in molecular embryology. Church and Brown, studying hybridization, obviated this difficulty by using high-labeled DNA from cell cultures instead of labeled RNA from embryos (Church and Brown, 1972; Brown and Church, 1971, 1972). The use of high-labeled DNA and long-term (up to two-week) hybridization in solution enabled the authors to investigate repetitive and single-copy sequences separately in the RNA preparations not only from the postembryonic, but also from the embryonic stages of the mouse. And although the utilization of such a scarce material as mouse embryos strongly restricted the number of the stages investigated, Church and Brown's data yielded quite new information on the nature of gene transcription in development (see Church and Brown, 1972; Church and Schultz, 1974).

The mouse genome is represented, by 10%, by satellite, highly repetitive (10^6 copies) DNA which consists of AT-rich sequences and is predominantly localized in the region of centromeres and terminal sections of chromosomes (Pardue and Gall,. 1970). This DNA is probably never transcribed. By 20-25%, the genome consists of intermediate repetitions (10^1-10^5). With conventional hybridization methods (on filters, with annealing for 1-2 days) the composition of RNA's transcribed only from this repetitive part of the genome can be investigated. And, finally, 70% of the DNA consists of single-copy sequences (genes), whose transcription in development is of special interest and whose investigation has been started in recent years.

Church and Brown's studies of competitive hybridization of RNA's from different tissues of an adult organism have confirmed the earlier results: about 50% of the read-out repetitive DNA sequences were found to be common to such different tissues as brain and liver or brain and kidney, while the remaining 50% are tissue-specific. If the read-out repetitions represent only the regulatory regions of the DNA, it is obvious that their activation differs by 50% in the above tissues, and their participation in development is of some interest. Therefore the data on RNA-DNA competitive hybridization obtained on sea urchins, fishes, and amphibians will not lose their significance even after they are supplemented by investigations into the activity of single-copy genes.

As shown by Church and Brown (1972), in the course of embryonic development of the mouse the hybridization level, i.e. the share of active repetitive genes, changes hardly at all. The hybridization increases about 2-fold (from 2.5 to 5%) in the earliest development from the 8-16-cell to the morula stage, then it decreases slightly at the beginning of differentiation (from the 4th to the 7th day of embryonic development) and remains at a constant level of about 4%, which is equal to 13-16% of all the repetitive DNA except satellites. In this sense mammals differ markedly from, say, amphibians where the share of repeats hybridization increases by a factor of 4 (from 5 to 20%) in the course of embryo and tadpole development.

At the same time, the activity of single-copy genes changes quite considerably in the course of mouse development. It remains very small (about 1%) for the whole organism up to the beginning of differentiation (blastocyst stage, fourth day of embryogenesis) but then quickly increases, reaching 5% within the next three days, 8.5% after two weeks, and almost 10% by the moment of birth. The subsequent changes are slight, and the total number of active single-copy sequences in the whole organism, thus determined, probably does not exceed 11-12% (Fig. 33).

The ratio between the repetitive and single-copy gene transcripts changes sharply in the course of embryonic development of mammals. For the mouse, up to the beginning of visible differentiation this ratio is 5 (2.5-5% for repeats and 0.5-1% for single-copy sequences), on the 7th day of development it is 1 (4% for either), and at birth it falls off to 0.4 (4% for repeats and 10% for single-copy sequences). A similar phenomenon was observed in differentiating cultured chicken myoblasts; in dividing myoblasts more repetitive sequences are transcribed, whereas during the formation of myotubes there is an increase in the share of transcribed single-copy sequences (Man and Cole, 1974).

Perhaps the most valuable part of Church and Brown's work con-

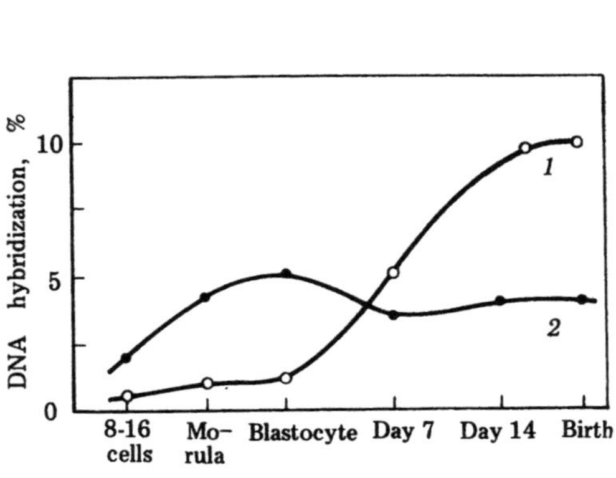

Fig. 33. Ratio of the number of RNA species transcribed from single-copy and repetitive sequences during the embryonic development of the mouse. RNA obtained at different stages of mouse development was hybridized with single-copy (*1*) and repetitive (*2*) labeled DNA fractions. Transcription of repetitive sequences changes only slightly, whereas transcription of single-copy sequences increases rapidly (Church and Schultz, 1974).

cerns the transcription of single-copy sequences in individual organs in the period of embryonic and postembryonic development. In a 14-day embryo the morphological differentiation of the organs is practically completed but cytodifferentiation still continues in a number of tissues and especially in the brain. This is reflected by the hybridization curves of single-copy sequences. In the liver, this value does not exceed 2-4%, whereas in the brain tissues it increases from less than 3% in the embryo to 5% in the newborn and almost to 10% in the adult mouse. It may be assumed that this growth is basically due to the increase in the diversity of nerve cell types, rather than to the increase in the number of active genes in each of them.

While the brain tissue RNA is characterized by a gradual increase in hybridization level and then retention of a constant high level, the picture is different for other tissues. Hybridization of RNA with single-copy DNA sequences in the liver, which may reach 2.5% in a newborn mouse, constitutes 4.5% in the first two weeks (3.5% in the kidneys, 3.2% in the spleen). In an adult animal, however, all these values decrease to 2%. This situation was analyzed by Brown and Church using the method of additional hybridization.

The traditional technique of competition between labeled and unlabeled RNA was inapplicable because in their experiments hybridization of single-copy genes was determined by the binding of labeled DNA rather than RNA. The authors mixed together equal quantities of RNA's from two sources (two different tissues or a single tissue at two stages of development) and compared the hybridization percentage obtained with the sum of percentages for the DNA's binding with the same RNA's, which were, however, hybridized separately. If the percentage of hybridization with a mixture of the two RNA preparations was equal to that of hybridization with one of the components (the one, obviously, which hybridizes with a higher DNA percentage than the other), this meant that complete overlapping was achieved. Consequently, the transcription of the more hybridizable RNA preparation involves all the DNA sequences which are also active in the synthesis of the other, less hybridizable, RNA preparation. Thus, RNA from the brain of a newborn mouse hybridizes with 5.2% of the DNA and that from the brain of an adult mouse, with 8.2%. A mixture of the two RNA's hybridizes to practically the same extent as the RNA from the brain of an adult mouse alone, namely by 8.4%. This means that during the postembryonic development of the mouse brain all the genes active at the moment of birth remain active, but some new genes are added to them.

In another version it may turn out that the percentage of hybridization with a mixture is much higher than with each of the components separately (at most, the hybridization percentage for the mixture is equal to the sum of those for each individual component). This would mean that there is no overlapping or that it is small, and that a large part of the active DNA sequences in the two cell types are different. This was practically the case when RNA's obtained from a mouse liver at the moment of birth (2.5%) and 6 weeks later (2%) were compared. A mixture of equal amounts of the two RNA preparations hybridized with 3.5% of the DNA, i.e. to a greater extent than each of the components (Church and Brown, 1972). It is easy to calculate that single-copy sequences constituting 40% of the active part of the liver genome in a newborn mouse and 50% in an adult one coincide at these two stages, whereas 60 and 50% of the active genes, respectively, are different. In other words, in the course of postembryonic development more than half of all the transcribable sequences are replaced. This replacement can be ascribed to a sharp change in liver function. In the embryo it mainly plays the role of a blood-forming organ. After birth, this function dies out gradually and is replaced by the usual liver function: the storage of carbohydrates, the secretion of bile, etc. (at this time the percentage of hybridization increases to 4.5%). In an adult mouse the liver performs only this principal function, which is quite different from its former one.

However, certain similarity is also observed when comparing other organs, for instance, such dissimilar ones as liver and brain. In a newborn mouse, a mixture of the brain RNA (5.2%) and the liver RNA (2.5%) hybridizes with 6.5% of the DNA, i.e. the overlapping is 6.5 — 5.2 = 1.3%. This means that 20% of the active brain sequences and more than 50% of the active part of the liver genome are common to both. In an adult animal, the brain RNA hybridizes with 8.2%, the liver RNA, with 2.0%, and their mixture, with 8.8% of the DNA. This means that the overlapping is only 8.8 — 8.2 = 0.6%, i.e. only 7.3% of the active brain genes and 30% of the active liver genes are common to both. It may be assumed that many metabolic links in these organs remain common. These data, however, are evidently insufficient to indicate the similarity or difference in the work of the genome and to correlate it with the function of the organ.

A more advanced methodological approach was used in the laboratory of Paul, where poly(A) mRNA was isolated from the polysomes of 14-day embryos and the adult brain and liver of the mouse, then complementary DNA (cDNA) was synthesized on it, and the kinetics of backward hybridization with its RNA or that of other tissues was studied (Young *et al.*, 1976). The authors found that the complexity of the poly(A) mRNA in the embryos, brain, and liver was 7×10^9, 15×10^9, and 7×10^9 daltons, which corresponds to 0.7%, 1.7%, and 0.8%, respectively, of the total genome. The lower values than those found by Church and Brown are evidently due to the fact that only polysomal and only poly(A) mRNA's were investigated here. It was also shown that about half of the active genes (55%) were the same in different tissues (Young *et al.*, 1976).

Other Species

Comparison of RNA complexity was made between *Drosophila* oocytes and embryos (Arthur *et al.*, 1979). Mature oocytes contain 1.45×10^4 species of RNA sequences, of which 103 are represented by 2×10^7 copies each. By gastrula, the number of RNA species in the oocytes decreases by 2.5%, and by late embryogenesis, by 5.3%. On the other hand, these stages reveal new RNA sequences, which increase their total number by 7.7 and 11%, respectively. It should be noted that the values obtained by these investigators seem overestimated, as compared with those of Davidson, Britten *et al.*, (see above), and they result in the minimum gene number values for *Drosophila* being 2-3 times as high as those obtained by other techniques.

We can now summarize the investigation by hybridization methods into the synthesis of RNA's in the course of organism development.

1. It has been shown on all the animals studied that the number of RNA species, i.e. of transcribed DNA sequences, or, what is presumably the same, the number of active genes, increases in the course of development. At least partly, this is due to the appearance of new cell types and probably, to some extent, also to an increase in diversity of the active genes within each cell.
2. The increase is achieved principally by adding new transcribed sequences. However, for some of them replacement of active genes, i.e. synthesis of stage-specific RNA's, has been shown convincingly. This is manifested most vividly in amphibians, in which active repetitive sequences are quite different in oogenesis and in embryogenesis. Synthesis of stage-specific RNA's has also been shown during gastrulation. Sea urchins and fishes, on the contrary, show a considerable overlapping, i.e. the same repetitive genes are active in oogenesis, in embryonic, and in larval development.
3. In the course of development of mammals, the share of active repetitive sequences steadily decreases due to the considerable increase in the diversity of active single-copy sequences. This relationship is probably typical of other animals as well. The organ and stage specificity for some of the single-copy sequences has been shown. This probably suggests an important and specific role of this part of the genome in development.
4. At all the stages of RNA transition from transcription to translation it undergoes partial transformation and selection, which change its composition. For instance, the share of repetitive sequences increases in the course of early development. At the late stages of development and in the adult organism, the opposite relationship seems to prevail: a decrease in the share of repeats in the course

of processing and selection. At the early stages, repetitive sequences in the cytoplasm RNA may also reflect the high intensity of transcription and translation of repetitive histone genes.

5. Despite a number of limitations in the interpretation of the data obtained by the hybridization method, it has already yielded definite, though not-too-unexpected, relationships. A further progress in this area is possible by separate investigation of single-copy and repetitive sequences both on whole embryos and on homogeneous cell populations.

Replacement of Histone mRNA's in Early Development

It has been shown in a number of papers that early development of sea urchin embryos is attended by consistent replacement of histone H1, H2a, and H2b subfractions. At the cleavage and blastula stages, the chromatin contains only two subfractions of these proteins (H1α, H2aα, and H2bα), as shown by electrophoresis, while beginning with the mesenchyme blastula stage they are gradually replaced by subfractions β, γ, and δ. More commonly used are other designations for H1: H1m for the cleavage stage (m — morula) and H1g for the later stages (g — gastrula). (For more information see Vol. II, Chap. 3.)

This process has aroused an appreciable interest and has been investigated in a number of laboratories because the replacement of histones is a case of protein synthesis regulation in early development by replacement of active genes which is accessible to study. Finally, the discovery of the great diversity of histones may lead to new concepts of the functional organization of chromatin.

The opportunities for studying the transcription of histone genes greatly increased after the histone-encoding DNA sequences were isolated and cloned.

The very first works on the problem (Kedes and Gross, 1969; Kedes *et al.*, 1969a, b) already showed that it is largely nuclear proteins that are synthesized in early development. The labeled mRNA's are dominated by the 9S fraction corresponding to histone mRNA's, while in low-molecular-weight polysomes the incorporation of lysine, which is abundantly represented in the histones, greatly exceeds that of tryptophan, which they do not contain. This led the authors to conclude that transcription of histone genes (Fig. 34) and translation of histone mRNA's are very prominent at the early stages. Later on, despite the contrary views (Johnson and Hnilica, 1971), this was demonstrated directly by identification of proteins synthesized *in vitro* on templates stored in the sea urchin egg (Gross *et al.*, 1973b; Ruderman and Gross, 1973; Skoultchi and Gross, 1973) (Fig. 35).

Fig. 34. Changes in the RNA species synthesized during early development of the sea urchin. Sea urchin embryos at the 32-blastomere stage were incubated with ^{3}H-uridine, and those at late blastula, with ^{14}C-uridine. The isolated RNA was analyzed by joint centrifugation in a sucrose gradient. *1*, Radioactivity of the ^{3}H-RNA labeled at the 32-blastomere stage; *2*, radioactivity of ^{14}C-RNA labeled at late blastula; *3*, quantity of RNA's ($E_{260\ nm}$). At the morula stage, 20S RNA and 9-10S RNA (encoding nuclear proteins) and at late blastula, a wide range of mainly high-molecular-weight RNA are synthesized (Kedes *et al.*, 1969a).

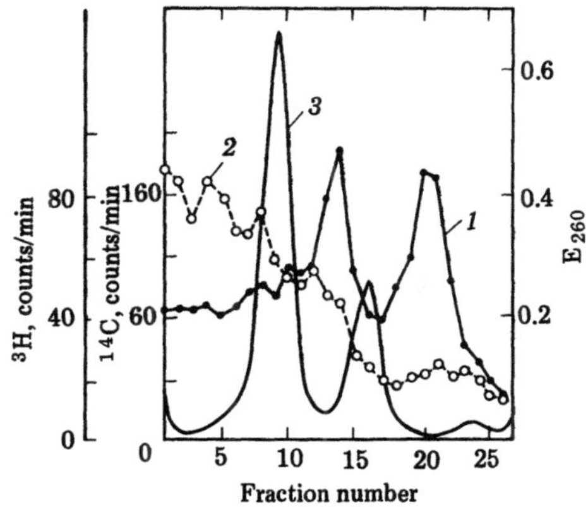

According to Nemer, 85% of the histone synthesis during the early development of the sea urchin occurs on templates stored in oogenesis, but the fraction of histone mRNA's transcribed during cleavage and at early blastula reaches about 60% of the total mRNA (Nemer, 1975). This was confirmed by electrophoretic separation of mRNA's isolated from the polysomes of early embryos and by their testing in the *in vitro* translation system. Three (out of eight) predominating fractions coded for histones (Levy *et al.*, 1975). Then individual histone mRNA's and corresponding cDNA's were obtained by means of reverse transcription; these cDNA's were used for testing individual histone mRNA's of all the five classes (Lifton and Kedes, 1976). Lifton and Kedes succeeded in showing that histone RNA's stored in oogenesis do not differ from those transcribed from histone genes in early sea urchin development. However, no reserve of histone mRNA's could be found in the eggs of *Spisula* molluscs (Gabrielli and Baglioni, 1975).

Tubulin synthesis is similar to that of histones: the bulk of the necessary templates has been stored in oogenesis (Raff *et al.*, 1971, 1972, 1975). They are contained in the polysomes of the early sea urchin embryo and can be detected by their specific binding with colchicine and precipitation by vinblastine (Hirama and Mano, 1974).

At the next stage of investigation it was important to establish whether the replacement of histone subfractions is controlled at the level of mRNA, or whether the small differences between these subfractions are achieved by modification of histone components, i.e. at the posttranscriptional level. It was first of all shown that subfraction H1m is replaced by subfraction H1g at gastrula and that this replacement is preceded by the transcription of specific mRNA for this later subfraction (Arceci *et al.*, 1976).

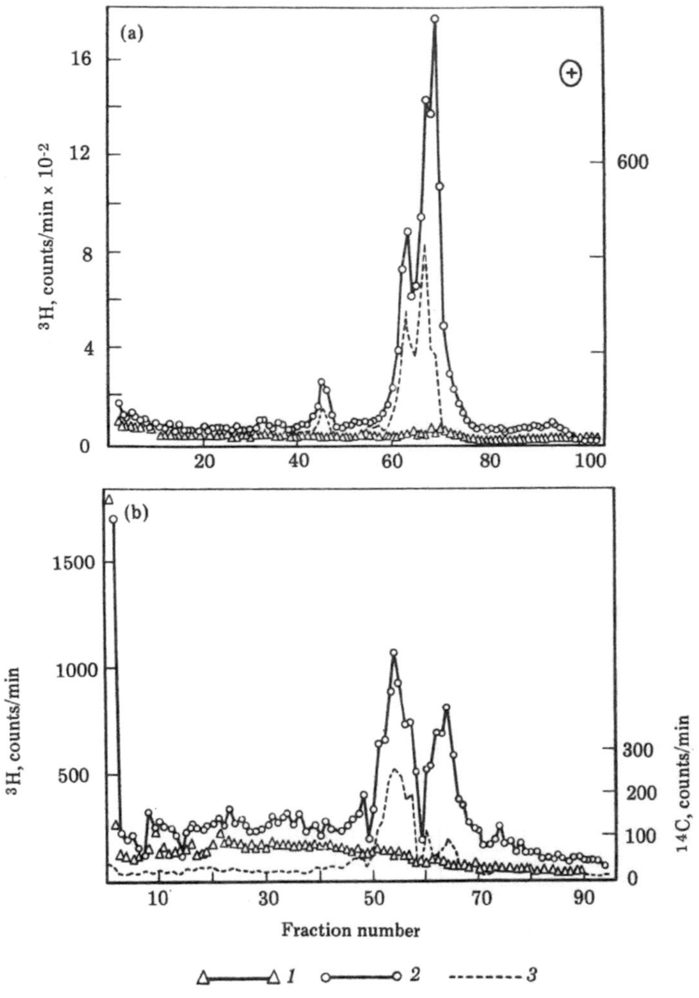

Fig. 35. Identification of histone mRNA's from sea urchin embryos. RNA was isolated in embryos of *Lytechinus pictus* at early blastula, then it was fractionated in a sucrose gradient, and the fraction in the 9S region was translated in cell-free protein-synthesizing system from Krebs' ascite cells with ^{3}H-lysine. The proteins were mixed with the sea urchin tracer histones labeled with ^{14}C-leucine and analyzed electrophoretically in SDS-polyacrylamide gel at a neutral (*a*) or an acid (*b*) pH. *1*, The protein synthesizing system without 9S RNA (control); *2*, the protein-synthesizing system with 9S RNA added; *3*, tracer histones. A considerable part of the 9S RNA-induced synthesis occurs in the histone region (Gross *et al.*, 1973b).

Besides, there exists another small histone subfraction H1x, which can be resolved electrophoretically from H1m. The amount of this subfraction is small at the beginning of development but increases appreciably within one cell cycle from 8 to 16 blastomeres. During the same period, *in vitro* translation of isolated templates shows that the brief increase in the synthesis of H1x is due to the increased quantity of mRNA for H1x (Senger *et al.*, 1978). At the 16-blastomere stage the micromeres contain a greater amount of subfraction H1x than do the meso- and macromeres. The temporary increase in H1x synthesis coincides with the appearance of RNA for some nonhistone proteins as well.

A similar replacement of templates (at the mesenchyme blastula stage) was also shown for subfractions of H2a and H2b. Histone mRNA species coding for early subfractions (H1α(m), H2aα, and H2bα) are the only ones stored in the egg and translated up to the blastula stage. Transcription of new species of mRNA's, on which new subfractions are translated — H1β, γ (or H1g); H2aβ, γ, δ; H2bγ, δ — begins from mesenchyme blastula (Newrock *et al.*, 1978).

The difference between the "early" and "late" (those transcribed and translated at mesenchyme blastula) histone mRNA's has been shown by their hybridization with cloned histone genes (Kunkel and Weinberg, 1978). The early mRNA's hybridize with cloned histone DNA's better than do the late ones (hybrid complexes melt at a higher temperature and are less sensitive to ribonuclease). This indicates that they differ in the nucleotide sequence and, consequently, are transcribed from different histone genes. Early mRNA's are apparently transcribed from the main histone genes which are represented in the genome by large repeats clusters. Late mRNA's are transcribed from other genes which constitute just a part of the total number of histone genes and are scattered through the genome (Overton and Weinberg, 1978).

The general picture of replacement of histone subfractions and their mRNA's in sea urchin development was presented in the paper of the Kedes group (Childs *et al.*, 1979). Using electrophoresis of histones and hybridization of their mRNA's with cloned histone DNA the authors succeeded in distinguishing in five classes of histones of sea urchin embryos over 20 histone subfractions and about 20 types of RNA's encoding these proteins. Histones H3 and H4 are the most homogeneous but some difference in their mRNA's, which are translated at the early and late stages of development, can be revealed as well. Histones H1, H2a, and H2b have subfractions detectable only during early cleavage. Subfractions α_1 and α_2 for H1, H2a, and H2b function only in early embryos (before gastrula) and their mRNA's, stored in oogenesis and transcribed from the embryonic genome, are present at these stages, accordingly. Subfractions β, γ, and δ appear at mesenchyme blastula and the synthesis of the respective mRNA's begins already at morula-early blastula and intensifies with further development. Switching from the early histone genes to the late ones is not sudden but occurs by gradual decrease in transcription on some genes and its intensification on others (Childs *et al.*, 1979).

Investigations on the relationship between the histone translation on the oogenesis-stored and newly synthesized mRNA have been carried out recently using interspecific hybrids of sea urchins and amphibians whose histones differed electrophoretically, and the species-specific identity of the RNA could be determined by hybridization with cloned histone DNA's.

In hybrids of sea urchins (*S. purpuratus* × *L. pictus*) transcription of histone genes is detected already at the two-blastomere stage and their RNA's enter the polysomes not later than the 16-cell stage. The paternal and maternal genomes of the hybrids function with equal efficiency (Maxon and Egrie, 1980). Experiments with sea urchin hybrids have shown that translation of histone H1 occurs on stored and on newly synthesized RNA's, and their ratio undergoes characteristic changes in the course of development. For instance, at the 16- to 32-blastomere stage 80% of histone H1 is synthesized on the new templates and only 20% on the stored, maternal, ones.

The involvement of newly synthesized RNA of histone H1 decreases to 74% at early blastula, to 54% during hatching, and to 40% at mesenchyme blastula (24 h of development) (the fraction of stored templates increases from 26 to 60%, respectively). During the subsequent period of development, however, the ratio changes drastically and at mid-blastula already 94% of histone H1 is translated on newly synthesized RNA's and only 6%, on the remaining histone RNA's of the oocyte (Easton and Whiteley, 1979).

The high contribution of the newly synthesized RNA to the synthesis of histones, found in this work, does not agree with earlier data (Nemer, 1975): the differences are apparently due to the fact that here the experiments were conducted on stage-16 to 32 embryos and only histone H1 was studied.

A similar investigation on amphibian hybrids (*X. laevis* × *X. borealis*) demonstrated that newly synthesized paternal templates for histone H1 appear only from mid-blastula (when the RNA synthesis in the nuclei is sharply intensified), and translation of the oogenesis-stored maternal RNA's for this histone continues only up to early gastrula (Woodland *et al.*, 1979).

The fraction of histone templates in total mRNA changes considerably in the course of sea urchin development (Kunkel and Weinberg, 1978). In the two species studied (*S. purpuratus* and *L. pictus*) they constituted 9-12 and 19-20% during cleavage and only 0.5-0.6 and 1.4% at mesenchyme blastula, respectively. This reduction was partly due to the intensification of transcription of nonhistone genes and, to a large extent, also to the reduction in the total quantity of mRNA's transcribed per unit time per cell and per embryo (Kunkel and Weinberg, 1978). The quantity of histone mRNA's synthesized in one cell of *S. purpuratus* from blastula (140 cells) to mesenchyme blastula (420 cells) drops from 1.45×10^{-3} to 3×10^{-5} pg per min, and in the entire embryo, from 2×10^{2} to 1.4×10^{-2} pg. Such a drastic decrease may be attributed to transition in histone transcription from multiple genes of the early type to a much smaller number of genes of the late type, rather than to the differences in the transcription rate of each gene.

Histone mRNA's are characterized by very intensive translation, which exceeds that of other templates (Newrock *et al.*, 1978).

Accumulation of mRNA's Encoding Specific Proteins in Differentiating Cells

While the presently available methods enable one to estimate only the diversity of mRNA species when considering transcription at the embryonic stages, in the case of differentiated cells one can already speak of transcription, accumulation, and translation of individual mRNA's. These data are particularly important because they help evaluate the true importance for cell differentiation of such events as switching on of genes or increase in the rate of their transcription, accumulation, and storage of mRNA and, finally, the efficiency of translation of specific proteins.

Until recently, the appearance and life span of individual mRNA's could be estimated only indirectly, by the synthesis of the corresponding proteins. In a number of cases the sedimentation constant is a sufficient characteristic of mRNA's: 9S RNA in reticulocytes corresponds to globin mRNA, 9S RNA in sea urchin embryos, to the mRNA of histones (Nemer, 1975); 26S poly(A) RNA in myoblasts is the largest and corresponds to mRNA for the heavy myosin chain (Buckingham *et al.*, 1974a). The content of specific mRNA's in preparations can be assessed by the synthesis of the corresponding proteins in a protein-synthesizing system *in vitro* or in the oocyte after an injection of these RNA's (see, for instance, Levy *et al.*, 1975). In recent years, however, the method for determining the quantity of individual mRNA's by hybridization with complementary DNA's (cDNA's), i.e. with copies synthesized by reversed transcription of mRNA, has been developed and increasingly applied. This method has proved much more sensitive than all the others (O'Malley *et al.*, 1975). At present, several models of cell differentiation are known, in which the number of molecules of mRNA encoding the specific protein responsible for its differentiation has been determined.

When the lens epithelium is placed in a serum-containing culture medium the cells begin to elongate and differentiate into lens fibers. δ-Crystallin is then produced in them (Milstone and Piatigorsky, 1975). ^{3}H-DNA which encodes δ-crystallins was synthesized on purified 17S mRNA. It helped determine the number of δ-crystallin mRNA molecules in the cells of the lens epithelium in the course of its differentiation into lens fibers (Milstone *et al.*, 1976).

The initial epithelium cells contain about 6×10^3 molecules of crystallin mRNA per cell, and this number remains unaltered 5 h after the beginning of serum stimulation of differentiation. But it increases almost three-fold within the subsequent 19 h. Since the quantity of the crystallin mRNA in the nuclei also increases several-fold, the increase in the number of templates is the result of enhancement of their transcription.

In the chicken lens, the epithelium cells contain 1.2×10^4 molecules of δ-crystallin mRNA, and the fiber cells 4 times as many (Milstone *et al.*, 1976).

The composition of the lens mRNA was studied by comparing the translation products in cell-free systems (Thomson *et al.*, 1978). It transpired that the templates for the synthesis of the main crystallins (α, β, and δ) were located both in poly$(A)^-$ and poly$(A)^+$ fractions. The compositions of crystallin RNA and of the translated proteins showed good agreement: the ratio of mRNA for α-, β-, and δ-crystallins in the lens epithelium and fibers and at different stages of development reflects the synthesis of the corresponding proteins in these cells (Thomson *et al.*, 1978). When the embryonic retina and pigment epithelium are cultivated for a long time, the former undergoes a transdifferentiation into lens cells and the amount of crystallin mRNA's, which were initially represented by single copies, increases by one or two orders of magnitude (Thomson *et al.*, 1979).

An analysis of complexity of the lens mRNA carried out by the same research group (Jackson *et al.*, 1978) showed the presence in the lens of a one-day old chick, of $6\text{-}7\times10^3$ different species of sequences, of which only four were represented abundantly, about 200 species, by a much smaller number of copies, and the others, by single mRNA molecules. The abundant mRNA's correspond to crystallin mRNA's, they are very sparse in the retina and the segment epithelium and are absent altogether outside the eye.

Another example of differentiation, where the number of mRNA molecules was measured, is the stimulation of the ovalbumin synthesis in the oviduct gland cells of young chickens by estradiol injections. With daily hormone administration the oviduct gland cells differentiate and the quantity of mRNA, as determined by hybridization with cDNA, increases from 0 to 5×10^4 molecules per cell within 18 days (the oviduct cells of an adult chicken contain 1.4×10^5 molecules each). After the cessation of estradiol administration the quantity of specific mRNA's drops to zero again. But a renewed estradiol administration leads to the appearance of the first ovalbumin mRNA molecules already within 30 min (O'Malley *et al.*, 1975). Three hours later their number reaches 10^3 and 24 h later, 4×10^3.

Palmiter (1975) gives somewhat different figures: on second ovalbumin synthesis stimulation the quantity of the ovalbumin mRNA's after 4 days reaches 7×10^4 as compared to 10^5 per cell in laying hens. According to him, the steady state achieved on the 5th day corresponds to the synthesis (and decay) of 34 ovalbumin mRNA molecules per minute per cell, with a half-life of these molecules of 24 h.

Induction of transcription of ovalbumin genes by estradiol is strictly specific, since ovalbumin mRNA's are not detected in other tissues or in the liver of uninduced chicks (Ono and Getz, 1980).

Myogenesis is another model used to study the accumulation of mRNA's specific to terminal differentiation. Several authors (Buckingham *et al.*, 1974a; Dym *et al.*, 1979; Robbins and Heywood, 1978) have shown that myosin mRNA can be detected in myoblasts before they fuse together into myotubes.

Synthesis of most of the mRNA's specific to terminal differentiation of muscles apparently takes place about as early, a few hours before myoblast fusion and synthesis of muscle proteins.

Zevin-Sonkin and Jaffe (1980) demonstrated this by means of hybridization of cDNA enriched in myotube-specific sequences, with mRNA's of various differentiation stages. They found that after several hours of incubation the proliferating myoblasts revealed accumulation of a large number of mRNA sequences absent in the liver.

A quantitative estimate of mRNA accumulation (Robbins and Heywood, 1978) showed that the number of myosin mRNA molecules in the myoblast culture increases from 5×10^2 at 3.5 h to 4.5×10^3 at 24 h and 1.2×10^4 at 48 h of incubation. The myoblasts begin to fuse together into myotubes only after 24 h and the myosin synthesis is at a low level. By 67 h, at the moment of myosin synthesis activation, the quantity of the mRNA decreases five-fold (down to 2.2×10^3 molecules per cell).

Besides myogenesis, other examples of differentiation are described, when the mRNA's of tissue-specific proteins accumulate beforehand and are not translated, remaining in the form of RNP complexes (informosomes). For example, transcription of casein mRNA in the mammary gland cells constitutes 15% of the total mRNA already in the middle of pregnancy, and by the beginning of lactation and casein synthesis the quantity of casein mRNA exceeds that of all the other mRNA's taken together (Takemoto *et al.*, 1980). In trout spermatocytes there is a transcription of protamine RNA's, whose synthesis begins only in the spermatides (Iatron *et al.*, 1978).

It is obvious that the quantity of each mRNA species in the cell is determined by the ratio between the rates of their synthesis and decay. But at the beginning of differentiation, when the quantity of mRNA of a given species in the cells is still small, the synthesis rate exceeds the degradation rate, which is a function of the number of molecules. Accumulation of mRNA molecules apparently continues until the number of molecules hydrolyzed becomes equal to that synthesized per unit time. An increased stability of mRNA, however, which can occur at a definite stage of differentiation, disturbs this equilibrium and leads to a new increase in the number of mRNA molecules in the cell. This increase in mRNA half-life has been described in a number of cases for crystallins (Yoshida and Katoh, 1972) and for myosin (Buckingham *et al.*, 1974a).

Investigations into the synthesis and accumulation of RNA in differentiating cells, where tens and hundreds of proteins are synthesized, have led to elaboration of a new technique, when a library of cloned cDNA's from given cells is first established, and then the synthesis and quantity of many RNA species are studied with its aid. Such an investigation was started by Kafatos and his coworkers on follicular cells of the silk moth, which synthesize the proteins of egg chorion (Sim *et al.*, 1979). The chorion of these eggs contains over 100 proteins, and the spectrum of their synthesis varies consistently for two days at the end of oogenesis. During this time the concentration of certain RNA species increases or decreases tens of times within a few hours, and the composition and relative concentration of the proteins synthesized change correspondingly. The two-day process of chorion formation falls into three periods according to the composition of the alternating transcribed RNA's, and for many mRNA's their respective chorion proteins translated during this period have been identified. The clones obtained correspond exclusively to the RNA's represented by a large number of copies, while the expression of most of the genes in the follicular cells has not yet been investigated (Sim *et al.*, 1979). This remarkable series of works by the Kafatos group on the synthesis of chorion proteins reveals the problems facing the researcher in studying the RNA composition even in specialized cells. The study of transcription at embryonic stages will apparently require new approaches.

RNA PROCESSING AND TRANSPORT INTO THE CYTOPLASM

The mechanisms of splitting and selective degradation of parts of giant RNA molecules (maturation, processing) and the mechanisms of selection and transport of the RNA into the cytoplasm after processing are not sufficiently understood. It is obvious that they involve specific proteins described by Georgiev (informofers), which form RNP complexes in the nuclei, and also specific ribonucleases and methylases. The involvement of proteins, however, does not eliminate the problem of identifying the various RNA sequences determining the points of splitting or methylation, the choice of sequences to be destroyed, and selection upon the transport out of the nucleus. Lately, this has been associatied with poly(U) and poly(A) sequences, the creation of the secondary RNA structure, and, in particular, with the formation of "hairpins" — loops stabilized by hydrogen bonds between complementary nucleotides. It is quite probable that the stages of proccessing of giant RNA molecules and of the mRNA transport are a single process, in which each of the subsequent links — methylation, splicing, splitting, degradation, polyadenylation, selection, and transport — is determined unambiguously by the preceding one. Therefore the division of the whole process into two — processing and transport — may prove purely tentative. On the basis of what is now known with respect to differentiating cells, all the posttranscription changes, such as the decrease in the molecular weight of the RNA after its synthesis, its methylation, and other changes specific to the translated RNA, may be regarded as processing.

The course of processing is described most comprehensively for rRNA, where this phenomenon consists of successive processes of methylation of the giant precursor, the splitting off and degradation of redundant fragments, producing 28S and 18S RNA, which actually constitute the ribosomal high-molecular-weight RNA's. It is assumed that methylation points are determined by the primary RNA structure, and the methylated nucleotides, in turn, protect the 28S and 18S rRNA during the splitting off of the noncoding sequences from the precursor. Both stages are controlled by methylases and nucleases. For methylases, it is known, mainly from the data on selective methylation of prokaryotic DNA, that they are capable of recognizing definite nucleotide sequences (see Bogdanov and Ledneva, 1975; Murray, 1975). The splitting off and degradation of RNA are controlled by specific endo- and exonucleases.

The processing of messenger RNA's is much more complicated. It includes, as the first stage, polyadenylation of most of the pre-mRNA's from the 3′-end with the aid of polyadenylate polymerase. What is the biological significance of polyadenylation and why some mRNA's are polyadenylated and others are not, is unclear so far. To the 5′-end of the mRNA molecule is added a cap formed by 7-methylguanosine which is bound with one or two methylated nucleotides via a triphosphate bond: $m^7GpppNmNm$ (Furuichi *et al.*, 1975). It is believed that the cap serves as a signal for mRNA recognition by a specific center of the ribosome and for the formation of the initiating complex (Hickey *et al.*, 1976).

The most complicated and intensely studied process is splicing, i. e. the cutting out of nonencoding sections of pre-mRNA complementary to introns and joining the encoding fragments complementary to exons. It is assumed that the enzymes which accomplish splicing are RNP complexes and contain low-molecular-weight RNA's, which determine the specificity of splicing. At the 5′-end of one of such RNA's (U_1-fraction) are three to eight nucleotides which proved to be complementary to the 5′-end of 43 introns of different genes. A bit farther (9th to 11th and, possibly, 12th to 20th nucleotides) follows a sequence complementary to the 3′-end of the same introns (Lerner *et al.*, 1980). It is supposed that the RNA of the splicing enzyme draws the intron ends together, as it were (bringing the neighboring exons closer to one another), while the protein portion of the enzyme (apoenzyme) cuts pre-mRNA along the boundaries of the encoding and nonencoding sections and joins the neighboring encoding portions, thereby discarding the noninformative part of the molecule.

Processing of Messenger RNA's

The processing of messenger RNA's in development may be indicated by the data on the differences in the RNA composition in the nuclei and the cytoplasm. But one cannot ignore the fact that part of the nuclear RNA's are not mRNA precursors. The differences between nuclear RNA's and cytoplasmic mRNA's are evident from their size distribution, in the relative content of repetitive sequences, and, finally, in their diversity (complexity), i. e. in how many types of sequence exist and by what number of copies they are represented in the two populations.

A comparison of the nuclear and cytoplasmic RNA's distribution in the sucrose gradient by just their optical density, i. e. by the absolute quantity, indicates the existence in the nuclei of heavy RNA's (>28S), which are practically absent in the cytoplasm. These differences are particularly prominent when one compares labeled, i. e. newly synthesized, RNA's. In loach embryos at the gastrula stage a considerable part of the label in the nuclei is found in the range of 30S to 60S, whereas in the cytoplasm most of the labeled RNA sediments in the range of 2S to 14S, with a peak at 8S to 10S (Rachkus *et al.*, 1971). It is obvious that the processing of most of the mRNA's consists in shortening the pre-mRNA's from 30-50S to 8-10S mRNA's, i. e. by about 15 to 25 times. In sea urchin embryos these differences seem to be not so pronounced: the size of the nuclear RNA is $0.5\text{-}3\times10^6$ daltons, and that of the cytoplasmic RNA, $0.1\text{-}2\times10^6$ daltons (Kung, 1974). The author assumes that in the evolution, as the organization increases, so does the size of nuclear RNA's and, consequently, the degree of degradation of the pre-mRNA's in the course of processing.

A comparison of hybridization with DNA's shows that nuclear RNA's contain no more than 20% of repetitive sequences, whereas the cytoplasmic RNA's hybridize with them by 40-50% at these early stages, i. e. they contain a considerable number of repeats. Later on, during organogenesis, the difference between the nuclear and cytoplasmic RNA's decreases and the content of repeats in the nuclear RNA's drops to 15%, and in the cytoplasm, to 8-12% (Rachkus *et al.*, 1969a). It may be assumed that in the course of development the nature of the processing changes somewhat and at the late stages a larger number of single-copy sequences go into the cytoplasm than at the early stages. The RNA transcribed from repetitive sequences has recently also been found in the cytoplasm of a frog embryo (Dobner and Flickinger, 1976).

In similar experiments on sea urchins at the blastula and gastrula stages (Volfson and Vorobyov, 1972) it was also shown that at the successive posttranscriptional stages the share of repetitive hybrid-forming sequences gradually increases. For chromosomal and nucleolar RNA this value is 21%, for the total nuclear RNA, 35%, for the total cytoplasmic RNA, 57%, and for the polysomal mRNA, 60%. Of special interest were the differences between chromosomal and nucleolar newly transcribed RNA (extracted with hot phenol) and the total RNA of the nuclei which included the readily extractable fraction. From these data it follows that the increase in the fraction of repeats is achieved precisely by the processing of the initial transcripts into the cytoplasmic RNA.

As shown in the same study (Volfson, 1972; Volfson and Vorobyov, 1972), stage-specific differences increase in the course of processing: while chromosomal and nucleolar RNA's (which have not gone through processing) practically do not differ from each other at the blastula and gastrula stages as regards competitive hybridization these differences are considerable for cytoplasmic RNA's. It is more probable that stage-specific differences do not arise in the course of processing and selection but are only revealed owing to these phenomena: identical RNA fragments degrade in the nuclei and a small part of the RNA which contains sequences characteristic of different stages of development passes into the cytoplasm.

The change in the nature of processing or, more precisely, in the share of nuclear RNA's split off in the course of the sea urchin development, has also been shown by comparing the intensity of RNA synthesis in the nuclei with the share of the RNA transported into the cytoplasm at the successive stages from cleavage to pluteus (Kijima and Wilt, 1969). While the RNA synthesis rate per embryo increases continuously, the same parameter calculated per nucleus decreases. In the period from early blastula to gastrula the synthesis rate per nucleus drops 5-fold. At the same time the transport of the RNA into the cytoplasm even increases during the same time. Consequently, the share of the RNA degraded in the nuclei is particularly high at the beginning of development and the percentage of the RNA entering the cytoplasm increases several-fold in the course of development. These data, however, fail to indicate whether larger portions of the pre-mRNA decay at the early stages of development or, what is more probable, a larger number of the nuclear RNA's are completely hydrolyzed.

A comparison of single-copy RNA sequences at the gastrula stage in the sea urchin showed that nuclei contain 10 times more species of sequences than the polysomal mRNA of the cytoplasm. In this case, too, it is difficult to know which part of this excess of nuclear RNA species is contained in the degrading fragments of pre-mRNA's, and which in the specifically nuclear RNA's degrading completely (Kijima and Wilt, 1969).

The large differences in the number of RNA species (RNA complexity) in the nuclei and in the polysomes and the considerable quantity of RNA's completely decaying in the nuclei pose the question of the role of the processing and selection in the control of protein synthesis. It is possible that a larger number of species or copies of pre-mRNA's than that of mRNA's transported into the cytoplasm are transcribed in the nucleus. If this assumption were true, the role of the processing would consist not merely in the transformation of pre-mRNA's to mRNA's, but also in the selection of the pre-mRNA's species to be processed. Then the processing could be regarded as a new level of regulation substantially affecting not only the transport rate, but also the composition of the mRNA's in the cell and their quantity. This problem will evidently be resolved when the composition of nuclear RNA's, their function, and mechanisms of their degradation are clarified.

Polyadenylation and Capping

Polyadenylation is an important stage of the processing (see review by Brawerman, 1976). After its discovery it was assumed that all mRNA's contain poly(A) fragments. The relative simplicity of determining and isolating the poly(A)-containing molecules (on a cellulose column bound with poly(U) or oligo-dT sequences) would have seemed to provide a reliable method for determining mRNA. It turned out, however, that histone mRNA's contain no poly(A) fragments, and then it was found that the share of mRNA's already being translated on polysomes and devoid of poly(A) can also be considerable (Nemer, 1975; Fromson and Verma, 1976; Sonenshein *et al.*, 1976).

Polyadenylation is one of the first stages of the processing and presumably occurs already at the chromosomal level, i. e. immediately after the termination of transcription (Bonaldo *et al.*, 1979).

It has already been mentioned that in the course of oogenesis not all the mRNA's are subjected to polyadenylation, and that a considerable part of them in sea urchin eggs is adenylated in the cytoplasm only after fertilization (Wilt, 1973, 1977; Slater and Slater, 1974, 1979; Brandhorst and Bannet, 1978; Slater *et al.*, 1978) or egg activation (Wilt and Mazia, 1974). This phenomenon, however, was not observed in mammalian embryos (Schultz, 1975), though it was shown in the mammalian tumor cell culture that the completion of the poly(A) fragment may occur in the cytoplasm as well (Brawerman and Diez, 1975). The activity of the enzyme responsible for polyadenylation (polyadenylate polymerase) remains unaltered in the course of development, but its localization changes: it moves from cytoplasm into nuclei (Egrie and Wilt, 1979). It is possible that intensive polyadenylation in the cytoplasm at the very beginning of development is simply due to the enzyme localization.

In *Urechis caupo*, the quantity of poly(A) increases 5-fold within the first hour after fertilization (Davis and Davis, 1978), which indicates that during oogenesis the bulk of mRNA contains no poly(A) and is polyadenylated in the cytoplasm after fertilization.

In the course of embryonic development the share of poly(A)-containing mRNA's translatable on polysomes changes; in the mollusc *Ilyanassa* embryos, for instance, it falls off in the course of development (Collier, 1975). According to Nemer, in sea urchin embryos, at the earlier stages 60% of all the templates are histone mRNA's, most of which contain no poly(A) fragments, and among the remaining ones only about 10% of the mRNA's do contain them. At the hatching (mid-blastula) stage, the quantity of histone mRNA's in the cytoplasm diminishes as well and they are replaced with poly(A)-containing mRNA's whose share then increases. The nonhistone mRNA's lacking poly(A) constitute about 30% of all the mRNA's throughout the development period investigated (up to early pluteus) (Nemer, 1975) (Fig. 36). Nemer has also established that poly(A)-free mRNA's bind fewer ribosomes in the polysome than mRNA's containing poly(A) fragments. And, finally, it is of principal importance that poly(A)-containing mRNA's almost do not compete with poly(A)-free mRNA's in the hybridization reaction with DNA, i. e. they encode different proteins (Nemer, 1975). This conclusion was confirmed in a subsequent investigation, where it was shown that at the mesenchyme blastula stage of the sea urchin the poly(A)-containing and poly(A)-free mRNA's have identical size distributions (with a maximum at 26S) and are trans-

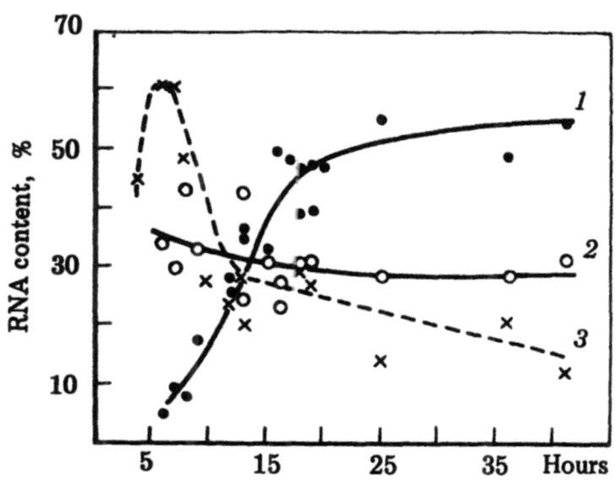

Fig. 36. Relative content of poly(A) mRNA in sea urchin embryos. Sea urchin embryos at different stages of development were incubated for 1 h with ^{3}H-uridine, after which mRNA was isolated from the polysomes. According to the ability for binding with oligo(dT)-cellulose, the mRNA was fractionated into poly(A)$^+$mRNA (containing a poly(A) fragment) (*1*), nonpolyadenylated poly(A)$^-$mRNA (*2*), and 9S histone mRNA subsequently obtained from poly(A)$^-$mRNA (*3*). The relative content of these fractions changes in the course of embryogenesis (Nemer, 1975).

lated in an *in vitro* system with similar effectiveness, but differ in the composition of the proteins encoded by them (Fromson and Verma, 1976). Later, however, in a more careful investigation using two-dimensional electrophoresis, the same authors failed to detect any differences in the proteins translated *in vitro* on poly(A)$^+$ and poly(A)$^-$ mRNA (Brandhorst *et al.*, 1979). Owing to the differences between nuclear high-molecular-weight RNA's as to the length of the poly (A) fragment, it was possible to divide them into three classes: α — those containing an oligo(A) fragment (25 nucleotides), β — those with long poly(A) fragments (175 nucleotides), and γ — poly(A)- and oligo(A)-free fragments. In the course of development class γ decreases (76% at early blastula, 63% at mesenchyme blastula), while class α and, particularly, β increase (Dubroff and Nemer, 1976).

In *Xenopus* oogenesis mRNA's with two size classes of poly(A) are synthesized. Long-lived mRNA's with relatively short (15 to 30 adenylate residues) poly(A) fragments are formed in early oocytes and mRNA's with a shorter life span and longer (40 to 80 adenylate residues) poly(A) fragments at later stages (Cabada *et al.*, 1977). After fertilization the number of these classes increases and there appear mRNA poly(A) fragments having a length of 100 adenylate residues (Sagata *et al.*, 1980).

In amphibians, as in the sea urchin, polyadenylation occurs partly in the cytoplasm, at the cleavage stages, when the new RNA synthesis is low. Owing to this the total number of poly(A) fragments in the oocytes and eggs changes, decreasing by deadenylation in the course of oocyte maturation and increasing after fertilization. Then the quantity of poly(A)$^+$ RNA increases considerably in the course of development from gastrula to larva (Sagata *et al.*, 1980).

Thus, at the stage of polyadenylation some sorting out of pre-mRNA's takes place, as a result of which some types of template receive poly(A) fragments, while the others do not. Later on, this may affect the translation efficiency (Sonenshein *et al.*, 1976) and the life span of the mRNA's.

Poly(A) fragments are bound by specific proteins whose role is not yet known. In early sea urchin development, at stages from fertilization to the 8-cell embryo, the quantity of poly (A) fragments bound with these proteins increases 2.5-fold and reaches 60% of the total poly(A)'s. These proteins contain two polypeptides with a molecular weight of 8.7×10^4 and 1.3×10^5 daltons, which form complexes with poly(A)'s sedimenting in the range of 8S to 12S and 15S to 20S, respectively. It is assumed that these proteins are involved in the translation control (Peters and Jeffery, 1978).

Mitochondrial RNA's are also subjected to polyadenylation (Avadhani, 1979).

The mRNA's stored in eggs and those synthesized in embryos of the sea urchin are practically fully capped. This follows from the fact that the egg and embryo mRNA preparations show template activity in the wheat germ cell-free translation system. Addition of a methylation inhibitor S-adenosylhomocysteine or of S-adenosylmethionine does not affect the template activity, while removal of the cap reduces it considerably (Pirrone *et al.*, 1977).

Processing of Ribosomal RNA's

The complicated and consistent picture of splitting and decay of giant RNA's was investigated earliest of all and most comprehensively on mammalian ribosomal RNA's, where out of a precursor with a molecular weight of 4.2×10^6 daltons (45S) there remain two molecules with a molecular weight of 1.5×10^6 daltons (28S) and 0.75×10^6 daltons (18S), while the other segments degrade already in the nucleolus. In many eukaryotes this process is less complicated, the precursor has a lower weight of about 3×10^6 daltons (39-40S), and only 20%, instead of 50%, of the initially transcribed molecules degrade as a result of the processing (Perry *et al.*, 1970; Hadjiolov, 1977).

In the mollusc *Urechis*, the rRNA precursor has a size of 38S. It is quickly processed to 30S and migrates in this form from the fibrillar part of the nucleolus into the granular cortex, where it slowly transforms into mature rRNA's (Das *et al.*, 1970).

In the sea urchin, the rRNA precursor is still smaller: 33S (molecular weight of 2.6×10^6 daltons). It disintegrates, in the course of processing, into 28S and 18S, then 28S splits off one more fragment and turns into mature 26S rRNA. This second stage is much slower than the first (Sconzo *et al.*, 1971).

The contradiction between Giudice's data and those of Emerson and Humphreys as to the beginning of rRNA synthesis seemed to be attributable to the delay in the processing, i.e. to the fact that the synthesis of pre-rRNA's begins in the sea urchin already at the blastula stage, and the formation of mature rRNA's and their incorporation into the ribosomes, only at gastrula. Indeed, the synthesis of heavy RNA's was detected both at blastula and gastrula (Giudice and Mutolo, 1969). But the authors failed to show that the rRNA precursor is actually present among these RNA's at the blastula stage (Giudice, 1973).

And yet, a change in the processing rate during the sea urchin development is actually observed. Since 18S RNA's are released already at the first stage of the processing, labeled 18S RNA's appear in the cytoplasm earlier than 26S rRNA's, and their molar ratio in the first few hours is not equal to unity. It was found that these differences, i.e. the delay in the release of 26S rRNA's, are greater at the gastrula than at the prism stage (Giudice and Mutolo, 1967).

The processing rate is different not only at different stages of development but also in different parts of the embryo. Thus, when an early embryo is divided into macro-, micro-, and mesomeres with their subsequent cultivation as aggregates (Hynes *et al.*, 1972b), the 26S rRNA : 18S rRNA radioactivity ratio after 4 h of incubation at a stage equivalent to early gastrula is 0.25 for mesomeres and 0.51 for micromeres. At a later stage, equivalent to early pluteus, these ratios increase to 1.45 and 1.22, respectively, but still do not reach the molecular weight ratio between 26S and 18S rRNA's, i.e. 2.0.

In trout embryos, the processing rate decreases if the blastoderm is incubated separately from the yolk. After a 20-hour incubation with labeled precursors at 4°C, most of the label in intact eggs is found in the peaks of 28S and 18S, whereas in an isolated blastoderm it is still predominantly in the precursor (40S) (Melnikova *et al.*, 1972).

It might be thought that here the delay in the processing is due to the lack of methionine, the methyl group donor. Such a delay mechanism was revealed on incubation of isolated salivary glands of *Drosophila* (Greenberg, 1969). However, addition of methionine to the trout blastoderm culture failed to speed up the processing (Melnikova *et al.*, 1972).

In amphibians, the transport of 18S rRNA's into the cytoplasm during embryonic development also precedes that of 28S rRNA's. On dissociated cells cultured at different stages in the presence of ^{3}H-uridine it was shown that the molar ratio between 18S and 28S rRNA's at early gastrula is 1.1 to 1.2, decreasing to 1.0 only in tadpoles and further to 0.9 in adult animals (Abe and Yamana, 1971).

In the course of early development of mammals the synthesis and processing of rRNA's are rapid enough, and mature molecules of 28S and 18S rRNA are detected at the 4- and even 2-blastomere stage (Knowland and Graham, 1972). In the mouse blastula (4-day embryo), after 30-minute incubation with ^{3}H-uridine, almost the entire label is observed in the region corresponding to the heavy precursor, but already after 60 min it is found predominantly in the 28S and 18S rRNA peaks (Pikó, 1970, 1975). However, no detailed comparison between the processing rates at different stages of early mammalian development has been made.

At a later stage of embryonic development of the mouse, the rRNA processing in differentiating erythroid cells was investigated in the yolk sack of 10- and 13-day embryos (Fantoni *et al.*, 1972). The authors showed that although the synthesis rate of pre-rRNA's in the erythroblasts of early embryos is much higher, their processing occurs differently. The formation of 28S rRNA's in early embryos is delayed, and 18S rRNA's decay on entering the cytoplasm. At a later stage (13 days) practically the whole pre-rRNA undergoes processing and forms mature 28S and 18S rRNA's.

Finally, in the differentiating cells of the lens the processing of the rRNA precursor occurs differently in terminally differentiated cells of the lens epithelium and lens fibers and in the dividing and differentiating cells of the equatorial zone, where the cells proliferate and turn into lens fibers. In differentiated cells, 45S pre-rRNA is synthesized and goes through the first stage of its transformation to 30S molecules. But practically no mature 28S and 18S rRNA's form in these cells. At the same time, in the lens epithelium zone where the cells divide, elongate, and turn into lens fibers, a rapid processing of the pre-rRNA's into 28S and 18S rRNA's takes place (Papaconstantinou and Yulku, 1968).

Thus, in various groups of animals rRNA processing, even if it is not a regulatory process, always depends on the stage of development and the nature of differentiation.

Processing of Transfer RNA's

The synthesis of tRNA's, as well as that of rRNA's, proceeds with the formation of a high-molecular-weight precursor whose processing consists in removal of noncoding sequences from the 5′- and 3′-ends of the precursor, in methylation, and other modifications of the polynucleotide chain, and, in some cases, in splitting out introns and adding the ACC-end. Specific enzymes (endonucleases, methylases, splicing enzymes, etc.) operate at each stage (see Burdon, 1971; Altman, 1978; Venkstern, 1980).

The most successful technique for studying the tRNA processing in eukaryotes is microinjection of various cloned tRNA genes into the nuclei of the *Xenopus laevis* oocytes. In these experiments it was shown that the processing enzymes are localized in the nucleus (Melton *et al.*, 1980).

In our work (Solovyova *et al.*, 1973b; Timofeeva and Solovyova, 1973) the synthesis and processing of tRNA were investigated at the early stages of loach development. Incubation of embryos with $^{14}CO_2$ or of blastoderm with ^{3}H-uridine or ^{3}H-cytidine at the 6-hour stage (mid-blastula) and later resulted in intensive label incorporation into the 4S peak. Hence, at these stages the tRNA synthesis is already sufficiently intensive. But during incubation of earlier embryos (4-5 h, early blastula) with a labeled precursor a moderate yet distinct label incorporation in the region shifted to the heavier side from the 4S peak was detected. In an actinomycin chase for 2.5 h all the label shifts to the 4S region, i.e. the pre-tRNA's undergo a slow processing into mature tRNA's. Methylation of tRNA's with the methyl group of methionine proceeds actively at the 6-hour stage and later. Not only 4S RNA's but also larger molecules (the heavy shoulder of the 4S peak) are methylated. The activity of methylases in the loach blastoderm extract at the morula-blastula stages is very low, but it increases by the gastrula stage (Osterman *et al.*, 1974).

All these data show that in loach embryos a pre-tRNA synthesis begins already at early blastula (4 h), even before the beginning of an active mRNA synthesis (6 h), but its processing at these stages is retarded. The processing is activated later, at the 6-hour stage. Then the tRNA synthesis is enhanced, and the relative fraction of the label incorporated in the pre-tRNA sharply diminishes (Solovyova *et al.*, 1973b).

Thus, two slightly time-shifted processes — activation of the pre-tRNA synthesis and activation of its processing — can be distinguished in the formation of tRNA during the early development of the loach. As in the case of rRNA, partial methylation is probably necessary to trigger the tRNA processing. But after the activation of the splitting off of noncoding nucleotides and the attainment by the tRNA molecule of its final size (4S) it undergoes further methylation, which ensures the buildup of the secondary structure.

In concluding this section it can be said that processing is as typical for embryogenesis as it is for differentiated cells. It is observed at all the stages of development, beginning with the earliest and ending with terminal differentiations (erythropoiesis). Its role in developing systems, however, is more complicated. In tissues with a stable set of active genes and proteins the processing is a necessary stage of posttranscription events resulting in the formation, from nascent giant RNA molecules, of the functionally active RNA molecules of the cytoplasm: rRNA's, mRNA's, and tRNA's. Because of the processing, a certain period of time, constant for each RNA species, elapses between the formation of these molecules and their transport into the cytoplasm.

In differentiating cells, the processing acquires an additional function: it controls the timing of the events occurring in the embryo. It delays the *de novo* formation of ribosomes in the embryos of the sea urchin and mammals and the appearance of new tRNA's in the loach. Differences in the processing also determine the different fate of the RNA's in the nuclei of bird erythroblasts and reticulocytes.

So far we have little ground to believe that the processing plays an active role, i.e. changes the time of RNA transfer from the nucleus, and thereby changes the course of events in the cytoplasm. Based on erythropoiesis events, it would seem more probable that the change in the set of active genes is the primary factor, and that the different types of processing only reflect the differences in the composition of the RNA's synthesized. However, at the early stages of development the delay in the rRNA or tRNA processing may be due not to the primary structure of their precursors, but, say, to the methylase or nuclease activity.

Anyway, it is obvious that the data on the features of the processing at different stages and for different RNA's are a necessary component of molecular developmental biology.

Transport of RNA's from Nucleus into Cytoplasm

Study of RNA transport from the nucleus into the cytoplasm is a difficult problem which cannot be fully resolved as yet. Four processes occur concomitantly in the cell: RNA synthesis, RNA decay in the nucleus, RNA transport from the nucleus into the cytoplasm, and RNA decay in the cytoplasm. There are data indicating that RNA transport from nucleus into cytoplasm requires energy of ATP or ADP hydrolysis (Clawson *et al.*, 1978). The RNA transport can be evaluated by the kinetics of label incorporation into the nuclear and cytoplasmic RNA. Since part of the material is inevitably lost in obtaining pure preparations of nuclear RNA, it is necessary, along with determination of the RNA specific activities, to find out independently the RNA content in the nucleus and cytoplasm of the cells investigated.

To evaluate the RNA transport, one must also either exclude the RNA degradation in the nucleus and cytoplasm and the synthesis of new RNA's, or take them into account when processing the experimental results. Overlooking these processes introduces an error, which is the larger the longer the experiment lasts. Under certain conditions this error can be estimated by determining the RNA half-life in the cell. To eliminate the error associated with the continued RNA synthesis use is made of a chase technique consisting of the shortest RNA labeling and the following up of the fate of the labeled molecules. The purpose of such experiments is to ensure that the RNA synthesis continues, owing to the unlabeled nucleoside triphosphates, forming the internal pool of the cell, after all of the labeled precursor is removed. Then the change in label localization helps trace the fate of the RNA after its transcription.

To carry out the chase, the synthesis of the labeled RNA's must be rapidly replaced by the synthesis of the unlabeled molecules. To achieve this, the pool of the precursors must be small and promptly replaceable by their intake from the medium. However, the embryonic cells usually contain a large pool of precursors, so when radioactive nucleosides penetrate the cell they are diluted in the internal pool. As a result, if the labeled precursor is simply removed from the medium, the synthesis of the labeled RNA's goes on for a considerable period. Even if the labeled compound in the medium is replaced by a large excess of the same unlabeled compound, the intracellular radioactive pool will only be diluted but not eliminated altogether.

A complete chase can be achieved only by stopping all RNA synthesis following the formation of the labeled RNA fraction, for instance, with the aid of actinomycin D. Then, in the course of further incubation, it becomes possible to trace the transport of the labeled RNA and even to determine its degradation. However, a serious disruption of the cell operation caused by the blocking of the RNA synthesis cannot but affect the processes occurring in it. For instance, a number of papers report a delay in RNA transport in the presence of actinomycin D. This is probably why one can hardly find any reports in the literature which would present a complete quantitative analysis of the fate of the RNA's transcribed.

Considerable changes in the rate of transfer of newly synthesized

10.3ак.1447

RNA's from nucleus to cytoplasm in the course of the sea urchin embryonic development have been established by Singh (1968). By comparing the 18- and 36-hour stages of development (before the beginning of gastrulation and at gastrula) he showed that the flow of short-labeled RNA from nucleus into cytoplasm is accelerated 10- to 15-fold within this time. Since the RNA synthesis rate per nucleus is higher at the early stages than at the late ones (see Neyfakh, 1971), the slow transport of the RNA at the early stages would have meant its accumulation in the nuclei at the beginning of development. It is, however, obvious that a considerable quantity of the RNA synthesized in the nuclei decays before passing into the cytoplasm. According to some calculations (Aronson and Wilt, 1969), the cytoplasm receives only 6% of the RNA synthesized in the nuclei per unit time. The fact that the specific activity of nuclear RNA's reaches a plateau already after 15 min and then remains constant for several hours suggests that the average life span of nuclear RNA's is very short. At the same time, relatively long-lived RNA's find their way into the cytoplasm (Aronson and Wilt, 1969). The transport of the RNA's from nuclei into cytoplasm also continues when the protein synthesis is blocked (Hogan and Gross, 1972). According to Brandhorst and Humphreys (1972a, b), 85% of all the nuclear RNA synthesized in the sea urchin embryos decays and only 15% is transported into the cytoplasm and incorporated into the polysomes. They found that the average half-life of nuclear RNA's is 7 min, whereas in the cytoplasm half of the mRNA's are hydrolyzed within 75 min.

In loach embryos, a considerable synthesis of nuclear RNA's is detected beginning with the mid-blastula stage (6 h). But the RNA transfer into the cytoplasm begins later, not before the 8th hour, i.e. shortly before gastrulation (Neyfakh *et al.*, 1972). The kinetics of RNA transport from nuclei into cytoplasm in later loach embryos is rather complicated. Within the first hour the cytoplasm receives about 30% of all the labeled RNA. In the subsequent hours the transfer of the previously synthesized RNA continues, so that a certain portion of the RNA leaves the nuclei even 5-6 h after its synthesis.

In amphibians, the fraction of the RNA transported from nuclei into cytoplasm falls off from 60% at early gastrula to 37% at tail bud. The average size of the heterogeneous nuclear RNA increases at the same stages (Shepherd and Flickinger, 1979).

The transport of various RNA species from nucleus to cytoplasm has been described for *Xenopus* cells dissociated at blastula (Shiokawa *et al.*, 1979) and neurula (Shiokawa *et al.*, 1977). At these stages, tRNA's and poly(A)$^+$ RNA's appear in the cytoplasm immediately following their synthesis. However, the transport of 5S RNA, whose synthesis begins at blastula, is delayed until the onset of the synthesis and transport from the nucleus of 28S RNA. In cells dissociated at blastula the synthesis of large rRNA's begins after 5 h of incubation and 18S rRNA appears in the cytoplasm right away, while the transport of 28S rRNA begins 2 h later. The transport of 28S rRNA at other stages of *Xenopus* development shows a similar delay. The low-molecular-weight nuclear RNA's, whose synthesis begins at blastula, are not transported to the cytoplasm (Shiokawa *et al.*, 1977, 1979).

No information is yet available on whether the composition of the RNA's transported into the cytoplasm changes depending on the duration of their residence in the nucleus. But it is clear that during the early development the rate of RNA transport into the cytoplasm is an additional factor in the control of protein synthesis on newly formed templates. This applies primarily to the protein synthesis intensity but probably also to the order of translation of templates encoding different proteins.

Migration of Nuclear RNA's during Mitosis

The behavior of RNA during mitosis is a feature of mitosis as essential as the behavior of the chromosomes, the nucleolus, the spindle proteins, etc. But until recently the fate of nuclear RNA during mitosis has been given very little space in the literature. Slightly more attention has been attracted by the fate of the nucleolus: its disappearance at the early prophase stage and its spreading along the spiralizing chromosomes, which is detectable by methods of classic histology. The fate of the nuclear RNA after the disappearance of the nuclear envelope concerns not only the problem of mitosis. It is of no less interest from the standpoint of the general problem of posttranscriptional changes in the RNA: the mechanisms of the RNA processing and transport from nucleus into cytoplasm. Mitosis in this sense is a unique Nature's experiment for the study of relationships between chromatin and RNA in the nucleus. Indeed, the division of the eukaryotic cell is an event which sharply disturbs the normal cell structure and the interrelationship of its components; the condensation of chromosomes, the degradation of the nuclear membrane, the mixing of nucleoplasm with cytoplasm obliterating all the differences in their chemical composition, the cessation of the RNA synthesis, and a considerable reduction of the protein synthesis, all this might be the price for cell reproduction. It is not then surprising that in many cases of pronounced cell differentiation, such as the formation of nerve cells, red blood cells, osteocytes, etc., mitosis becomes incompatible with preservation of the normal cell structure.

Following the division of nucleus and cytoplasm the mitosis is completed by the restoration of the disturbed cell organization and its normal functions: chromosome decondensation, formation of the nucleolus and the nuclear membrane, resumption of RNA synthesis, etc. This is naturally attended by the restoration of the differences in the chemical composition of the nucleus and cytoplasm. At present, when so much is known about the differences between the nuclear and cytoplasmic RNA's, the problem of restoration of the normal distribution fully refers to this type of macromolecule as well.

There are three theoretical possibilities for nuclear RNA's fate during mitosis: (1) Nuclear RNA's remain bound with the chromosomes; this possibility is supported by observations of the fate of the nucleolus and by some biochemical data (see Chentsov, 1966). (2) Nuclear RNA's mix with the cytoplasm and remain there. Mitosis thus accelerates, as it were, the natural transport of the RNA but, on the other hand, it gives rise to the problem of the fate of those giant pre-RNA molecules which were synthesized immediately before mitosis and had no time to go through the processing. The *de novo* synthesis of nuclear RNA's must apparently take place already in the daughter cell nuclei. (3) A considerable part of the nuclear RNA's enter the cytoplasm for the duration of mitosis and spread throughout the cell, but after the nuclei of the daughter cells are formed they return into these nuclei, thereby restoring the normal distribution of the RNA's in the cell. Our data have shown that the third possibility is the most realistic (Neyfakh, 1970).

The numerous autoradiographic data show that at the beginning of mitosis almost all of newly synthesized RNA leaves the nuclear region and spreads throughout the cytoplasm. Gould (1969b), for instance, notes that in uridine-labeled metaphase cells of *Urechis* embryos the whole label distributes over the cell. The author assumes that this RNA then decays in the cytoplasm.

The assumption about the release of nuclear RNA's during mitosis was not shared primarily by those researchers who attempted to isolate the chromosomes and determine their chemical composition. Depending on the isolation method, the isolated chromosomes revealed, along with DNA and protein, a greater or lesser quantity of RNA, including labeled one (if the cells were incubated with a radioactive precursor prior to mitosis). Thus, in one of such works (Fan and Penman, 1971) it was shown that a considerable part of the total labeled RNA is bound with isolated chromosomes. At the same time it is obvious that chromosomes are capable of adsorbing RNA during their isolation. This is shown particularly clearly in the investigation of Salzman *et al.* (1966), in which unlabeled chromosomes were incubated with labeled rRNA. Here, the label became firmly bound to the chromosomes. RNA adsorption onto chromosomes evidently depends on the isolation conditions (ionic strength, pH, etc.), which can hardly be compared with the conditions in a living cell.

Despite the fact that autoradiographic investigations are usually semiquantitative and fail to determine the RNA type, they produce more reliable information. Obviously, on instantaneous fixation of the cell substantial redistributions of the RNA between chromosomes and cytoplasm are hardly possible. These data indicate unambiguously that a considerable part of the nuclear RNA's synthesized prior to mitosis find themselves outside the chromosomes during mitosis and spread throughout the cell, although part of the label remains bound with the chromosomes themselves.

Autoradiographic preparations of early loach embryos incubated with ^{3}H-uridine showed that in all the interphase cells the label is almost completely concentrated in the nuclei (transport into the cytoplasm within these periods was still insignificant) (Neyfakh and Kostomarova, 1971). At the same time, in many dividing cells the label was scattered throughout the cytoplasm with no preference for the metaphase plate region. During incubation for 1 to 2 h a considerable part of the cells had undoubtedly passed through mitosis (at the early stages the mitotic index is rather high), but no interphase cells in which the label would be scattered throughout the cytoplasm were found on the preparations. These observations would have enabled one to assert that after mitosis the labeled RNA returns to the nuclei, but for one objection. The experimental conditions used do not rule out the possibility that the RNA which is transported into the cytoplasm during mitosis will decay rapidly, and a synthesis of new RNA will occur in the nuclei immediately upon division, using both the labeled decay products and the radioactive precursor pool which remained after the incubation with ^{3}H-uridine.

This possibility was eliminated in specially designed experiments. Isolated loach blastoderms were partly synchronized by cooling (Neyfakh and Rott, 1959), short-incubated with ^{3}H-uridine, and then placed into a medium with actinomycin D where a significant part of them divided during the first hour. It was shown by autoradiography that under these conditions the dividing cells have most of the newly synthesized RNA's (84%) in the cytoplasm. In the cells already divided, almost the whole label was located over the daughter nuclei. Since no synthesis occurs in the presence of actinomycin D, most of the label in the nucleus belongs to the RNA which went into the cytoplasm for the duration of mitosis and returned to the nuclei after mitosis.

The next step in this study was made on a culture of mammalian cells, the Chinese hamster fibroblasts (Neyfakh *et al.*, 1971) (Fig. 37). First ^{3}H-uridine and 10 min later colcemide, which arrests mitosis at the metaphase stage, were added to the synchronous fibroblast culture. Then metaphase and interphase cells were separated (Stubblefield *et al.*, 1967) and incubated in the presence of actinomycin D. In the interphase cells the label passes gradually from nucleus into cytoplasm. In the dividing cells, the label distributes uniformly throughout the cytoplasm and returns to the daughter nuclei after mitosis. As compared with interphase cells, mitosis does not accelerate but, on the contrary, decelerates, or even arrests, the normal transfer of RNA out of the nuclei.

Whether or not this RNA retains the properties of the nuclear RNA becomes clear from our data (Abramova and Neyfakh, 1973). The cytoplasm (mixed with the nucleoplasm) and chromosomes were isolated from the metaphase cells, and the cytoplasm and nuclei, from the interphase cells. Then RNA's were obtained from these fractions and analyzed in the sucrose gradient. Figure 38 demonstrates the results of the experiment. In the fractions of the interphase cells the label distribution

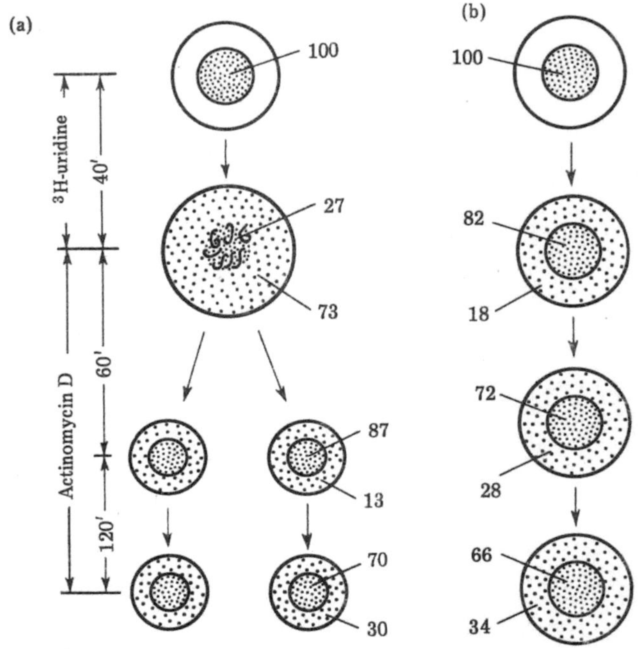

Fig. 37. Migration of newly synthesized RNA during mitosis. Schematic representation of autoradiographic data. Cultured Chinese hamster fibroblasts were short-labeled with ^{3}H-uridine in the presence of colcemide. The metaphase (*a*) and interphase (*b*) cells were separated and incubated in a medium with actinomycin D (2 μg/ml) where the metaphase cells completed their division. The distribution of the labeled RNA above the nucleus and cytoplasm was determined autoradiographically and expressed in percent. In the interphase cells, the newly synthesized RNA gradually passes into the cytoplasm. At metaphase, the entire RNA is distributed uniformly over the cell, and after the completion of mitosis it returns into the daughter nuclei (Neyfakh *et al.*, 1971).

is usual: the cytoplasm contains predominantly ribosomal RNA's (28S and 18S) and also transfer RNA's (4S). No labeled RNA's with a sedimentation constant exceeding 28S are detected in the interphase cell cytoplasm. The nuclei contain, along with some RNA of the same species, heavy RNA's (30-45S), corresponding to the rRNA precursors, and also giant RNA's (50-60S) which have not yet gone through the processing.

The label distribution among the fractions of the metaphase cells is different. The chromosomes contain different species of RNA, including heavy ones. As has been mentioned above, it is difficult to determine to what extent these species of RNA are actually bound with the chromosomes in the dividing cell and to what extent they were adsorbed in the course of chromosome homogenization and precipitation. Unlike the

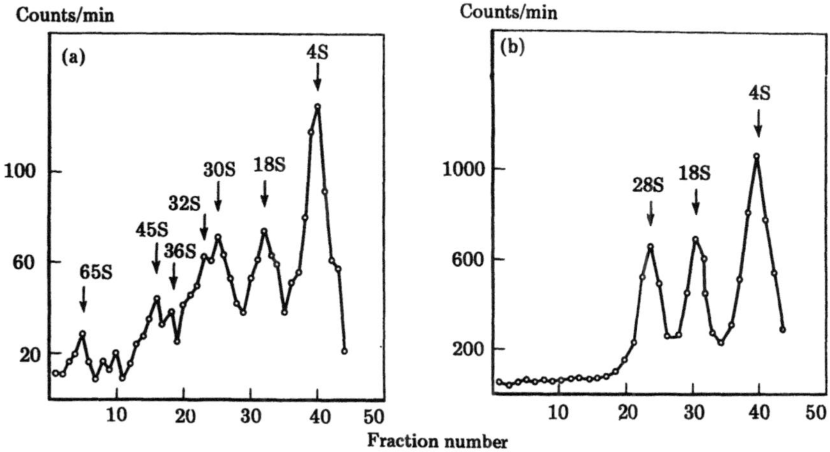

Fig. 38. Giant RNA in the "cytoplasm" of metaphase cells. Metaphase cells short-labeled with ^{3}H-uridine were used to obtain a "cytoplasmic" fraction (4000*g* supernatant), then RNA was isolated and analyzed by centrifugation in a sucrose gradient (*a*). Interphase cells were likewise treated for control (*b*). Curve *a* (metaphase "cytoplasm") shows peaks of giant RNA (30-65S), usually localized in the nucleus and absent in the cytoplasm of the interphase cells (Abramova and Neyfakh, 1973).

interphase cytoplasm, the "cytoplasm" of the metaphase cells has revealed heavy RNA's typical of nuclei and sedimenting at 30-60S. Thus, it has been shown that during mitosis the cytoplasm contains nuclear RNA's which retain their typical properties there, and first of all their high molecular weight.

Similar data on the behavior of RNA in mitosis were obtained on the amoeba, one of the primitive eukaryotes. Rao and Prescott (1970) showed that, in an amoeba with a long-term labeled RNA, during mitosis the whole label is found in the cytoplasm, and after the mitosis, in the nuclei again. Since in these amoebae the nuclei and cytoplasm were labeled uniformly, one cannot assert that the same RNA which had run out into the cytoplasm at the beginning of mitosis returns to the nuclei. We have reproduced these experiments (Yudin and Neyfakh, 1973) with a slight modification: amoebae were incubated for a relatively short time (3 h) with high-labeled ^{3}H-uracyl, which penetrates the amoebae and is incorporated into the RNA. The label in the amoebae does not manage to pass into the cytoplasm within this time. Then the amoebae were transferred into an actinomycin D-containing medium, the just-divided cells were withdrawn and their nuclei isolated. The radioactivity of both daughter nuclei, as determined by autoradiography, agreed well with that of the maternal cell nucleus. We thus confirmed the data of Prescott and Rao, and also reinforced the conclusion that the

RNA returning to the nuclei was the same as the one which had left the nuclei at the beginning of mitosis.

A very important phenomenon has been established in the same species (Goldstein *et al.*, 1969). If an amoeba nucleus with labeled RNA is transplanted into another amoeba, after some time part of the transplanted nucleus labeled RNA will be found in the nucleus of the recipient. The authors showed that part of the nuclear RNA can migrate from nucleus to cytoplasm and back and find its way into the other nucleus, if there are two of them in the cell. This means that a small part of the nuclear RNA resides in the cytoplasm all the time. The nuclear RNA's are evidently in some dynamic nucleus-cytoplasm equilibrium, shifted strongly toward the nucleus. This shift can be attributed to the chemical affinity between the nuclear RNA and the chromatin, which, however, retains not all the nuclear RNA in the amoeba, but only 98%, while 2% is in the cytoplasm (according to Prescott and Rao, the labeled RNA from the transplanted nucleus is distributed in a ratio of 50:1 between nucleus and cytoplasm). It should be recalled that here we deal not with all the RNA's — the cytoplasm has many more of them than the nucleus — but only with the nuclear ones, i.e. those which have not yet left the nucleus or reside in it permanently.

In conclusion we shall consider an embryological example in which the RNA migration during mitosis apparently does not simply serve for restoring the premitotic distribution of the various RNA's between nucleus and cytoplasm.

In some of their experiments Davidson and Hough (1969) used the South-American frog *Engystomops pustulosis*, which has a relatively short oogenesis (of about 3-4 months). Labeled uridine was injected to this frog at the age when the oocytes were at the lampbrush stage. Three months later, when the frog shed eggs and these were fertilized, labeled RNA was found in the nuclei at the early blastula stage. The authors believe that this is RNA synthesized at the lampbrush stage. It is obvious that during oocyte maturation this RNA was in the cytoplasm and then returned to the nuclei.

In our investigation, ^{3}H-uridine was injected to loach females concomitantly with choriogonin, a hormone inducing ovulation and stimulating an RNA synthesis in the germinal vesicle during maturation. It has been shown previously that during maturation the RNA synthesized in the germinal vesicle spreads, on its rupture, throughout the egg more or less uniformly. It was established that it finds its way (returns) into the blastomere nuclei. In this instance one does not necessarily deal with unprocessed mRNA, since many hours elapse between ovulation and cleavage. One may assume it to be specific nuclear RNA which fulfills some special functions in the nuclei. An alternative possibility is that this RNA is the pre-mRNA stored in the nuclei and remaining unprocessed for a long time. Since there is no RNA synthesis in the nuclei at the early stages of development, it is natural to suppose that this

RNA is stored in oogenesis (at the lampbrush stage and during maturation), and then enters the blastomere nuclei. It is to be emphasized that the germinal vesicle RNA, being diluted in the whole egg cytoplasm, is capable of leaving the nuclei and returning to them several times during maturation and on each cleavage. The mechanisms of restoring the nuclear structure after mitosis in this case evidently perform specific functions of RNA storage in oogenesis and during early development.

Now we can try to explain the data on RNA migration during mitosis and the general problem of RNA transport from nucleus into cytoplasm. There is a chemical affinity between nuclear RNA's and chromatin. Recently the RNA-chromatin association was demonstrated directly (Kimmel *et al.*, 1976). The RNA-chromatin bond can be broken by treating the cells with hot phenol. This serves as a basis for the method of thermal fractionation of nuclear and cytoplasmic RNA's after Georgiev: cold phenol extracts readily extractable RNA's, i.e. cytoplasmic RNA's and a small fraction of nuclear RNA's apparently no longer bound with chromatin and dissolved in the nucleoplasm. The hot phenol treatment enables one to extract the chromatin-bound RNA as well. This RNA has presumably not gone through the processing as yet. It may be assumed that it is not simply the large size of the molecules, but some other properties that determine the affinity of nuclear RNA's for chromatin. Maybe it is the presence of special sequences responsible for the bond with chromatin. According to the data obtained on the amoeba this affinity is not too strong, and the dissociation constant is sufficiently high, because a certain part of the nuclear RNA's is capable of moving rapidly into the cytoplasm and back and migrating into other nuclei. It is possible that, in distinction to the amoeba, this affinity is much higher in the cells of higher eukaryotes and the nuclear RNA's are retained more firmly in the nucleus.

In the interphase cell, many (if not all) nuclear RNA's are subjected to changes, i.e. to the processing. In the course of it they are modified, cut into parts, and only some of the segments thus formed are transported into the cytoplasm. The size of these parts may constitute 80% of the whole molecule synthesized (e.g. the cold-blooded animals rRNA), 50% (e.g. the warm-blooded animals rRNA), and even less (as is the case with the globin mRNA). Their transfer out of the nucleus means that they have lost their affinity for chromatin and have first entered the readily extractable fraction of the nucleoplasm, and then passed through the nuclear membrane, which evidently is not an obstacle to RNA, and into the cytoplasm. Thus, the processing and transport of RNA from the nucleus are explained in the same way, and the complicated problem of RNA selection in the nucleus can be reduced to a simpler one, i.e. the presence or absence of regions (sequences) showing an affinity for chromatin.

Thus, transport of RNA from the nucleus in the interphase as a result of the processing is due to the fact that the RNA in the chromatin-

RNA complex undergoes a change. However, during mitosis considerable changes occur in the chromatin, the second constituent of the complex. It is condensed into chromosomes and thus apparently loses, partly or completely, its affinity for nuclear RNA's. Besides, the nuclear membrane breaks down at the same time. In this period, the nuclear RNA molecules not bound with chromosomes diffuse freely through the cell. There is no processing in the cytoplasm since the former is evidently determined by nuclear enzymes. Then the telophase sets in, the chromatin undergoes decondensation and is surrounded by a nuclear membrane. About this time nuclear RNA's reassemble in the nucleus. The impression is that the RNA returns only to the already formed nuclei surrounded by a nuclear membrane. If this is the case, it confirms once more that the nuclear membrane does not play the decisive part in the RNA transport between nucleus and cytoplasm. Apparently, when the chromatin is decondensed, its affinity for nuclear RNA's is restored. It is hard to imagine a mechanism which would transfer the nuclear RNA molecules straight into the nuclei of daughter cells. It is much more natural to visualize this occurring by diffusion: the RNA's enter the nucleus and bind with the chromatin, which "traps" them. In this way the normal interphase distribution of the RNA's between nucleus and cytoplasm is re-established. The processing is then resumed in the nucleus and the processed RNA's return into the cytoplasm.

The scheme of the chromatin-RNA relationship presented here is highly hypothetical in its details. But it rests on three basic groups of data: (1) the existence, in nuclei, of a large, hard-to-extract fraction of RNA which practically represents all the nuclear RNA's; (2) the relationship between the RNA processing and transport in the interphase nuclei; (3) the ability of nuclear RNA's to return to the daughter cell nuclei. The proposed hypothesis of the RNA-chromatin affinity, which is disrupted by the RNA processing in the interphase and by DNA condensation in mitosis, offers a plausible and unified explanation for the entire combination of problems.

RNA IN THE CYTOPLASM

The traditional concepts of three types of RNA — mRNA, rRNA, and tRNA — are valid only for those RNA's which have gone through the processing and were transported from nucleus into cytoplasm. Their function in the protein-synthesizing apparatus involves many other molecules, primarily enzymes and structural proteins. Only tRNA is in a free, not protein-bound, state, although it interacts with aminoacyl-tRNA-synthetases and with the ribosomal polypeptide and proteins.

The ribosomal RNA's, which, according to the current notions, play in the ribosomes the part of a skeleton ensuring their buildup from dozens of different specific protein molecules (see Spirin and Gavrilova, 1971; Bogdanov and Ledneva, 1975), are obviously protein bound, although they also fulfill a functional role by interacting with tRNA and mRNA.

The role of the proteins bound with mRNA is somewhat vague, although they appear to be of major importance in the storage of mRNA and the regulation of its translation.

Ribosomes

The formation of ribosomes in oogenesis has been discussed above. By the time of ovulation the eggs of all animal species contain large ribosome stockpiles ready for protein synthesis.

In most animals (fishes, amphibians, echinoderms), at the early stages of development — from fertilization to gastrula and later — the protein synthesis takes place only on ribosomes stored up earlier in oogenesis. Then, when new ribosomes begin to form, the old ones are gradually displaced (degrade), so that the total quantity of rRNA remains almost unaltered for a long time.

In fishes, and possibly in other animals with incomplete cleavage, a considerable part of the ribosomes resides among the yolk granules and does not find its way into the cytoplasmic prominence of the egg from which first the blastomeres and then all the embryo cells are formed. But at mid-late blastula the ribosomes move from the yolk part of the egg to the cells. In loach embryos this is observed between the 5th and 9th hours of development and manifests itself as a sharp increase in RNA content in the blastoderm (with the total egg quantity of RNA unchanged) (Aitkhozhin *et al.*, 1964). A similar process was shown on trout embryos (Hagenmaier, 1969). The mechanism of this transfer is not clear. It may consist in active migration of ribosomes into the cells adjoining the yolk (Aitkhozhin *et al.*, 1964; Thomas, 1968) or simply in incorporation of the upper layer of the yolk portion of the egg into the periblast cells during their division. It is also unclear how these ribosomes are distributed at a later stage: whether they remain only in cells of the periblast region or whether they penetrate the higher-lying cells. But at late blastula there is a gradient in rRNA distribution from high concentrations in the basal layer to low ones in the outer cells of the blastula (Kostomarova and Nechayeva, 1970). This gradient seems to determine a similar gradient in protein synthesis intensity (Krigsgaber *et al.*, 1971).

At present there are two groups of animals — *Xenopus* and the sea urchin — for which data on the life span of old (maternal) egg ribosomes have been obtained. We have already mentioned a mutation in *Xenopus*, when the nucleolar organizer is absent altogether. In frogs heterozygous with respect to this mutation the cells contain one nucleolus each, which is sufficient to ensure the normal number of ribosomes throughout the lifetime. In other words, judging by the number of ribosomes rather than by the number of nucleoli, this mutation is recessive. When these heterozygous mutants are crossed, 25% of the embryos are homozygous with respect to this mutation and completely devoid of nucleoli. Nevertheless, they develop normally up to the tail-bud stage and go on living until the swimming tadpole stage (Elsdale *et al.*, 1958). At the same time it is well known that in amphibians the synthesis of new rRNA's begins already at the end of gastrulation, and since the total quantity of RNA's does not increase, this indicates that the formation of new ribosomes is attended by an equivalent decay of the old ones.

Is it possible to determine the time of conservation of egg ribosomes in a normal embryo using anucleolate mutants as a development model? Repression of the protein synthesis terminates the development almost immediately. Hence, the protein synthesis in mutant embryos is close to normal up to the late stages of development, and the maternal ribosomes are capable of functioning. This may be achieved, to some extent, by compensatory intensification of the protein synthesis per one preserved ribosome. Although no direct determinations of the number of ribosomes in homozygous anucleolate mutants have been made, it is likely that the degradation of the egg ribosomes in them is retarded as compared to the control. In other words, ribosomes are able to function during development for a longer period than under normal conditions, when they are "displaced" by the new ribosomes of the embryo itself. If the rRNA synthesis begins at gastrula, then the gradual replacement of the old ribosomes by new ones during normal development begins somewhere at the neurula stage.

When studying sea urchin embryos Nemer found that the ribosomes stored in the egg differ from the newly formed embryo ribosomes. When they were heated to 60°C and RNA was isolated, it turned out that the 18S rRNA from the oocyte ribosomes decomposed into two segments sedimenting at 13S, whereas the 18S rRNA from new ribosomes was heat-resistant. It turned out later that this property is inherent only in some specimens of the sea urchin *Strongylocentrotus purpuratus*, and the differences between the maternal and embryonic ribosomes are due to their genetic dissimilarity. In any case, these differences enabled Nemer to show quantitatively how the new ribosomes replace the oogenesis-stored ones during development. This process begins at the mesenchyme blastula stage (20 h). The replacement of the old ribosomes by new ones reaches 50% after 50 h (prism), proceeds rather uniformly, and ends in a complete replacement at the pluteus stage after 70-80 h of development (Nemer and Infante, 1967). It is likely that the two processes are interrelated, namely, the appearance of new ribosomes is caused by the decay of the old ones, or the old ribosomes decay because they are displaced by new ones.

Ribosome degradation may be the manifestation of cell differentiation if it results in the formation of cells with a low-level protein synthesis.

This was observed, for instance, during the differentiation of lens fibers (Zapisek and Papaconstantinou, 1973).

The ribosomes in the cells may be in a free state or they may be bound with intracellular membranes, the endoplasmic reticulum. As a rule, undifferentiated cells have almost no reticulum and its amount increases with differentiation. The fixation of ribosomes on the endoplasmic reticulum probably occurs at the same time.

Polysomes

A polysome is a set of ribosomes involved in the translation of one mRNA molecule. Since the distance between the ribosomes along the mRNA chain is relatively constant, the polysome size is determined by the length of the mRNA and correlates with the molecular weight of the protein synthesized. This correlation may not be very strict, because part of the mRNA's consists of untranslatable poly(A) sequences and sequences preceding the initiating codon. For instance, the poly(A) fragment of the globin mRNA contains 150 adenylate nucleotides, i.e. constitutes about 20% of the entire length of the mRNA.

Sedimentation analysis of the polysomes shows a broad distribution corresponding to values from 100S to 800S, or from 2 to 50 ribosomes per polysome. According to this criterion, all particles heavier than ribosomes are usually assumed to be polysomes. But this is actually not quite true, because, if a particle is identified by its radioactivity rather than by its optical density, the informosomes will fall into the same category (see below). A more reliable criterion is the buoyant density, which is close to 1.55, 1.51, and 1.40 g/cm^3 for ribosomes, polysomes, and informosomes, respectively. These values have been obtained on loach embryos (Spirin *et al.*, 1964; Ovchinnikov *et al.*, 1969b; Spirin, 1969). For other preparations the deviation does not exceed ± 0.01 g/cm^3. That polysomes are reliably lighter than ribosomes points to a higher (by about 30%) protein content. This cannot be attributed to the addition of synthesizable polypeptide chains alone and indicates that in the course of formation and functioning a polysome acquires additional proteins which become tightly bound either with the ribosomes themselves or with the mRNA. It has been shown at Glišin's laboratory that the polysomal mRNA in sea urchin embryos has a form of RNP which contains 60% protein (Ruzdijič and Glisin, 1972).

Protein is practically not synthesized on single ribosomes (monosomes) in the cell. Therefore, the fraction of ribosomes included in polysomes may serve as an indirect indication of the protein synthesis intensity or, more precisely, of the degree of loading of the protein-synthesizing apparatus. When the protein synthesis proceeds very slowly, for instance, in unfertilized sea urchin eggs or under the effect of protein synthesis inhibitors, almost no polysomes are formed. But even with the most intensive protein synthesis in embryonic cells the share of ribosomes contained in the polysomes usually does not exceed 60%. This is due to the fact that a certain optimum concentration of free ribosomes must be maintained in the cytoplasm, to ensure a continuous initiation of protein synthesis.

A good example of correlation between the protein synthesis and the number of polysomes is the early development of sea urchin eggs. In unfertilized eggs the protein synthesis is hardly noticeable, and the polysome content does not exceed 2% of the total ribosome content. After fertilization the protein synthesis increases sharply almost immediately and the share of the ribosomes contained in the polysomes increases and reaches a maximum within several hours (from 40 to 60% according to different authors) (Kedes and Gross, 1969; MacKintosh and Bell, 1969b; Finamore and Clegg, 1969; Mano, 1971a). At this time the protein synthesis is also at a maximum.

Knowing that the mRNA fraction in the total RNA of the polysomes is about 2%, it is possible to calculate the quantity of the mRNA involved in translation. According to Humphreys (1971), 2 h after fertilization of sea urchin eggs the polysome fraction is 20%, which corresponds to 6×10^{-12} g of mRNA, or to just 0.4% of the total egg RNA, i.e. is much less than the quantity of potential mRNA, as determined from its template activity.

As mentioned above, the polysome size reflects the molecular weight of the proteins synthesized. At the late stages of development, when most diverse proteins are synthesized, the polysome size varies widely. The variation in polysome size at the early stages of development of the sea urchin was investigated by Infante and Nemer (1967) (Fig. 39). They showed that after fertilization a rather broadened peak of large polysomes with a sedimentation constant fluctuating in the region of 300S appears in the eggs and begins to increase. At later stages (8 to 16 blastomeres) a new class of polysomes with a maximum at about 200S appears. In the presence of actinomycin D a 300S peak appears and increases in the same way as in the norm, while there is no small-polysome peak. These data show that larger proteins are synthesized on the templates stored up in oogenesis. On new templates, transcribed already in the embryo itself (their appearance can be suppressed by actinomycin D), a class of proteins with a lower molecular weight is translated.

These investigations were continued at Gross' laboratory. If labeled RNA precursors are used to detect polysomes, those polysomes can be detected on which newly synthesized templates alone are translated. New polysomes at the earliest stages are very small and include no more than 2 to 7 ribosomes (Kedes *et al.*, 1969a). This corresponds to the synthesis of proteins with a molecular weight of about 8×10^3 to 25×10^3. These proteins are histones, which provide the protein component of the quickly increasing mass of nuclei and are intensively synthesized during cleavage. Thus, the data on polysome size helped estab-

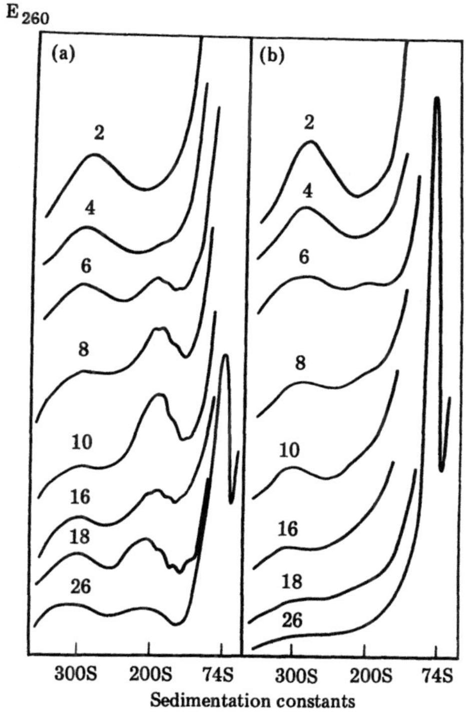

Fig. 39. Changes in the size of polysomes during development of the sea urchin. Embryos of *Strongylocentrotus purpuratus* developed for different time periods under normal conditions (*a*) and in the presence of actinomycin D (*b*); they were used to obtain an amitochondrial extract which was analyzed by centrifugation in a sucrose gradient. The numerals at the curves denote the stages of embryonic development in hours. In the course of normal development (*a*) the heavy-polysome peak (300S) slightly decreases, and a light-polysome peak (200S) appears. The polysomes gradually disappear in the presence of actinomycin D (*b*). The maternal templates are apparently translated on 300S, and the newly transcribed ones, on 200S polysomes (Infante and Nemer, 1967).

lish (although by an indirect method) that in the course of early development histones are synthesized on new templates or, conversely, that the bulk of the mRNA's synthesized at the earliest stages of development in the sea urchin serves as templates for the nuclear protein synthesis (Kedes *et al.*, 1969a, b). It was shown recently that the number of ribosomes per mRNA unit length in sea urchin embryos may vary in different polysomes. Thus, mRNA's at the blastula stage are bound with a smaller number of ribosomes than at later stages. For poly(A)-containing mRNA's the loading with ribosomes increases to a greater extent and reaches a maximum, while for mRNA's lacking poly(A) fragments this increase is insignificant (Nemer, 1975).

In *Drosophila* oogenesis and embryogenesis the ribosomal fraction of polysomes is relatively constant and makes up 56%, although the absolute number of ribosomes grows rapidly in oogenesis and remains unaltered in embryogenesis. But if mature oocytes are delayed in the ovaries of starved flies, they pass from stage 13 to stage 14 and, accordingly, the percentage of ribosomes in the polysomes decreases to 10% within 24 h (Mermod and Crippa, 1978).

The polysome size acquires one or several discrete values on terminal differentiation of cells resulting in a preferential synthesis of only a few proteins. Thus, reticulocytes usually have a single polysome peak containing 9-10S mRNA which serves as a template for the globin polypeptide synthesis. In the chick embryonic muscles, three polysome peaks are prominent: large polysomes containing up to 50 ribosomes each, medium, and small polysomes. They correspond to three types of muscle proteins whose synthesis dominates in these cells: myosin, actin, and tropomyosin (Heywood and Rich, 1968).

Thus, in a number of cases a polysome analysis may provide some information on the sizes of the proteins synthesized, the relative intensity of their synthesis, its transcription dependence, etc.

Transfer RNA's

There is a certain optimum ratio between the quantity of transfer RNA's and the ribosomes, which ensures the maximum operating efficiency of the protein-synthesizing apparatus. Based on the figures obtained for many differentiated tissues, this ratio is about 10-15 tRNA molecules per ribosome or, if we consider only polysomes, 20-30 tRNA molecules per functioning ribosome, i.e. approximately one tRNA molecule for each type of amino acid.

As we have already said, in a mature amphibian oocyte this ratio is one order of magnitude lower, i.e. about 2-3 tRNA molecules per ribosome. Consequently, only 1/10 of all the ribosomes can function at a normal rate in the oocyte. This is not surprising, since it is obvious that most of the oocyte ribosomes are inactive and are only stored until embryogenesis. In amphibian embryogenesis the quantity of tRNA begins to increase from early gastrula, then increases in parallel with the number of cells for some time, which, with a constant number of ribosomes, leads to the normal ratio of 15 tRNA molecules per ribosome already at the neurula stage (Brown and Littna, 1966). Quantitatively, this means that the mass of the tRNA constitutes about 10% of the rRNA.

Unfertilized eggs and early embryos of the loach contain about the same relative quantity of tRNA as those of amphibians (Solovyova *et al.*, 1973a). With an egg or an early embryo having $5\text{-}9 \times 10^{11}$ tRNA molecules, there are 0.8-1.4 tRNA molecules per ribosome. In the loach, the tRNA synthesis begins earlier than in the frog, at the early blastula stage, but the relative quantity of tRNA does not change substantially up to early gastrula. It may be assumed that at these stages the synthesis of tRNA's only compensates for their losses, while the increase in the number of tRNA molecules per ribosome in the loach occurs at later stages.

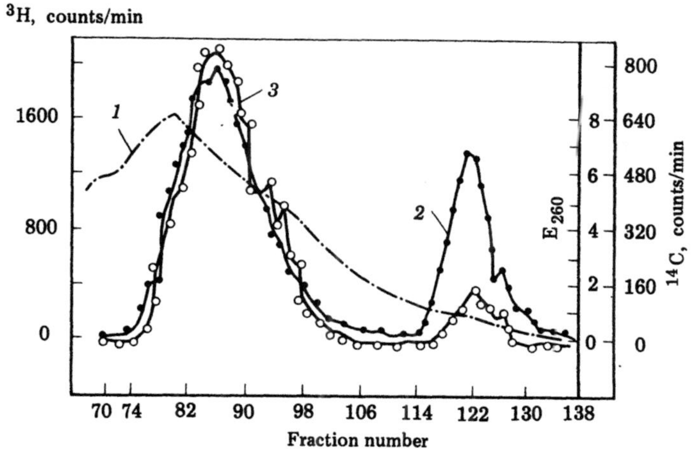

Fig. 40. Changes in tRNA composition during the early development of the sea urchin. The tRNA's from eggs and from blastulae of *Strongylocentrotus purpuratus* were aminoacylated with ^3H- and ^{14}C-leucine, respectively, and jointly chromatographed on a MAK column. *1*, Optical density ($E_{260\ nm}$); *2*, ^3H-leucyl-tRNA; *3*, ^{14}C-leucyl-tRNA. The relative content of the two isoacceptor leucine tRNA's changes in the course of development (Zeikus *et. al.*, 1969).

The tRNA spectrum in embryonic development has been investigated several times, and it has been shown that although the embryo contains the same species of tRNA at different stages of development, their quantitative ratio changes somewhat; thus, changes in the composition of some isoacceptor tRNA's have been detected in the development of insects (White *et al.*, 1973; Smith and Forrest, 1973; Araya *et al.*, 1975), sea urchin (Zeikus *et al.*, 1969) (Fig. 40), early bird embryos (Portugal, 1972), mammalian embryos (Clandinin and Schultz, 1975), and in their postnatal development.

These data are insufficient to visualize the overall pattern of changes in tRNA composition and to understand the meaning of these changes. The amino acid composition of the proteins synthesized may change slightly in the course of development, for instance at those stages where the basic nuclear proteins constitute a considerable part of the newly synthesized proteins, the requirement for lysine and arginine tRNA's may increase somewhat. The change in the ratio of different tRNA's may also be due to the change in the pool of free amino acids: a decrease in the concentration of one of them may be partly offset by an increase in the concentration of the corresponding tRNA, so that a sufficient number of aminoacyl-tRNA molecules will still be present. Finally, a change in tRNA spectrum in the course of differentiation may serve as a mechanism for protein synthesis control (see Osterman, 1977, 1979).

Masked mRNA

Of greatest interest are, of course, events occurring to messenger RNA in the cytoplasm. Sufficient evidence has been accumulated by now to reject the idea that as soon as mRNA appears in the cytoplasm it becomes fully incorporated into the polysomes and is translated until it degrades. This gives rise to two questions: (1) is there a nontranslatable (masked) mRNA in the cytoplasm? and (2) what role does the mRNA life span in the cytoplasm play in the regulation of protein synthesis and by what mechanisms is this life span determined?

The term "masked RNA" was proposed by Spirin (1966) and Tyler (1967) and was abbreviated to mmRNA (masked messenger RNA). It refers to mRNA which is in the cytoplasm, is not being translated at the moment but becomes involved in translation at a later date. The best known example of this form of RNA is mRNA stored in oogenesis. Indeed, the existence of this RNA has been convincingly demonstrated on unfertilized sea urchin eggs. Its quantity could be estimated from the *in vitro* template activity (Slater and Spiegelman, 1966). The very low level of protein synthesis in sea urchin eggs indicates that almost none of these mRNA's are used before fertilization.

It is more difficult to obtain irrefutable proofs of the existence of masked mRNA in amphibian and fish eggs, because the protein synthesis in them continues at all the stages of development and it can hardly be asserted that not all the mRNA's take part in translation. These species have, however, an advantage over the sea urchin in showing practically no mRNA transcription throughout their early development. At the same time, the nature of the translated RNA changes in the course of early development, as shown by experiments on competitive hybridization of polysomal mRNA's, i.e. some masked RNA's become involved in translation at the earliest stages and others at later stages (Crippa and Gross, 1969).

One more example of new protein synthesis beginning in the absence of that of new mRNA's is spermatogenesis. The RNA synthesis in differentiating spermatozoa ceases very early, prior to the second meiotic division or immediately after it. This was shown on *Drosophila* (Hess, 1968; Hennig, 1968) and other arthropods (Brasiello, 1968) and repeatedly demonstrated on various mammals (Monesi, 1965; MacLaughlin and Terner, 1973, Kierszenbaum and Tres, 1975; see the review by Auriault, 1975). After the cessation of the RNA synthesis, several days elapse during which a second-order spermatocyte gradually differentiates into a spermatozoon. As has been shown definitely on *Drosophila*, mRNA templates for this process are synthesized on the Y-chromosome prior to meiosis. Throughout the spermatogenesis, proteins are synthesized and some enzymes are replaced by others (Turkington and Majumder, 1974). It cannot be asserted in this case that the templates for all the indicated syntheses are not translated at the earlier stages of spermatogenesis but are at first stored and then gradually enter the polysomes. A number of enzymes, however, appear only in the acrosome (Turkington and Majumder, 1974), arginine-rich histones begin to synthesize and replace lysine-rich histones in the spermatozoon head only at the very end of spermatogenesis (Das *et al.*, 1964; Marushige and Marushige, 1975). It can then be stated that the templates for these histones were synthesized not later than the premeiotic stage and were stored in a masked form throughout spermatogenesis (about 8 days).

Finally, it was shown directly that in the trout there is protamine mRNA transcription already in spermatocytes. This mRNA is not translated but is accumulated in the form of cytoplasmic RNP particles which enter the polysomes and begin to be translated only at the spermatide stage (Iatrou *et al.*, 1978). Similar accumulation of casein mRNA in the mammary glands of the mouse occurs already during pregnancy, i.e. before the onset of translation (Takemoto *et al.*, 1980), and that of myosin mRNA in the myoblasts before their fusion into myotubes and the *en masse* synthesis of muscle proteins (Dym *et al.*, 1979).

One of the forms of masked mRNA may be RNP complexes (particles). Such complexes were detected and named informosomes (Spirin *et al.*, 1964). They were recently studied very thoroughly in sea urchin eggs (see next section).

Informosomes

Spirin and coworkers investigated the cytoplasmic fraction of ^{14}C-uridine-labeled loach embryo cells in sucrose gradients (Spirin *et al.*, 1964; Spirin, 1966, 1969). They detected in the postribosomal region of the gradients (particles lighter than 80S) a number of radioactivity peaks which did not correspond either to the ribosomes or to their subunits, or to the ribosomal RNA as regards their sedimentation (Fig. 41). Similar peaks were detected after incubation with labeled amino acids. All this suggested that a new type of ribonucleoprotein particles containing protein and mRNA had been discovered in the cytoplasm. They were called informosomes. A distinguishing feature of informosomes was their buoyant density (ρ) in the CsCl gradient, which proved to be a constant (1.40 g/cm^3) for all classes of informosome. Considering that the buoyant density of protein is 1.35 g/cm^3 and that of RNA, 1.95 g/cm^3, this means that the RNA : protein ratio in informosomes is 1 : 3, which is much less than in ribosomes (1 : 1.5) and polysomes (1 : 2) (Spirin, 1969; Ovchinnikov *et al.*, 1969b) (Fig. 42). By using this property as a criterion informosomes were also found in the preribosomal (heavier-than-80S) zone, where they could not be distinguished from polysomes in the sucrose gradient.

The sedimentation characteristics of informosomes vary from 20S to 100-200S, which means that at a constant RNA : protein ratio they differ from each other in the size of RNA molecules. Indeed, after deproteinization they yielded a number of size classes of RNA, typical of

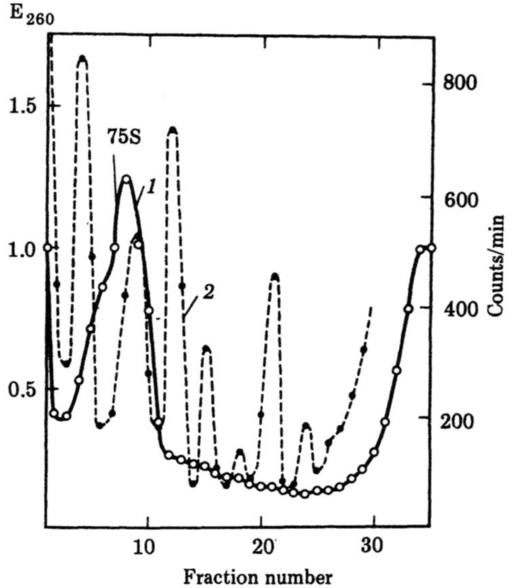

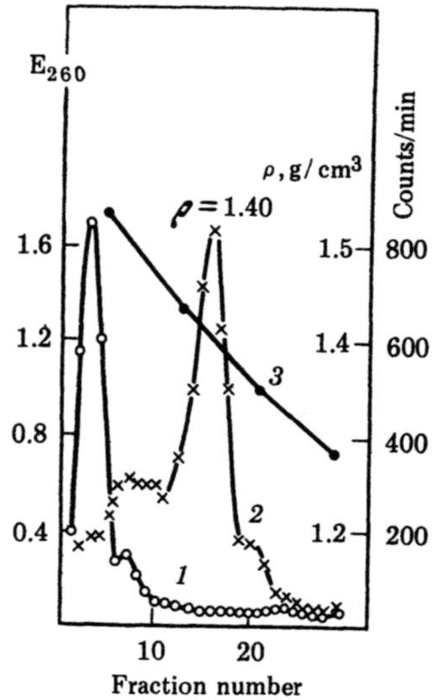

Fig. 41 (*left*). Informosomes of loach embryos. Loach embryos at the blastula stage were incubated for 6 h with ^{14}C-adenine and ^{14}C-uridine. The cytoplasmic (anucleate) extract was analyzed by centrifugation in a sucrose gradient. *1*, Optical density ($E_{260\,nm}$); *2*, radioactivity. The radioactive peaks of the informosomes are located in the heavier and lighter regions with respect to the ribosome peak (75S) (Spirin, 1966).

Fig. 42 (*right*). Buoyant density of informosomes and ribosomes. Loach embryos were incubated with ^{14}C-uridine from 13 to 16 h of development, a cytoplasmic extract was obtained, the ribonucleoprotein complexes were fixed with formaldehyde and their density distribution investigated in a CsCl gradient. *1*, Optical density ($E_{260\,nm}$); *2*, RNA radioactivity (counts/min); *3*, buoyant density (g/cm^3). The optically revealed ribosomes have a buoyant density of 1.55, and the informosomes forming the radioactivity peak, 1.40 (Ovchinnikov *et al.*, 1969a).

mRNA. The content of informosomes in the cell is very small, and they could not, so far, be isolated in quantities measurable by their optical density. Therefore they can be detected only by their radioactivity (usually incorporated into RNA). Later on, informosomes were found in some other types of animal cell (see Spirin, 1969, 1972), where their buoyant density was also close to 1.4 g/cm^3, and also in germinating plant seeds (Aitkhozhin *et al.*, 1973). Informosomes have been described by Nemer, Spirin, and Infante (Spirin and Nemer, 1965; Nemer and Infante, 1965; Infante and Nemer, 1967) in sea urchin embryos. At a later date Mano and Nagano (1970) established that unfertilized sea urchin eggs contain RNP's which sediment at 300-400S and include 20-60S RNA molecules. After fertilization, probably as a result of proteolysis, these particles disintegrate and the RNA is incorporated into the polysomes.

New data have been obtained on a model of differentiating cultured myoblasts which turn into muscle fibers. The synthesis of myosin mRNA begins already in the myoblasts, and prior to the fusion of myoblasts into myotubes, i.e. before the beginning of visible differentiation, it is found in the cytoplasm in the form of RNP complexes and according to its characteristics can be classified as the informosomes (Buckingham *et al.*, 1974a; Buckingham and Gross, 1975; Jain and Sarkar, 1979). In the embryonic retina, ^{3}H-uridine is first incorporated into RNP particles containing about 60% of protein, then enters the polysomes (Sarkar and Moscona, 1971a). This is probably one of the few cases where RNA migration from informosomes to polysomes could be shown.

Already in the first years after informosomes were detected, a number of hypotheses were advanced concerning their important role in the protein synthesis control. Informosomes were regarded as an intermediate step in mRNA migration from nucleus to polysomes (Spirin, 1966). They became the center of attention. In some communications, informosomes were regarded as an artefact, unstable RNP complexes, arising nonspecifically during cell homogenization (Girard and Baltimore, 1966). Indeed, it was shown that informosome-like complexes arise on mixing any types of RNA (e.g., rRNA from bacteria) with a cell extract (Ovchinnikov *et al.*, 1968).

Nevertheless, by using mainly embryonic loach cells some evidence on informosome specificity was also found. Thus, "artificial informosomes," resulting from incubation of the loach embryonic cell cytosol with various RNA's, turned out to be more stable if they were formed with loach mRNA at temperatures optimal for the loach embryo (Stepanov and Voronina, 1972). Among the cytosol proteins, the informosome complex with RNA can be formed only by a very small protein fraction (about 0.3% of all the proteins) with definite physicochemical properties (molecular weight about 2×10^{5} daltons) (Stepanov *et al.*, 1972).

The existence of informosomes in intact cells was proved in the following ingenious experiment. Extract of cell cytosol was mixed with labeled mRNA in the presence of an excess of unlabeled rRNA. No informosomes with $\rho=1.4$ g/cm^{3} were formed. This was probably because the amount of informosome-forming protein was insufficient and it was distributed throughout the RNA molecules, as a result of which the RNA : protein ratio exceeded 1 : 3. But when, instead of cytosol with labeled RNA, whole cells were homogenized in the presence of an excess of unlabeled rRNA, ordinary informosomes were detected. The possible explanation was that the rRNA which was added to the existing informosomes at the moment of homogenization could not compete for their protein component, which was tightly bound with the loach mRNA (Ovchinnikov *et al.*, 1969a).

It was found in Georgiev's laboratory that nuclei of many cell types contained RNP complexes whose protein was then named informofer (Samarina *et al.*, 1965). Unlike the informosomes, these RNP complexes are present in nuclei in readily measurable quantities, and therefore it was possible to obtain their image with an electron microscope. The buoyant density of nuclear RNP complexes is also close to 1.4 g/ cm³. Later these RNP complexes were named nuclear informosomes. Nuclear informosomes differ from cytoplasmic ones in their protein component and probably in their functional role. The nuclei and cytoplasm of sea urchin embryos revealed RNP complexes differing in RNA : protein ratio (Alfageme and Infante, 1975).

According to the scheme proposed in Georgiev's laboratory (Samarina, 1975), the role of nuclear RNP complexes amounts to the following. Immediately after the transcription of pre-mRNA, RNP complexes with informatin (informofer protein) are formed, and as a result the RNA chain becomes bound with several informofer globules. It is assumed that this complex protects the pre-mRNA from the nonspecific action of nucleases. The pre-mRNA in the RNP complex undergoes the processing, i. e. is subjected to restriction, is polyadenylated, and becomes mature mRNA. After the mRNA passes from nuclei into cytoplasm, its bond with the informofers seems to be lost, and in the cytoplasm it joins the ribosomes directly or forms RNP complexes (informosomes) with the cytoplasmic proteins.

The role informosomes play in the cell cytoplasm is not clear. In the cytoplasm of loach embryonic cells up to 80% of the synthesized RNA is bound with these particles and only 20%, with polysomes (Spirin, 1966). At the same time it has been shown on other cells, both adult and embryonic, that mRNA enters the polysomes very quickly and is translated (Kedes *et al.*, 1969a, b; Dworkin and Infante, 1976; Brandhorst, 1980).

If informosomes are only an intermediate step, it was natural to expect that in short-label experiments with a subsequent chase, or simply on prolonged incubation with a radioactive precursor, the share of the informosome label must decrease and that of the polysome label, increase. This, however, has not been observed.

It might be thought that there is a dynamic equilibrium between polysomes and informosomes and that a constant mRNA ratio between them (for instance, 20% and 80%) means that only a small part of all the templates are translated at each moment. This is indicated, for instance, by the fact that mRNP complexes with a buoyant density $\rho = 1.4$ g/ cm³, i. e. similar to informosomes, rather than free mRNA's, are released on polysome dissociation. This, together with the data on the protein excess in polysomes, suggests that polysomes form from ribosomes and informosomes, rather than from free mRNA's and ribosomes.

This assumption is confirmed by more recent data of Spirin and his coworkers obtained on vegetal (Vlasik *et al.*, 1978) and animal (Ovchinnikov *et al.*, 1978) RNA-binding proteins. It was found that these proteins include the initiation and translation factors. It may be assumed (Spirin, 1978) that these factors are bound with the RNA in the cell-forming mRNP complexes which are included into polysomes. Thus, the initiating complex has "at hand" the necessary translation factors, and its function is less dependent on the relatively low concentration of these factors in the eukaryotic cell (Spirin, 1978).

Dworkin and Infante (1976) studied nonpolysomal free RNP particles in sea urchin embryos. They showed that these RNP's contain the same types of RNA as do the polysomes but differ by a lower degree of polyadenylation at the blastula stage (40-50% of RNA in the polysomes and about 30% of RNA in the free RNP's have poly(A)). In the course of development the content of free RNP's in the cytoplasmic newly synthesized RNA's decreases from 50% during cleavage to 20% at the mesenchyme blastula stage. A comparison of the synthesis rates and the life spans of RNA in nuclei, polysomes, and RNP's shows that mRNA from free RNP's can be a precursor of only a small part of polysomal templates and that these RNP particles must be regarded as an excess of transcribed mRNA's used for the control of translation.

Similar conclusions were drawn from experiments on *Drosophila* embryos (Goldstein, 1978). The composition of RNA translated on polysomes and stored in RNP complexes proved to be identical; therefore, these complexes can be regarded as a reserve for regulation of translation intensity.

Raff and coworkers (Kaumeyer *et al.*, 1978; Jenkins *et al.*, 1978; Young and Raff, 1979), using sea urchin eggs, isolated RNP particles which sedimented in a sucrose gradient at 60 to 65S. They contained approximately equal quantities of poly$(A)^+$ RNA (about 10^6 daltons). All mRNA's in these particles were capped (Pirrone *et al.*, 1977). The protein content in the RNP complexes (and, consequently, their buoyant density) depended markedly on the composition of the saline medium during isolation (Kaumeyer *et al.*, 1978). The greatest amount of protein (1.4×10^6 daltons) in the RNP complexes is maintained in 0.35*M* Na^+ in the presence of 5m*M* Mg^{++}, and the smallest amount (0.25×10^6 daltons), in the absence of magnesium and on addition of EDTA (5mM). RNP particles isolated in a K^+- and Mg^{++}-containing medium are almost incapable of translation *in vitro*, although the mRNA isolated from them is readily translated under the same conditions. However, isolation of RNP's in the presence of sodium ions has a destabilizing effect and they begin to be translated. According to other data (Ilan and Ilan, 1978), translation of RNP's from sea urchin eggs in an *in vitro* system requires a very high concentration of $MgCl_2$ (12mM), whereas translation of RNA, as low as 3m*M* of $MgCl_2$. The high magnesium concentrations apparently cause a release of mRNA from RNP.

RNP particles stored in sea urchin oogenesis differ from those synthesized in early embryogenesis (Young and Raff, 1979). The egg RNP's have a buoyant density of 1.45 g/ cm^3, corresponding to an approximately equal RNA and protein content. On the other hand, the newly synthesized RNA of the embryo is contained in particles that have a buoyant density of 1.55 to 1.66 g/ cm^3. These particles contain no more than 10% of protein.

Irrespective of the specific mechanisms of informosome formation and their actual functional role, the very idea of mRNA-protein complexes proved to be extremely fruitful and is discussed in dozens of papers published in the last few years. The solution of the problem of the RNP complexes structure and function might give the clue to many problems of protein synthesis regulation in the cell at the translation level.

Life Span of Messenger RNA

The life span of the various RNA's after synthesis is an important parameter, which, together with their transcription rate, determines the quantity of RNA's in the cell. This parameter can serve as a measure determining the ratio of different mRNA's and the changes in the level of the same mRNA in the course of development. This issue will be discussed in more detail in Volume II, in relation to mechanisms of translation control. Here, we shall just dwell very briefly on some investigations which established the mRNA half-life ($t_{1/2}$, the period of decay of half of simultaneously synthesized molecules).

The simplest method for determining the RNA life span is tracing the fate of labeled RNA after repression of transcription by actinomycin D. However, the use of this inhibitor may affect the results. According to some data, actinomycin D increases the life span, while according to others, it shortens it (Singer and Penman, 1973). Less invasive methods are also used, such as tracing the short-labeled ("pulsed") RNA and analysis of a double label, etc.

In bacteria, most of the mRNA exists for minutes, but there is also a long-lived fraction which persists for tens of minutes (Davidson, 1969). In eukaryotic cells, at least three groups of mRNA's can be distinguished according to their half-life. Short-lived RNA's have a $t_{1/2}$ of around 20 to 60 min. In the lymphocytes, the average $t_{1/2}$ of this fraction is 17 min (Berger and Cooper, 1975). For the mRNA of the δ-aminolevulinate synthetase $t_{1/2}$ is about 1 h (Tschudy *et al.*, 1965). In sea urchin embryos, the average $t_{1/2}$ of the short-lived RNA's in the cytoplasm proved to be 23 min (Grainger and Wilt, 1976) and 75 min (Brandhorst and Humphreys, 1972a, b). For nuclear, informosomal, and polysomal RNA these values were 12, 40 and 65 min, respectively (Dworkin and Infante, 1976). In the cultured embryonic cells of *Drosophila* the short-lived fraction of mRNA has a $t_{1/2}$ of 50 to 60 min (Lengyel and Penman, 1977) and even 20 min (Levis and Penman, 1977). In the mitochondria, part of the mRNA's have a $t_{1/2}$ of 45 min (Avadhani, 1979).

The second group consists of medium-lived mRNA's with a $t_{1/2}$ from 2-3 to 24 h. In the cell culture of *Drosophila* these mRNA's have a $t_{1/2}$ of 3 h (Levis and Penman, 1977) and 6 to 7 h (Lengyel and Penman, 1977), in HeLa cells 33% of mRNA's have a $t_{1/2}$ of 7 h and the other 66%, of 24 h (Singer and Penman, 1973), in myoblasts, the myosin mRNA have a $t_{1/2}$ of 10 h (Buckingham *et al.*, 1974a), and in the hatching glands of the silk moth the mRNA's encoding most of the proteins have a $t_{1/2}$ of 2.5 h (Kafatos, 1972). Part of the mitochondrial mRNA's have a half-life of 3.5 h (Avadhani, 1979). Halving of the protein synthesis in actinomycin D-treated sea urchin embryos also occurs within 10 to 12 h (Gross and Cousinou, 1964; Neyfakh and Krigsgaber, 1969; Krigsgaber and Neyfakh, 1972).

Finally, the third group consists of the longest-lived mRNA's, which usually determine the synthesis of specific proteins: globin in erythroblasts, fibroin in the silkworm, and crystallins in the lens (Milstone *et al.*, 1976). For instance, in the above-mentioned gland cells of the silk moth the cocoonase mRNA has a $t_{1/2}$ of about 100 h and the myosin mRNA in the myotubes (not the myoblasts), of 50 h. In the *Drosophila* cell culture, some mRNA's have particularly long $t_{1/2}$ of over 100 to 200 h, comparable with that of rRNA and exceeding that of tRNA (53 h) (Lengyel and Penman, 1977).

The prolonged preservation of mRNA in oocytes, which continues for months and even years, deserves special consideration.

An electrophoretic analysis of the proteins in the rabbit embryo at the 2- and 16-cell stages developing in the presence and absence of α-amanitine showed that protein synthesis at the 16-cell stage is determined by two classes of stable templates: mRNA's which are apparently present in a still unfertilized egg and mRNA's synthesized after fertilization at the 2-cell stage and later (Tucker and Schultz, 1978).

In the section on regulation of translation we shall cite examples of variation in $t_{1/2}$ in the course of development and the very scarce data on the proposed mechanisms determining the mRNA life span and the role of poly(A) fragments in this process.

Chapter Three

PROTEIN

The synthesis of the protein molecule completes the series of processes of transformation of hereditary information, which is usually regarded as the subject-matter of molecular biology. After the protein is formed, the next, much less investigated, stage begins: the development of functions and the appearance of new supramolecular, intracellular, and supracellular structures — processes constituting the essence of morphogenesis. If we restrict ourselves to the consideration of protein synthesis, the entire development can be described as the process of succession of proteins, different at different stages of development and in different parts of the embryo. It is obvious that this protein succession must reflect all the processes of morphogenesis and all the newly arising functions, which are controlled by the hereditary apparatus. Nevertheless, information on this protein succession is much scantier so far than could be expected.

In the course of development, new proteins appear, but the quantity of newly synthesized proteins is small compared to the mass of egg proteins, which are replaced slowly. In eggs with incomplete cleavage — in fishes and, to a still greater extent, in birds and reptilians — the embryo is separated from the reserve proteins of the yolk and its growth is a sufficiently convincing evidence of the formation of new proteins during development. Another manifestation of this process is the growth of the mammalian embryo.

Among the proteins synthesized one should distinguish those types which were present in the egg and are only renewed in the course of development, and those which appear for the first time at one stage or another. Such a division cannot be too strict, and many enzymes, which would seem to exist at all the stages of development, are actually replaced: their new isozymes are formed. It is evident for many proteins that they first appear in definite tissues and at definite stages. This primarily refers to such tissue-specific proteins as muscle myosin, hemoglobin, collagen, and lens crystallins. The appearance of the hatching enzyme, digestive enzymes of the pancreas, cholinesterase, etc., is associated with definite stages of development and with the beginning of specific functions. The total number of such new proteins, known at present, runs into the dozens.

At the same time it is obvious that according to the lowest possible estimates the number of structural genes functioning during development and, hence, the number of new proteins, must run into the thousands, or even tens of thousands. The information on the proteins which determine specifically the shape, movements, and mutual relations of cells, i.e. the morphogenesis processes, would be particularly valuable. That such proteins do exist is beyond any doubt: many mutations, the blocking of RNA or protein synthesis lead quite rapidly to disturbances or cessation of morphogenesis. But practically nothing is known yet about the morphogenesis proteins. The first candidates to be included into this category are tubulin, which forms microtubules in the cells, and the actin-like proteins which form microfilaments and affect the shape and movement of the cells. In recent years the number of identified structural proteins involved in the cell shape and movement control has increased.

Surprisingly, the meagerness of our knowledge concerning the "development proteins" coexists with the continuously developing techniques of protein separation and the advances in immunological procedures, which would seem to enable us to isolate and identify any proteins. The reason probably is that morphogenesis-controlling proteins are present in the embryo in insignificant quantities, hundreds of times less than the other proteins. Besides, these proteins are most probably insoluble under the normal conditions of extraction. And, finally, what is most important, we still do not know the specific tests for these proteins which would enable us to distinguish them from "nonmorphogenetic" proteins and from each other.

Investigation of protein synthesis seems simple if we just determine the incorporation of labeled amino acids into the total protein. The problem becomes much more complicated when individual embryonic proteins are the object of investigation. Obtaining them in sufficient quantities and freeing them from other labeled proteins involves great difficulties, and until now this approach has been used only in a few studies. An accurate comparison of the protein synthesis rates at different stages or in different parts of the embryo can only be achieved by solving the difficult problem of determining the specific radioactivity of the amino acid pool.

The immediate precursors of the protein molecule are activated amino acids (aminoacyl-tRNA's), but the measurement of their specific radioactivity is necessary only when the actual synthesis rate, i.e. the elongation of the polypeptide chain per unit time, is being determined.

Determining the actual synthesis rate is a special problem, which requires solution only on rare occasions. Much more often one has to compare the rate of protein synthesis at different stages of development, in different tissues, or in tissues exposed to various treatments. The comparison is valid only when one is quite confident that the specific activities of the activated amino acids in the tissues or cells being compared are identical.

The problem can be simplified by assuming that the rates of ami-

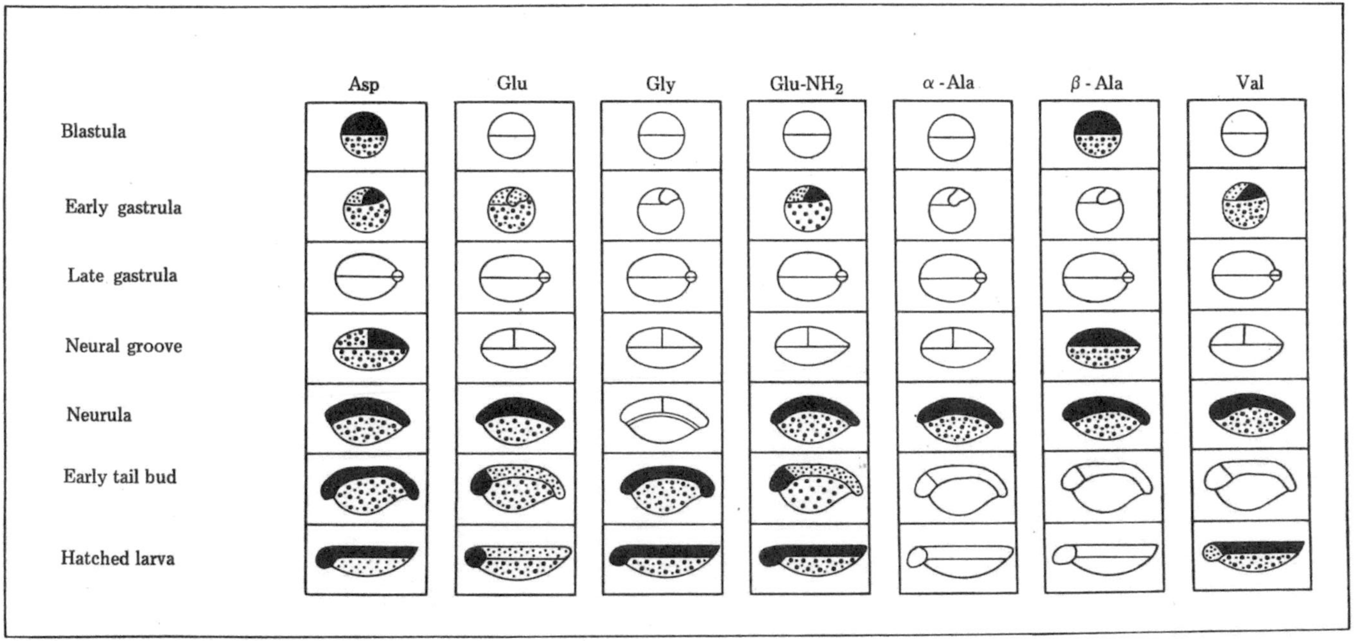

Fig. 43. Changes in the concentration and localization of free amino acids in *Xenopus* embryos. The solid areas denote a high concentration, the open ones, a low concentration, and the dots indicate an intermediate concentration (Deuchar, 1956).

noacyl-tRNA's formation, i.e. the activities of aminoacyl-tRNA-synthetases and the tRNA concentration at different stages or in different anlages are identical or high enough. Then the specific radioactivity of the aminoacyl-tRNA will be proportional to that of the free amino acids. However, the determination of the specific radioactivity of the amino acids in the whole embryo may prove insufficient because it is quite probable that the composition of the amino acid pool and its specific radioactivity in the different parts of the embryo and in different tissues are different. That this is possible can be seen from the data of Deuchar (1956; 1963a) (Fig. 43).

In some instances, when the rate of protein synthesis is compared at two closely spaced stages of development, for instance, before and immediately after fertilization, the so-called "preloading" technique is used. A prefertilized egg is incubated with a labeled amino acid, assuming that after penetration the acid will form a pool whose specific radioactivity and the rate of the aminoacyl-tRNA formation will change only slightly in the period of development investigated. It has actually been shown in a number of experiments that the radioactivity of the amino acid pool changes little, and that the errors associated with its possible change are not very large. Nevertheless, as has been noted above, when dealing with the RNA synthesis it is necessary to keep in mind the source of this error, and the more important or unexpected the obtained result is, the more attention should be given to the question of determining the specific radioactivity of the amino acid pool.

The appearance of new proteins or changes in their quantity can be detected not only by working with individual proteins obtained from a tissue, an organ, or a whole embryo. A simpler method consists in measuring the enzymatic activity. However, even this widespread method may lead to grave errors if one forgets that enzymes may change their intracellular localization, become activated or inactivated in the course of development.

The most important phenomenon in the formation of proteins during development is the regulation of their synthesis at the transcription and translation levels. Correlation between the synthesis of mRNA and that of protein is usually hard to establish, because the methods for testing and isolating individual mRNA's are much more complicated than those for detection and isolation of proteins. The use of genetic methods and selective inhibition of the RNA synthesis, however, reveals the protein synthesis regulation at the transcription level. Changes in the pattern of protein synthesis, which occur in the absence of genetic control or on its experimental suppression, may be regarded as the manifestation of regulation at the level of translation.

PROTEINS IN OOGENESIS

Formation of Yolk

The oocytes of different animals may differ in diameter by tens of thousands of times: 7 μm in parasitic hymenoptera and 22×10^4 μm in one of the shark species. With respect to volume these differences are 13 orders of magnitude. The egg size is mainly determined by the amount of stored substances, the yolk being the major one. Therefore, small eggs (those of ichneumon fly, mammals, echinoderms, and some other sea invertebrates) must differ but slightly from somatic cells in their overall chemical composition. On the other hand, particularly large eggs of birds, reptilians, and some fishes are nearly completely filled with yolk, and their overall chemical composition is close to that of yolk.

Apart from proteins, yolk contains considerable amounts of neutral fats and phospholipids. As a rule, it is contained in yolk granules (platelets, spherules), but in some fishes these granules fuse together into a single yolk mass (liquid yolk, ichthulin) at the end of vitellogenesis. There are three types of yolk (yolk granules): the fatty yolk (high fat content), the carbohydrate yolk (high polysaccharide content), and the protein yolk (high protein content). The formation of carbohydrate and fatty yolk precedes, as a rule, that of protein yolk (see Raven, 1961). The major protein constituents of protein yolk and, hence, of the large eggs protein, are lipovitellin and phosvitin. As follows from the name of these proteins, they contain considerable amounts of lipids and phosphorus.

The eggs of various animals differ not only in the amount of yolk, but also in the structure and arrangement of the yolk granules, which is described in detail in corresponding sections of embryological manuals.

A great many morphological investigations on the origin of yolk granules have been published, in which the formation of yolk granules is associated with a special yolk nucleus, mitochondria, the Golgi apparatus, etc. (Ward, 1962). The morphological relationship with these structures is beyond any doubt, but the yolk protein synthesis proper may proceed in the egg itself, or such proteins may enter the egg from the outside.

Three sources of yolk proteins in oogenesis may be distinguished: protein synthesis in the egg itself, protein synthesis in the cells surrounding the egg, and protein synthesis in remote parts of the organism with the subsequent protein transfer into the oocyte through the blood. The last pathway has now probably been investigated most comprehensively with respect to yolk lipovitellin and phosvitin of the eggs of amphibians and birds (see reviews by Wallace and Bergink, 1974; Tata, 1976). As has been shown for the frog (Rudack and Wallace,

1968; Wallace and Dumont, 1968) and for birds (Clavert, 1953), yolk proteins are synthesized in the liver. The synthesis of these proteins is accompanied by the binding to them of calcium, phosphate, and lipids brought into the liver by the bloodstream. The blood receives the precursor protein, vitellogenin (Ca-phospholipoprotein), with a molecular weight of 5×10^5 daltons containing 8-13% of lipids and 1.4% of phosphorus (Wallace and Dumont, 1968). Vitellogenin is accumulated by oocytes by means of pinocytosis; this is probably the only way of massive passage of the protein through the cellular membrane. In the oocyte, presumably near its surface, vitellogenin decomposes into two proteins: lipovitellin and phosvitin, which become incorporated into the yolk granules (Dehn and Wallace, 1973). When vitellogenin is injected into the oocyte, it degrades without turning into yolk, as has been shown in *Xenopus* oocytes (Dehn and Wallace, 1973).

The rapid entry of the vitellogenin into the oocytes is probably ensured by the specific receptor sites on the membrane because the other proteins penetrate into the oocyte many hundreds of times more slowly (Wallace and Jared, 1976). This process involves follicular cells and is controlled by the gonadotropic hormone (Follett *et al*., 1968).

Lipovitellin has a molecular weight of about 1.9×10^5 daltons and contains phospholipids (20% of the lipids). The other protein, phosvitin, has a low molecular weight of the order of 3.5×10^4 daltons and contains up to 10% phosphorus, i.e. there is one phosphate residue per three amino acid residues. Phosvitin contains 56% serine (Tata, 1976). In the frog yolk, the lipovitellin-phosvitin weight ratio is 10 : 2 and the molecular ratio, 1 : 1. In yolk granules they form a densely packed crystallin structure and constitute at least 98% of the content of the granule, which is surrounded by a thin lipoprotein membrane. The egg proteins have one more component, livetin, which was formerly identified with the blood serum proteins (Jared and Wallace, 1968).

Vitellogenin, which is readily determinable by the presence of alkali-labile phosphorus, is absent in the male blood and is present in insignificant amounts in the female blood. After estradiol induction it appears in males and sharply increases in females (Wallace and Dumont, 1968; Tata, 1976). On injection of labeled precursors — amino acids or ^{32}P-phosphate — the specific activity of the vitellogenin fraction in females becomes higher than in males because of its rapid uptake by the oocytes in the ovaries. Previously (see, for instance, Romanoff and Romanoff, 1949) several proteins were found in the egg yolk of birds; at present, it is believed that they largely consist of lipovitellin and phosvitin.

The synthesis of yolk proteins in the liver does not occur only in amphibians and birds, where this fact has been established experimentally. An increase in the size of the liver before vitellogenesis, followed by a decrease during oogenesis, has also been noted in fishes.

There are quite a lot of data on the hormonal regulation of the egg protein synthesis in the liver (Follett *et al.*, 1968; Wittliff and Kenney, 1972a, b) (Fig. 44).

Under the influence of environmental factors (temperature, day/night cycle), the hypothalamus activates synthesis of gonadotropic hormones in the pituitary, which in turn stimulate the secretion of estrogen in the ovary. The estrogen stimulates in the liver cells synthesis of vitellogenin, the yolk precursor. This process mobilizes the reserves of lipids in the fatty tissue and those of phosphorus and calcium in the bone. Under the effect of gonadotropic hormones vitellogenin moves into the oocyte by pinocytosis and decomposes there into yolk proteins: lipovitellin and phosvitin. Both the female and the male liver are competent to the synthesis of the egg proteins: 3 h after injection of choriogonin or 17-β-estradiol to males or females, a synthesis of tRNA's, rRNA's, and proteins begins in the liver. In the frog *Xenopus laevis* a lipovitellin synthesis in the liver begins 9-12 h after its induction with estrogen (Fig. 45). This is preceded by a synthesis of uridine-rich RNA's and then of rRNA's (Wittliff and Kenney, 1972b). As shown on rooster liver, lipovitellin synthesis takes place on very large polysomes (30 to 40 ribosomes) (Roskam *et al.*, 1976). The synthesized molecules of the yolk proteins enter the blood. In females, they pass into oocytes rather rapidly, while in males, with estrogen induction, they continue circulating in the blood for a long time (Wallace and Dumont, 1968; Follett *et al.*, 1968; Tata, 1976). The pinocytosis of the protein by the oocytes proceeds via the follicular cells and is controlled by the gonadotropic hormone.

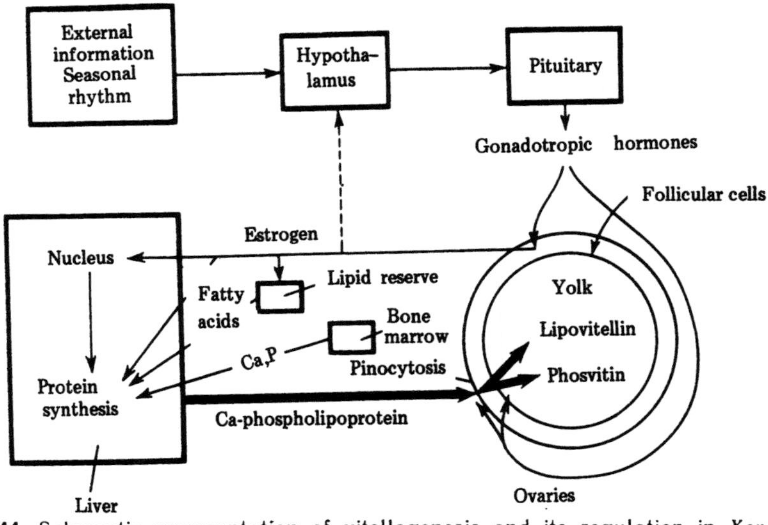

Fig. 44. Schematic representation of vitellogenesis and its regulation in *Xenopus* (Follett *et al.*, 1968).

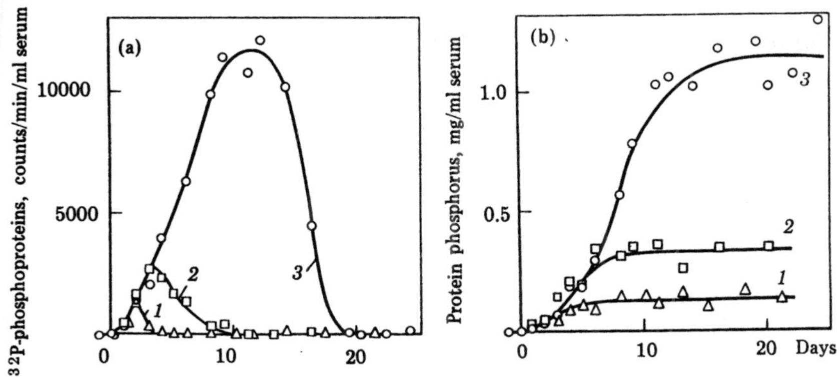

Fig. 45. Estradiol-induced synthesis of yolk phosphoproteins in the liver and their accumulation in the blood plasma of *Xenopus* males. The males were given single injections of 0.1 (*1*), 0.25 (*2*), or 1 (*3*) IU of 17β-estradiol, then the phosphoprotein synthesis intensity in the blood was determined at different times by incorporation of ^{32}P-phosphate into the plasma proteins (*a*), and by the amount of phosphoproteins in the plasma (*b*). The synthesis of phosphoproteins and their accumulation in the blood depended on the dose of the hormone (Wallace and Jared, 1968).

In many insects, the site of the extraoocyte protein synthesis is the fat body. Vitellogenin is produced there and is then transported into the oocytes by means of pinocytosis (Telfer and Melins, 1963; Roth and Porter, 1964; Telfer, 1965; Pan *et al.*, 1969; Bell and Barth, 1971; Hausman *et al.*, 1971; Warren *et al.*, 1979).

In distinction to the vertebrates, three types of vitellogenin molecules with a molecular weight of 4.4, 4.5, and 4.6×10^4 daltons form in the fat body of *Drosophila*. They are initially synthesized as previtellogenins (molecular weight of 4.5 and 4.6×10^4 daltons) with signal peptide sequences and subjected to two stages of the processing (Warren *et al.*, 1979).

The initiation of vitellogenin synthesis in the fat body of insects is controlled by the juvenile hormone, which is secreted by the corpus allatum (Bell and Barth, 1971). The juvenile hormone controls not only the secretion of vitellogenin, but also its selective uptake by the oocytes from the hemolymph. If ovaries are implanted to the roach *Eublaberus posticus* males with a ligature of the channel running from the corpus allatum, injections of yolk do not result in vitellogenesis in the oocytes. It is only injection of the juvenile hormone that causes an active accumulation of yolk (Bell and Barth, 1971).

Some insects (orders Neuroptera, Diptera, Lepidoptera, Hymenoptera, and Coleoptera) have the nutrimental type of oogenesis, in which the formation of oocytes involves the nurse cells. But here, too, the main source of yolk proteins is not the nurse cells, which are active only at the early stages of oocyte growth and degenerate by the beginning of intensive yolk formation in the oocyte (Raven, 1961).

In some species of *Protostomia* the nurse cells take part in yolk formation. Thus, the rotifer ovaries are divided into the vitellarium consisting of nurse cells (usually eight fused cells with large nuclei) and cells giving rise to oocytes proper. Oocytes grow using the content of the nurse cells. The relationships between nurse cells and oocytes in flat worms *Turbelaria* during the formation of yolk are even more interesting. The mature oocytes of this species have a very low yolk content. The yolk accumulates in the nurse cells localized in the vitellarium. During oviposition one or several oocytes and the nurse cells are enclosed in a common egg membrane. During the embryo development the yolk of the nurse cells is used in the same way as the oocyte yolk is used in other animals (Raven, 1961). In this instance the source of proteins for the nurse cell yolk remains unknown. It is not clear whether these proteins are synthesized in the nurse cells themselves or brought by the blood from some other source.

The role of the follicular cells in the formation of yolk is rather obscure. When vitellogenin is synthesized outside the oocyte (in the liver of amphibians and birds or in the fat body of insects), follicular cells take part only in the transport of proteins from blood to oocyte.

However, synthesis of yolk proteins seems sometimes to take place in the follicular cells themselves. In the house fly, for instance, when the ovaries are incubated outside the female body, the label incorporation into protein is detected in follicular cells after 30 min. After 1 h the label migrates into the oocyte yolk (Chia and Morrison, 1972). Finally, the oocyte cytoplasm proper may also be the site of yolk protein synthesis. This has been found in worms (Dumont, 1969) and is particularly prominent in some crustaceans, in particular in the river crayfish *Orconectes immunis* (Kessel, 1968; Ganion and Kessel, 1972). It was mentioned in early investigations that just before vitellogenesis there develops the endoplasmic reticulum in the crayfish oocytes. During yolk formation, granules are formed in the reticulum, which are at first localized in the reticular cisterns and then transported to the yolk granules (Beams and Kessel, 1962). Ganion and Kessel (1972) have shown, by autoradiography of ovaries incubated *in vivo* and *in vitro*, that incorporation of labeled amino acids occurs during vitellogenesis, and the label is first detected in the rough endoplasmic reticulum, then around the yolk granules and, finally, directly over them. At the same time, during the transplantation of unlabeled ovaries into the bloodstream of the crayfish with previously labeled blood proteins the label practically does not enter the oocytes. The very small label which is then detected in the oocyte is not localized over the yolk granules. Although the authors regard these results as evidence for the intraoocyte synthesis of the bulk of the yolk protein, they do not exclude an exogenous source of protein, such as follicular cells (Ganion and Kessel, 1972). Since it was the ovary tissues, and not the oocytes, that were used in these experiments, a supply of protein by the surrounding cells, in which an active protein synthesis occurred, cannot be excluded.

Protein synthesis inside the oocytes has also been shown in sea urchins (Piatigorsky *et al.*, 1967; Monroy, 1967; see Giudice, 1973; Rinaldi and Parente, 1976). But early morphological descriptions of oogenesis (Lindahl, 1932) mention that the sea urchin ovaries have amoeboid cells which phagocytize the residual oocytes. These cells find their way into the ovary cavity and form a syncytium filled with lipid droplets and granules. The syncytium attaches itself to the surface of the growing oocyte with the aid of pseudopodia. The content of the syncytium seems to be transferred into the oocyte because the number of lipid granules and droplets in the syncytium decreases as the number of oocytes increases.

Thus, in this case, too, the existence of two sources of the oocyte yolk protein may be assumed.

Protein Synthesis in the Oocyte

In addition to vitellin-type reserve proteins, a growing oocyte experiences an increase in the number of various structures and in the content of various functionally active proteins (Heideman *et al.*, 1975). At the earliest stages, a simultaneous growth of the oocyte and its nucleus is observed, resulting in the nucleus many times as large as an ordinary one. This nucleus — the germinal vesicle — contains chromosomes which take on a lampbrush shape in many animal species. Besides, in a number of cases there is either a large nucleolus which is many times larger than the nucleoli of somatic cells, or a multitude of small nucleoli containing amplified rDNA (for more details see Chapter 1).

The formation of a larger nucleolus or of additional nucleoli undoubtedly requires a considerable amount of nucleolar proteins which are to be synthesized at the earliest stages of oogenesis. The volume of the nucleoplasm also exceeds that of the nuclei of somatic cells. The protein content in the frog oocyte nucleus is $15\text{-}20 \times 10^{-7}$ g, and in the chromosomes, 3.5×10^{-7} g. The protein : DNA ratio in the nucleus is 1060, and in the chromosomes, 550 (Izawa *et. al.*, 1963b). These values exceed 150- to 300-fold those in the somatic cell nuclei. For instance, in the chromatin of the cattle liver the protein · DNA ratio is 3.

The synthesis of nuclear proteins in oogenesis is detected autoradiographically and biochemically. Thus, an active synthesis of proteins of the chromosomal apparatus is shown by Gall and Callan (1962). When ^{14}C-phenylalanine is injected to female frogs, labeled proteins appear in the chromosome loops, covering them uniformly, already 24 h after injection.

In the early oocyte, the incorporation of label into the protein is higher in the nucleus than in the cytoplasm (Ecker and Smith, 1971).

This does not prove, however, that the protein synthesis occurs in the nucleus, but can only indicate the high rate of formation of nuclear proteins and their relatively rapid migration into the nuclei (Feldherr, 1975; Bravo and Allende, 1976). But the oocyte cytoplasm contains a pool of nuclear proteins which enter the germinal vesicle gradually within several hours (Feldherr and Richmond, 1976). Cognetti *et al.* (1977) observed histone synthesis and accumulation in sea urchin oocytes.

The proteins accumulated in the nucleoplasm must contain specialized oocyte proteins along with the usual proteins of the nuclei. They include, for instance, the "cleavage factor," which is, most probably, a protein. In frog and star sturgeon eggs maturing without formation of the germinal vesicle, fertilization or activation is not followed by regular cleavage (Dettlaff *et al.*, 1964, 1965; Skoblina, 1970). Injection of the germinal vesicle content into an enucleated egg restores its ability for cleavage (Smith and Ecker, 1969).

The germinal vesicle of a large oocyte also contains a "DNA synthesis factor," which may possibly be a specific DNA endonuclease (Benbow and Ford, 1975).

Finally, data have been obtained indicating that the oocyte has, by the moment of maturation, some factors in the germinal vesicle which determine the course of the early differentiation (gastrulation, neurulation) by the embryo. Humphrey (1966) has established that eggs of the Mexican axolotl bearing o-gene mutation can develop only up to the gastrula stage. This mutation is recessive, i.e. the +/o heterozygotes show no deviations from normal development. But on crossing, 1/4 of the specimens obtained are o/o homozygotes. They develop to an adult state, and half of them are sterile males. The other half consists of females whose eggs develop, after fertilization with any type of sperm (+/+ or +/o), only to the stage of mid-late gastrula.

Consequently, the normal (wild) allele of the o-gene does function in oogenesis but determines the course of gastrulation and neurulation, rather than oogenesis itself. The products of this gene accumulate in the germinal vesicle. If the content of the germinal vesicle of normal specimens (+/+) is injected into an egg obtained from an o/o female, the embryos are capable of normal development (Briggs and Cassens, 1966).

According to some properties (heat and proteolysis sensitivity, precipitation by ammonium sulfate, etc.), this o-factor is a protein. It accumulates in the germinal vesicle (though is apparently synthesized in the cytoplasm) and when the latter disintegrates, on maturation, it gets into the ooplasm.

Thus, oogenesis is attended, aside from the synthesis of usual nuclear proteins (proteins of nucleoli, nucleoplasm, and chromosomes), by a synthesis of oocyte-specific proteins which are used in subsequent development.

In the cytoplasm of a growing oocyte, not only the total mass of the cytosol increases, but so does the number of cellular structures, such as mitochondria, ribosomes, endoplasmic reticulum, the Golgi apparatus. In fishes, for instance, the total amount of mitochondrial proteins per oocyte increases throughout oogenesis, but this process is particularly intensive in the vitellogenesis. During loach oogenesis, in the small-growth period, the rate of mitochondrial protein mass growth exceeds the rate of growth of the whole oocyte, so the mitochondria concentration increases. During vitellogenesis, the mitochondria mass increases by another 12 times, but the oocyte volume increases more rapidly due to yolk deposition, and by the time of maturation the mitochondria concentration decreases to the initial value (Ozernyuk, 1972; Ozernyuk and Palmbakh, 1975) (Fig. 46).

The mouse egg contains about 9×10^4 mitochondria (Pikó and Matsumoto, 1976), which, calculated per cytoplasm volume, is twice as many as in the cells of an adult liver. But since the egg mitochondria are much smaller, the volume occupied by them (2.5%) is 1/8 that of the mitochondria in the liver (20%). The egg mitochondria increase in size beginning with the 4- to 8-blastomere stage (Pikó and Chase, 1973).

The increase in the number of mitochondria is the result of their multiplication inside the oocyte. In insects, where nurse cells play an important role, the process of direct transfer of mitochondria from nurse cells to oocyte has also been described (Raven, 1961).

In the course of oogenesis, the mass of the ribosomes (see Chapter 2) and ribosomal proteins (Hallberg, 1974) increases markedly. In amphibians, the synthesis of ribosomal proteins constitutes 31.5% of the total protein synthesis by mid-oogenesis (Hallberg and Smith, 1975). In this period the synthesis of ribosomal RNA reaches a maximum (Scheer, 1973). By the end of oogenesis, there are 4 μg of ribo-

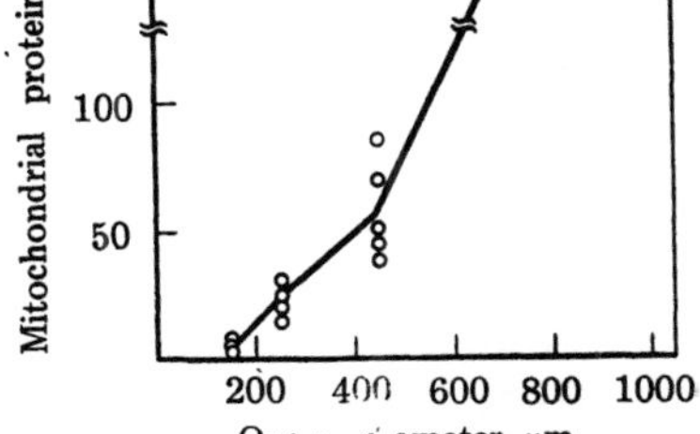

Fig. 46. Accumulation of mitochondrial protein during loach oogenesis. Mitochondria were isolated from oocytes of different size and the amount of protein in them was determined. The amount of mitochondrial protein in the oocytes increases by two orders of magnitude (Ozernyuk, 1972).

somal proteins in the *Xenopus* oocyte. Since in this period there is an active synthesis of all the other proteins, the relative fraction of the ribosomal proteins decreases to 1.5% (Hallberg and Smith, 1975).

A different picture was obtained for the ribosomal protein synthesis in the mouse oocytes (La Marca and Wasserman, 1979). Using two-dimensional electrophoresis, the authors traced the synthesis of 12 ribosomal proteins and measured their absolute synthesis rate at the stages of growth and meiotic maturation of the oocyte, in an unfertilized egg, and at the early stages of development. It has been shown that the synthesis of ribosomal proteins proceeds at all stages and its relative fraction in the total protein synthesis is very small: from 1 to 1.5%. Although the protein fractions studied are represented by equimolar quantities in the ribosomes, their synthesis rate may differ by an order of magnitude. In the mouse oocyte there accumulates 0.12 ng of ribosomal proteins. Since the volume of the mouse egg is 4×10^3 times less than that of *Xenopus*, the ribosome concentration in it is five times lower. This apparently explains the early beginning of formation of ribosomes in the mammals. During meiotic maturation one observes a 40% reduction in the synthesis rate of ribosomal proteins (from 620 to 370 fg/h per cell). The synthesis of the total protein and of such proteins as tubulin decreases only by 23% in this period. As a result the share of ribosomal protein synthesis falls off to 0.5% of the total proteins. At the same time, even in an unfertilized egg, the ribosomal protein synthesis continues in the absence of that of ribosomal RNA's, though at a lower rate (La Marca and Wasserman, 1979). The maximum of the intensity of ribosome formation falls on vitellogenesis (Mairy and Denis, 1972). At the small-growth stages, amphibian oocytes also accumulate 42S RNP particles which contain 5S rRNA and tRNA. The protein component of these particles does not belong to ribosomal proteins. Gradual utilization of 5S rRNA for production of new ribosomes leads to the decay of 42S RNP particles and probably to the disappearance of the protein they contain (Thomas, 1970; Mairy and Denis, 1972).

In addition to ubiquitous cellular structures — mitochondria and ribosomes — structures specific to the egg, such as cortical granules and porous membranes, also appear in the oocyte (Raven, 1961; Aizenshtadt, 1965; Aizenshtadt and Dettlaff, 1972). A substantial share of total protein synthesis in oogenesis belongs to tubulin. It has been demonstrated for sea urchin (4%; Cognetti *et al.*, 1977) and axolotl (Raff, 1977) oocytes.

The main source of synthesis of the whole mass of functionally active proteins and structures seems to be the oocyte cytoplasm. It is difficult to estimate accurately the ratio between proteins synthesized inside the oocyte and those arriving from the outside, because there are no reliable criteria for distinguishing proteins of different origin. Direct evidence for protein synthesis in the oocyte cytoplasm was given by

Gould (1969a). The oocytes of the echiuroid worm *Urechis caupo* were isolated and incubated *in vitro* with labeled amino acids in the absence of follicular and nurse cells. In these immature oocytes, an intensive protein synthesis has been shown by autoradiography.

It may be assumed that the reserve yolk proteins are exogenous, and the other oocyte proteins are endogenous, although, as mentioned above, there are animals, for instance, the river crayfish, in which vitellogenins are also synthesized in the cytoplasm of the oocyte itself (Ganion and Kessel, 1972).

The ability of the protein-synthesizing system of the amphibian oocyte for translation, including that of exogenous templates, is illustrated by numerous experiments with injection of various mRNA's (see Gurdon, 1974). The high translation efficiency was shown on such templates as mRNA of promelitine, a bee venom component (Kindas-Mügge *et al.*, 1974), this pointing to the absence of species and tissue specificity of ribosomes and of the tRNA set in the oocyte.

The share of exogenous proteins is evidently higher in large eggs and very small in eggs with practically no yolk. At the same time it is obvious that large eggs have not only more yolk, but also more other structures of clearly endogenous origin. Thus, for instance, the number of ribosomes, as well as that of mitochondria, is proportional to the egg size and, hence, many times more ribosomal and mitochondrial proteins are synthesized in the large eggs. The same is true for the germinal vesicle proteins. In small eggs, which are almost devoid of yolk, the share of endogenous proteins probably reaches 90-95%, but it is also hardly less than 10% in large eggs.

Experiments of Blackler (1970) on two species of *Xenopus* are very illustrative as to the role of the oocyte's own genome in the protein synthesis. He transplanted gonocyte-containing parts of *X. tropicals* embryo into the neurula of *X. laevis*. As a result, normal *X. laevis* specimens developed whose ovaries contained some oocytes with the genotype of *X. tropicals*. The entire somatic environment of such oocytes — follicular cells, the ovarian stroma, the endocrine glands, the yolk-synthesizing liver — evidently belonged to *X. laevis*. Nevertheless, these eggs retained all the traits of *X. tropicals* in their size (the eggs of *X. tropicals* are much smaller than those of *X. laevis*), pigmentation, and development pattern. An exception was their ability to be fertilized with the sperm of *X. laevis*. In the control, the eggs of *X. tropicals* grown in the female body of its own species were fertilized with foreign sperm only in 1% of the cases, whereas eggs of *X. tropicals* grown in the body of *X. laevis* were fertilized with the sperm of *X. laevis* by 95%. This, however, was probably due not to a change in the cortical layer of the egg cell, but to the properties of the gelatinous coat of the egg, which belongs completely to the recipient *X. laevis*, since it is formed on the egg surface owing to secretion by the oviduct cells during the passage of the egg. It is obvious that the yolk in these eggs also belongs to *X. laevis* since it has been synthesized in its liver, but the amount of the exogenous yolk is determined by the genotype of the oocyte, which is indicated by the small oocyte size characteristic of *X. tropicals*. It would be important to verify the species specificity of the remaining egg proteins, because the gonocyte transplantation method undoubtedly helps distinguish proteins of exogenous and endogenous origin, provided they are species-specific.

An approximate estimate of the number of types of oogenesis-synthesized proteins can be obtained from the data on the percent of active genes. The lowest estimate (see Chapter 2) is about 1%. Even if we assume that only 1/10 of the functioning genes are structural (this estimate is absolutely arbitrary), then at least 10^4 different types of protein must be synthesized during amphibian oogenesis. This figure should not seem overestimated if we recall that we know at present at least a thousand enzymes and most of them can be detected in the oocytes and eggs.

To check this estimate, one can also use the data on competitive hybridization, which show that about 1/3 of the RNA types synthesized in oocytes are completely utilized during oogenesis, i.e. serve as templates for the egg protein synthesis and do not participate in the post-fertilization protein synthesis (Crippa *et al.*, 1967).

In conclusion one could mention the data of Bakken (1973) obtained in *Drosophila*. He examined the manifestations of 98 autosomal mutations leading to female sterility, i.e. affecting the oogenesis or oviposition in one way or another. These mutations disturbed the ovarian structure, the yolk synthesis and deposition, the differentiation of the nurse cells, the formation of the chorion (egg membrane), the ovulation and oviposition, or they destroyed the embryo at the earliest stages of embryogenesis. Of the 98 mutations found by Bakken only part refer to genes controlling the metabolism in the oocyte itself. However, the diversity of their action shows that although a considerable part of the genes function outside the egg and determine the conditions of oogenesis (formation of follicular and nurse cells and the yolk synthesis) most of them determine the protein synthesis in the egg itself.

The different proteins accumulate in the egg unevenly. Thus, the ratio of the activities of glycolytic enzymes changes substantially during loach oogenesis (Milman and Yurovitsky, 1973). The thyrosinase activity in the *Xenopus* oocyte is induced by choriogonin, but this induction is particularly effective in oocytes of stages 3 and early 4, and less so at stages 2 and 5 (Eppig and Dumont, 1974).

In the course of oogenesis the protein distribution in the egg is also uneven. Thus, the antigen composition of the *Drosophila* egg is different in its anterior and posterior parts (Graziosi and Robertis, 1975). These differences apparently reflect the peculiarities of the germ plasm of the egg (Illmensee and Mahowald, 1976).

The protein synthesis in amphibian oocytes is retarded or ceases by

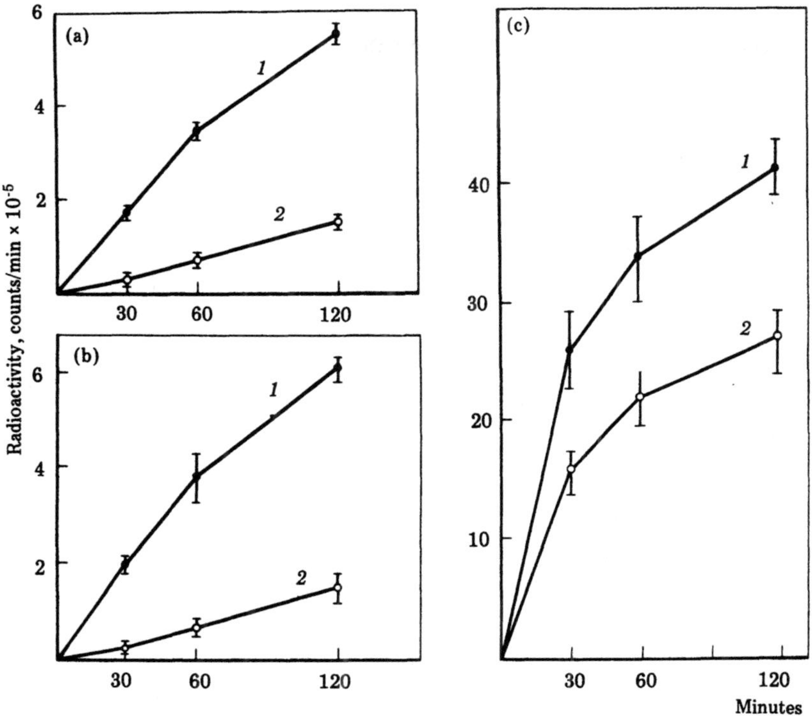

Fig. 47. Comparison of protein synthesis rates between oocytes and mature eggs of the sea urchin. Oocytes (*a*), unfertilized mature eggs (*b*), and embryos at the 16-blastomere stage (*c*) of *Paracentrotus lividus* were incubated with ^{3}H-leucine and the uptake of leucine by the embryo (*1*) and its incorporation into the protein (*2*) were determined. Leucine is taken up by the embryo and incorporated into the oocyte and egg proteins at an equal rate; after fertilization (*c*) the uptake and incorporation rate increase (Rinaldi and Parente, 1976).

the time of vitellogenesis completion at the end of the large-growth period. The synthesis is resumed upon maturation of the oocyte, and this process is activated by the gonadotropic hormone (Smith *et al.*, 1966; Ecker *et al.*, 1968). It was shown later, however, that protein synthesis in amphibian oocytes increases by only 70% during maturation and by another 50% after fertilization (Shih *et al.*, 1978). In the sea urchin the protein synthesis in the mature oocytes and in unfertilized eggs proceeds at the same rate, which increases sharply after fertilization (Rinaldi and Parente, 1976) (Fig. 47). In the mouse oocytes the protein synthesis rate slightly decreases at the beginning of maturation (from 43 to 33 ng/h per oocyte), but increases again after fertilization (Schultz *et al.*, 1979b).

The amount of proteins in the egg apparently increases insignificantly during the short maturation time, but these proteins are necessary for the maturation process. Protein synthesis inhibitors — puromycin or cyclohexymide — block the maturation of oocytes in amphibians (Dettlaff, 1966; Brachet, 1967; Dettlaff and Skoblina, 1969), echinoderms, and worms (Zampetti-Bosseler *et al.*, 1973).

The electrophoretic spectrum of proteins synthesized in the frog oocyte during maturation differs somewhat from the same spectrum in the egg after fertilization (Ecker and Smith, 1971). A considerable part of these proteins are concentrated in the germinal vesicle: the intensity of the autoradiographic label over it is 6-7 times higher than over the cytoplasm. Nuclear proteins synthesized during maturation remain in the egg even after the germinal vesicle rupture. As the number of nuclei increases in the course of cleavages, these proteins are concentrated in the embryo's nuclei. If nuclei from the blastula cells are transplanted to an egg, the proteins labeled during maturation rapidly migrate into these nuclei (Ecker and Smith, 1971). The nature of the synthesis and the spectrum of the proteins synthesized during oocyte maturation do not change, provided the germinal vesicle is removed beforehand. This means that the mRNA templates for the synthesis of these proteins are transcribed and enter the cytoplasm before the beginning of maturation, still in the growing oocyte (Ecker and Smith, 1968, 1971). This is also indicated by the fact that actinomycin D does not block the frog oocyte maturation *in vitro*, which is stimulated by progesterone (Masui, 1972). Maturation also proceeds in an oocyte deprived of the germinal vesicle (Skoblina, 1970).

However, actinomycin D represses maturation stimulated by a gonadotropic hormone (Dettlaff, 1966). The gonadotropic hormone acts through follicular cells, in which a nucleus-dependent synthesis and excretion of progesterone take place, the latter being directly responsible for the oocyte maturation (Masui, 1967; Masui, 1972). Similar results have been obtained on the oocytes of other amphibians (Thornton and Evernett, 1969) and fishes (Dettlaff and Skoblina, 1969).

Little is known about the composition of proteins synthesized during maturation. In the course of maturation, a factor causing the disintegration of the germinal vesicle membrane and inducing the onset of meiosis appears in the amphibian oocyte. This factor accumulates in the animal zone and is bound with the oocyte hyaloplasm. Its content sharply decreases by the moment of fertilization (Masui and Markert, 1971; Masui, 1972). Somewhat later, the oocyte begins to reveal a cytostatic factor which disappears within 90 min of fertilization by the beginning of the first cleavage (Masui, 1972). In the brown frog *Rana temporaria*, however, no cytostatic factor was detected in the oocyte (Dettlaff, personal communication).

In distinction to amphibians, in echinoderms the factor responsible for the disintegration of the germinal vesicle membrane appears long before the beginning of maturation, already in the first-order oocytes. In starfish and worms, among the proteins synthesized during maturation there are the proteins of the microtubules of the mitotic apparatus or the factors responsible for its assembly, since puromycin treatment of their oocytes causes abnormalities in the organization of the spindle and centrioles (Zampetti-Bosseler *et al.*, 1973).

In the mouse oocytes, the spectrum of synthesized proteins changes substantially at the beginning of maturation and remains unaltered after fertilization (Schultz *et al.*, 1979a).

Apparently, not all the proteins synthesized during maturation are maturation proteins proper in the sense of ensuring this process. A part of them, maybe the greater part, are used in embryonic development. Thus, in starfish and worms even a 50% inhibition of protein synthesis does not arrest the maturation process (Zampetti-Bosseler *et al.*, 1973).

Utilization of Yolk in the Course of Development

With complete cleavage of eggs, the entire yolk is found inside the cells; it is usually distributed nonuniformly: more yolk is located at the vegetal than at the animal pole of the blastomeres. In the course of development, a gradual resorption of yolk granules occurs, i.e. the hydrolysis of their proteins to amino acids and their utilization for the synthesis of the embryo protein and, partly, for energy needs. In amphibians, blastomeres overfilled with yolk become part of the endoderm and form the intestinal walls, with the yolk serving, at the later stages, as a stockpile of nutrients for these cells and for the tadpole.

With incomplete cleavage, the blastomeres are separated from the yolk, although in the course of cell divisions part of the yolk granules gradually enter into the cells adjoining the yolk (Thomas, 1968). Amino acids and other low-molecular-weight compounds which are formed as a result of the gradual yolk hydrolysis, are transported into the embryo cells. At the early stages of development this transport is promoted by cytoplasm cords which run through the yolk and in which an active motion of plasmic flows is observed (Tung and Tung, 1944; Oppenheimer, 1947; Svetlov *et al.*, 1962). At later stages, the yolk sack becomes enclosed by the circulatory system. Besides, in bird embryos, ovalbumin degradation occurs in extraembryonic parts of the blastoderm and the decay products enter the embryo through the circulatory system (Hassel and Klein, 1971).

The noncellular yolk includes a considerable portion of the proteolytic enzymes of the embryo (Neyfakh and Davydov, 1964), and also the glycolytic enzymes and mitochondria. The role of the proteases presumably amounts to hydrolysis of yolk proteins to amino acids. The function of the yolk energy system is less understood, although in loach embryos its power constitutes 1/3 to 2/3 of the total power of the egg energy system (Abramova and Neyfakh, 1965; Milman and Yurovitsky, 1973).

For mammalian embryos the principal source of amino acids is the maternal organism, and for parasitic insects, the host tissues in which the parasite's eggs are laid.

In animals whose embryos have no external sources of nutrition protein synthesis proceeds at all the stages of embryonic development, but the total amount of protein in the egg remains almost or completely unaltered throughout the embryonic development. This means that the synthesis of the embryo proteins is offset by the decay of the yolk proteins. The total protein content decreases at the late stages in amphibians and at the early larval stages in fishes. In fishes, for instance, the reduction runs to 40% of the initial egg protein. This value corresponds to the share of protein spent on energy needs. The remaining 60% is presumably converted into cell proteins of the embryo and the larva.

The yolk utilization kinetics was evaluated by Saito *et al.* (1965). They injected ^{14}C-leucine and ^{35}S-methionine to a teleost *Ozyrias latipes* female. Most of the injected label (90%) was found in the yolk proteins, another part (2%), in the proteins of the early embryo, while the remaining 8% were contained in the pool of free amino acids. This distribution remained practically unaltered until gastrulation, i.e. the yolk proteins were not used for building up the embryo proteins. Then the specific activity of the amino acid pool rapidly increased, reaching a maximum and then falling off again. It is to be noted that the dynamics of this process turned out to be different for leucine and methionine. The radioactivity of the embryo proteins increased simultaneously, and by the end of embryonic development 30% of the labeled methionine and 60% of the labeled leucine were incorporated into the embryo proteins. Thus, an appreciable consumption of yolk proteins was detected only after gastrulation and it proceeded by protein degradation to free amino acids and migration of the latter into the embryo cells. The difference in the metabolism of the two amino acids means that the utilization of yolk proteins proceeds in a differentiated manner. It is obvious that the yolk composition must also change gradually (Saito *et al.*, 1965).

The preferential utilization of yolk proteins was also shown indirectly by Deuchar (1956, 1963a, b, 1964), who determined the content of free amino acids in different parts of amphibian embryos at different stages of development. In the course of development the concentrations of amino acids change consistently relative to each other, and these changes are specific to the different parts of the embryo. In the course of yolk utilization some amino acids undergo mutual transformations. Thus, the labeled leucine, which was incorporated into the yolk during oogenesis of a *Xenopus*, in the embryo is partly converted into proline and incorporated into proteins in this form (Deuchar, 1963b).

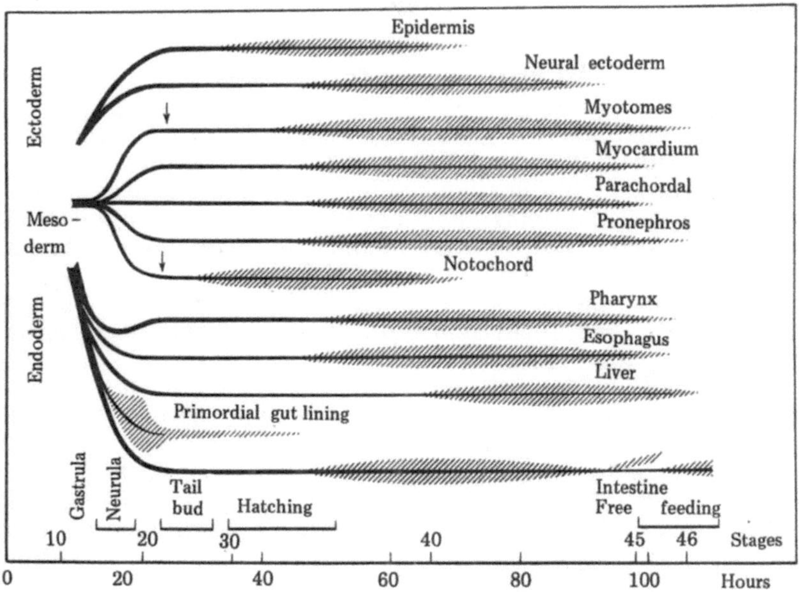

Fig. 48. Yolk uptake in various organs of *Xenopus* embryos and larvae. Histological preparations were stained with safranine, which revealed the proteolysis of the yolk granules. The shaded areas indicate the proteolysis period. Yolk utilization begins mainly after hatching and is completed only at the beginning of the feeding tadpole stage (Selman and Pawsey, 1965).

The utilization of the yolk granules can also be traced microscopically using safranine staining (Fig. 48). By using this method on *Xenopus* embryos it was shown that a perceptible proteolysis of the intracellular yolk begins from the neurulation stage, and only in the cells of the roof of the primordial gut. A little later, at the tail-bud stage, the yolk uptake is detected in the cells of the early differentiated notochord. In most of the tissues, however, the utilization of yolk granules does not begin until the tadpole is hatched (Selman and Pawsey, 1965).

For amphibians and other animals with intracellular yolk localization the maintenance of a definite concentration of free amino acids necessary for protein synthesis is provided by proteolysis control in the same cells. The mechanisms of this control are unknown but not difficult to conceive. More complicated is the stored protein proteolysis regulation in animals in which the yolk is located outside the cells, as are the proteolytic enzymes.

PROTEINS IN EMBRYOGENESIS

Protein Synthesis at the Beginning of Development

In most of the animals the protein synthesis does not cease at the end of egg maturation. Thus, in amphibians the protein synthesis, which is retarded in maturation-ready oocytes, increases following hormonal activation and the disintegration of the germinal vesicle and then remains constant until the moment of fertilization (Merriam, 1966; Smith *et al.*, 1966; Shih *et al.*, 1978).

After fertilization, the level of protein synthesis in amphibian embryos (Ecker and Smith, 1966, 1968; Shih *et al.*, 1978) begins to rise gradually during cleavage (Fig. 49).

In the sea mollusc *Ilyanassa* fertilization occurs at the germinal vesicle stage, i.e. formally before complete maturation. The protein synthesis rate starts rising 15 min after fertilization and within 2 h increases by a factor of 2.5 (Collier and Schwartz, 1969; Mirkes, 1970, 1972) (Fig. 50). Similar increase in protein synthesis after fertilization is observed in eggs of the echiuroid worm *Urechis caupo*

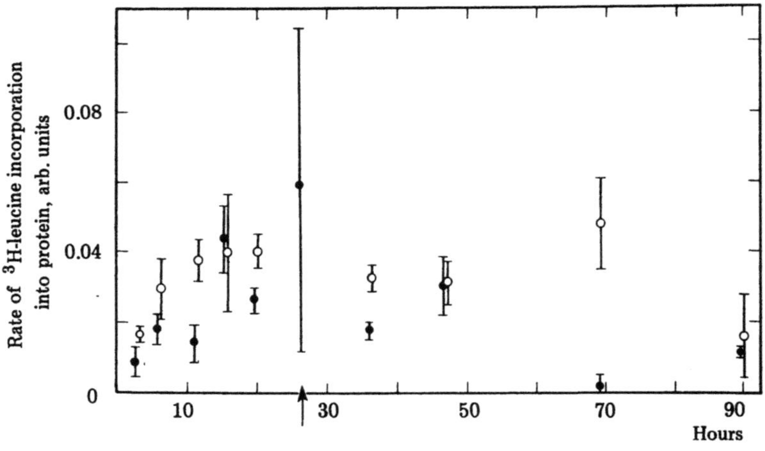

Fig. 49. Protein synthesis rate in *Xenopus* eggs during maturation and after fertilization. The maturation of oocytes was stimulated with progesterone. Part of the oocytes were previously enucleated. After 27 h the eggs were activated (arrow). ^{3}H-leucine was injected into them at different times after stimulation and the kinetics of its incorporation into the protein was determined. Solid circles indicate normal oocytes, open circles indicate enucleated oocytes. Abscissa: the time after progesterone treatment at 19.5°C. The measurement error is shown. After stimulation the protein synthesis increases up to the end of maturation (24 h), after which it remains at the level attained. The oocyte nucleus is not involved in the activation of protein synthesis (Smith and Ecker, 1970).

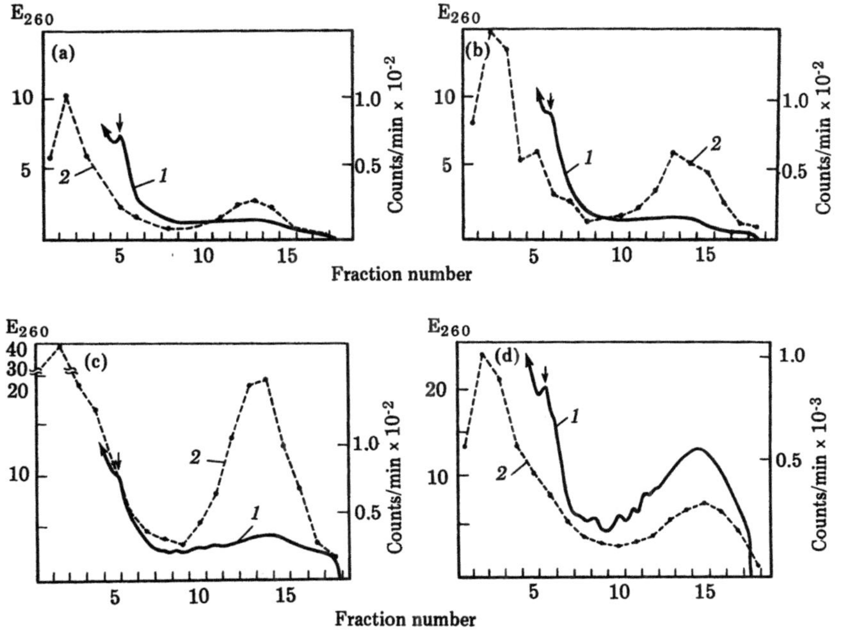

Fig. 50. Protein synthesis and polysome formation during the early development of the mollusc *Ilyanassa.* Eggs and embryos at different stages were incubated with ^{24}C-leucine (15 min, 20°C), then the homogenate was fractionated by centrifugation in a sucrose gradient (bottom right). *1,* Optical density (E_{260} nm); *2,* radioactivity; *a,* unfertilized eggs; *b,* stage of pronuclei fusion; *c,* 32- to 64-blastomere stage; *d,* larva (veliger). The protein synthesis is manifested as a radioactivity peak in the polysome region (fractions 10 to 17). It increases immediately after fertilization (Mirkes, 1972).

(Gould, 1969b), the bamboo worm *Sabellaria cementarium* (Wines-dorfer, 1965), and the mollusc *Spisula solidissima* (Bell and Reeder, 1967). In mammals, the protein synthesis remains at a low level after fertilization and increases only from the 8- to 16-cell stage (Epstein and Smith, 1973).

The eggs of echinoderms are a special case in this respect. A peculiarity of the sea urchin reproduction consists in the ability of mature water-spawned eggs to await fertilization for many hours until an accidental encounter with a spermatozoon. Before fertilization the eggs show very little metabolic activity, and although they retain the same synthesis rate as in oogenesis (Rinaldi and Parente, 1976), the protein synthesis in them constitutes a negligible fraction of that at the blastula stage. It begins to increase already 5-10 min after fertilization (depending on the temperature and the sea urchin species) and reaches a maximum after several hours, at the early blastula stage (Fig. 51). This increase in protein synthesis rate has been the object of numerous

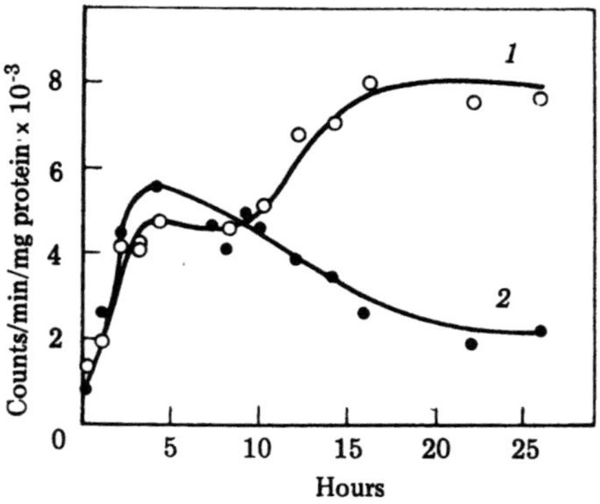

Fig. 51. Protein synthesis in early sea urchin embryos in the norm and in the presence of actinomycin D. Embryos developing under normal conditions (*1*) and in the presence of actinomycin D (20 μg/ml) (*2*) were placed for 20 min in a medium with ^{14}C-valine, then its incorporation into the protein was determined. During the first hours of development the protein synthesis increased rapidly both in the norm and upon transcription inactivation by actinomycin D (Gross and Cousinou, 1964).

investigations. Such an interest is not accidental, and it cannot be attributed solely to the simplicity of work with sea urchin eggs and the clarity of the effect, although these considerations have played their part. In the 1950s, when these investigations began, fertilized sea urchin eggs with a sharply increased protein synthesis rate were one of the few convenient models for studying the protein synthesis regulation mechanism. This mechanism is sufficiently interesting even now, but it should probably be regarded as a specific, rather than general, case of regulation, which is of greater interest to molecular embryology.

The beginning of the investigations is associated with the name of Hultin (Hultin, 1950; Hultin and Wessel, 1952). The main studies were conducted in the laboratories of Monroy, Gross, Nemer, Mano, and others, where the results which appear to determine the present level of understanding of this problem have been obtained (see also reviews by Nemer, 1967; Monroy and Gross, 1967; Giudice, 1973; Stearns, 1974).

The manifold increase in the rate of incorporation of labeled amino acids into the proteins of sea urchin eggs after fertilization (Hultin, 1950; Nakano and Monroy, 1958; Monroy and Vittorelli, 1962; Mano, 1970) reflects the actual increase in protein synthesis rate. It can be ascribed neither to the increase in the permeability of the egg membrane to amino acids, nor to the change in the intracellular amino acid pool. This is indicated by the consistency of the results obtained in work with different amino acids and other precursors, and also by direct experiments in which the rate of amino acid transport into the egg was measured and taken into account and, finally, by experiments with preloading of the egg with labeled amino acid.

After fertilization, amino acid transport actually increases (Tyler *et al.*, 1966; Silver and Comb, 1966; Fry and Gross, 1970a, b; Timourian and Watchmaker, 1970). According to Epel (1972), in unfertilized eggs the permeability to lysine and valine is sufficiently high. After fertilization it increases only by 8 and 21 times, respectively (Fig. 52). By contrast, unfertilized eggs show only slight permeability to alanine and especially to glycine. After fertilization the permeability to these amino acids increases by a factor of 290 (to alanine) and 1490 (to glycine).

When valine or lysine is used as labeled precursor, the increase in egg permeability to these amino acids is insufficient to explain the increase of label incorporation into the protein. The increase in permeability, for instance to valine, lasts about 2 h, whereas the increase in the rate of incorporation of labeled valine into the protein lasts much longer (Epel, 1972). Under natural conditions, in *Arbacia* eggs at early stages of development a constant pool of leucine (4×10^{-8} μmole per egg) is maintained, which is sufficient for only 5 min of protein synthesis. When this amino acid is added to the medium, the intracellular leucine pool quickly increases several-fold, but within an hour 80% of this excess transforms to glutamine and other amino acids (Fry and Gross, 1970a, b). All this greatly complicates calculations of the actual synthesis rate, but nevertheless it is obvious that the protein synthesis begins to increase 5-10 min after fertilization (Fry and Gross, 1970b; Epel, 1972) (Fig. 53).

In a later paper (Regier and Kafatos, 1977) the protein synthesis rate in unfertilized sea urchin eggs and in gastrulae was determined more accurately. In that case the authors took into account not only the specific radioactivity of the amino acid (leucine) in the intracellular pool, but also the specific activity of the immediate precursor of the protein synthesis, leucyl-tRNA. This provided more accurate data, which, however, differ only slightly from the values obtained by other researchers: 0.11 μg of protein per 10^6 eggs per minute are synthesized

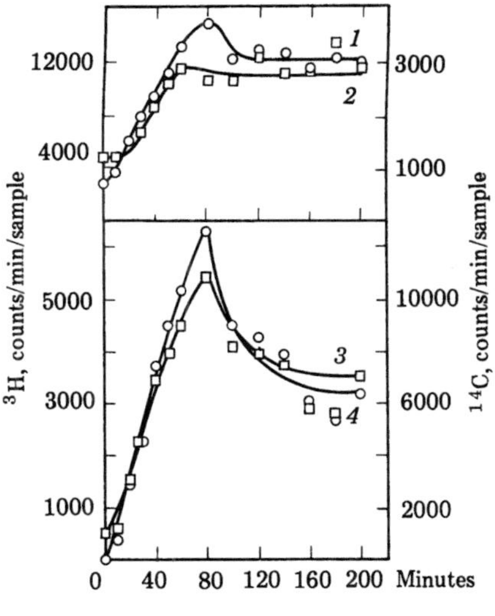

Fig. 52. Permeability of sea urchin eggs to amino acids after fertilization. At different times after fertilization, eggs of *Strongylocentrotus purpuratus* were incubated for 10 min with ^{14}C-phenylalanine (*1*), ^{3}H-lysine (*2*), ^{14}C-glycine (*3*), or ^{3}H-valine (*4*), after phenylalanine (*1*), ^{3}H-lysine (*2*), ^{14}C-glycine (*3*), or ^{3}H-valine (*4*), after termined. Permeability to lysine and phenylalanine increases several-fold, it grows tens of times for valine and many hundreds of times for glycine (Epel, 1972).

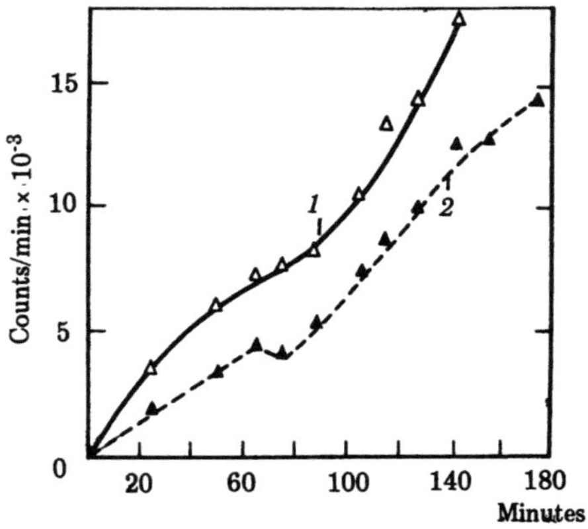

Fig. 53. Postfertilization changes in permeability of sea urchin eggs to leucine and in incorporation of leucine into protein. Eggs of *Strongylocentrotus purpuratus* were fertilized and incubated at 18°C. Then they were placed for 5 min into ^{14}C-leucine, after which the radioactivity of the acid-soluble fraction (^{14}C-leucine uptake by the egg) (*1*) and that of the protein (*2*) were determined. Permeability to leucine and incorporation into the protein increases, but this increase slows down at 60 to 90 min (first cleavage) (Fry and Gross, 1970a).

in unfertilized eggs and 12.4 μg of protein (i.e. 113 times more) in gastrulae (Regier and Kafatos, 1977). This is in good agreement with the observation that the number of ribosomes contained in the polysomes at these stages increases 90-fold (Galau *et al.*, 1974).

The augmentation of protein synthesis is accompanied by incorporation of free ribosomes into the polysomes. The formation of polysomes should be regarded as the expression of this increase. The data comparing the kinetics of the number of polysomes and the protein synthesis rate reported by different researchers are contradictory. Some authors found that the increase in the amount of polysomes lags behind the increase in protein synthesis, which suggested a fundamentally important conclusion that the enhancement of protein synthesis is also due to its efficiency per one active (polysomal) ribosome (Piatigorsky and Tyler, 1970; Mano, 1971a).

Humphreys arrived at the opposite conclusion. In his first paper (1969) he shows that the translation efficiency (time of addition of one amino acid to the polypeptide chain or the rate of synthesis per one ribosome contained in a polysome) is the same for unfertilized eggs and for developing embryos. However, from his subsequent data (Humphreys, 1971) it follows that the share of ribosomes incorporated into polysomes at the early stages increases from 0.75% in the unfertilized egg to 20% after 2 h, i.e. by 30 times, whereas the protein synthesis intensity increases only by 15 times within this period. The translation efficiency per ribosome is thus halved. It is possible that these different conclusions stem from the fact that the degree of preservation of polysomes on their isolation from the eggs or from the embryos is different in different works. Even slight differences in the quantity of polysomes in unfertilized eggs will lead to substantial divergencies in the results. Characteristically, according to Humphreys, the quantity of polysomes is, in general, smaller than according to other authors. The quantity of free ribosomal subunits in unfertilized eggs is about 3% of the total number of ribosomes, and after fertilization it decreases by 10-25%, i.e. changes only slightly (Jenkins and Denny, 1970).

In the early 1960s attention was focused on the role of the genetic apparatus of the egg, because it was assumed, by analogy with bacteria, that an increase in protein synthesis must be caused by the appearance of new (short-lived) RNA templates synthesized in the nucleus. Numerous and diverse experiments showed this assumption to be wrong. In the first experiments of this kind the protein synthesis was determined in the anucleate half-eggs activated by various agents causing parthenogenetic development. Irrespective of the presence or absence of nuclei, protein synthesis was practically nil before activation, while after activation it began and increased in both halves in the same way as in normally fertilized whole eggs (Brachet *et al.*, 1963; Denny and Tyler, 1964; Berg, 1965). A similar result was obtained on inactivation of nuclei with actinomycin D (Gross and Cousinou, 1964; Mal'kin *et al.*, 1964) or with high radiation doses (Neyfakh *et al.*, 1968). The protein synthesis intensity in the first hours of development was found to be the same as in the control, or slightly higher. This suggests that the sea urchin egg contains all the necessary components of the protein-synthesizing apparatus and that the protein synthesis activation after fertilization is due to translation regulation mechanisms.

Three components involved in protein synthesis were considered as possible synthesis-controlling factors: (1) mRNA, (2) ribosomes, and (3) components of cytoplasm, such as tRNA, aminoacyl-tRNA-synthetases, translation factors (T_1, T_2), HTP, and amino acids.

Some differences in the activity of aminoacyl-tRNA-synthetases, in the tRNA composition, and in translation factors between fertilized and unfertilized eggs have been revealed (Candelas and Inverson, 1966; Timourian, 1967; Castaneda, 1969; Metafora *et al.*, 1971; Sharma *et al.*, 1973). But these are insufficient to explain the activation of a protein synthesis after fertilization. Most of the authors find that the high-speed supernatant of the homogenate (cell sap) of unfertilized and fertilized eggs are equally efficient in maintaining the protein synthesis on natural or artificial (poly(U)) templates (Hultin, 1961; Wilt and Hultin, 1962; Stavy and Gross, 1967, 1969; Kedes and Stavy, 1969). The choice of the main regulatory factor is thus reduced to two possibilities: mRNA or ribosomes.

In a number of experiments with the use of ribosomes obtained from fertilized and unfertilized sea urchin eggs it was shown that there is no difference between them (Wilt and Hultin, 1962; Nemer and Bard, 1963; Stavy and Gross, 1967, 1969; Kedes and Stavy, 1969; Infante and Graves, 1971). This was mainly proved in *in vitro* experiments using Nierenberg's system, where polyuridylic acid (poly(U)) served as templates, and phenylalanine incorporation was measured.

Later, the identity of ribosomes from eggs and embryos was shown in a system which contained, aside from ribosomes, factors of translation from Krebs' ascite cells and rabbit globin mRNA as templates (Clegg and Denny, 1974). As a result, the conclusion was drawn that a special masked form of mRNA (mmRNA, or informosomes) is responsible for the control of protein synthesis in the sea urchin early development (Spirin and Nemer, 1965; Tyler, 1967; Spirin, 1966, 1969; Gross, 1967; Stavy and Gross, 1969; Kedes and Stavy, 1969; Humphreys, 1971). This remained the only possible assumption if it is true that all the other components of the protein-synthesizing apparatus do not differ from each other. It is also true that since mRNA is definitely present in unfertilized sea urchin eggs (Slater and Spiegelman, 1968) and since it does not take part in translation, it can be called masked mRNA (Spirin, 1966), or mmRNA (Tyler, 1967). According to Mano, in unfertilized eggs a considerable part of the active template RNA is in the granules precipitated in the presence of Ca^{2+} at 12,000*g*. Following fertilization, possibly as a result of proteolytic action, the RNA from these granules moves to the polysomal fraction (Mano and Nagano, 1970; Mano, 1971a, b). This protein-bound mRNA was detected in polysome complexes as well (Ruzdijič and Glišin, 1972).

As before, the main argument in favor of the assumption that the protein synthesis activation in sea urchin eggs after fertilization is due to demasking of the mRNA is the similarity of ribosomes isolated from unfertilized and from developing eggs. This finding has, however, been disputed in a number of works carried out in Monroy's laboratory (Maggio *et al.*, 1964, 1968; Monroy *et al.*, 1965; Rinaldi and Monroy, 1969; Metafora *et al.*, 1971; Gambino *et al.*, 1973). They showed, in particular, that trypsin treatment of unfertilized eggs activates protein synthesis in them (Monroy *et al.*, 1965). This, of course, is insufficient to prove the leading role of ribosomes, because trypsin may have affected not only the ribosomes, but also the masked mRNA. Moreover, according to other data, trypsin may activate unfertilized eggs to parthenogenetic development and stimulate the protein synthesis indirectly (Hand, 1971). But later Monroy and coworkers found some slight differences in ribosomes obtained from fertilized and unfertilized eggs. During prolonged incubation in a medium with a low magnesium concentration, the dissociation of ribosomes isolated from fertilized eggs occurs faster than that of ribosomes from unfertilized eggs. The differences disappear if both types of ribosomes are pretreated with trypsin. This suggested that ribosomes from unfertilized eggs have a special protein which binds their subunits and interferes with their function (Maggio *et al.*, 1968). During fertilization, a number of enzymes (see below), including proteases, become activated (see Epel, 1967; Epel *et al.*, 1969b; Vacquier *et al.*, 1973). Proteases destroy the binding protein, activate the ribosomes, and switch on the protein synthesis (Metafora *et al.*, 1971). To confirm this, a protein lacking in fertilized eggs and inhibiting the protein synthesis *in vitro* was isolated from ribosomes of unfertilized eggs (Gambino *et al.*, 1973).

Nevertheless, this hypothesis cannot be considered sufficiently substantiated: the effect of proteases on ribosomes and on their activation mechanism remains unconfirmed. The hypothesis has been disproved by subsequent investigations, where a careful comparison of the physicochemical properties of ribosomes did not reveal any differences between them (Kedes and Stavy, 1969; Infante and Graves, 1971).

The proponents of the hypothesis of "masked RNA" and of its demasking as the process triggering the protein synthesis at present have rather few data on the differences between the pre- and postfertilization state of mRNA (apart from the obvious fact that in the former case it is outside the ribosomes and in the latter it is contained in the polysomes).

From MacKintosh and Bell's experiments it follows that it is the rate of initiation that directly determines the activation of protein synthesis. These authors have found that the elongation inhibitor — cycloheximide — causes accumulation of polysomes in unfertilized eggs. Apparently, in unfertilized eggs the weak protein synthesis is limited by initiation, while in fertilized eggs the higher level of the synthesis is restricted by the rate of elongation (MacKintosh and Bell, 1969a, b). But these data, too, cannot clarify which of the two — the ribosomes or the mRNA — is responsible for the enhancement of initiation.

A new approach to the problem of protein synthesis activation is pioneered by investigations into events occurring immediately after fertilization or activation of the egg and preceding the stimulation of the protein synthesis. Epel (1967) records about 30 biochemical and morphological changes which take place in the sea urchin egg within the first 10 min following fertilization. The relationship between these changes and their bearing on the stimulation of the protein synthesis are unknown, as a rule, although proteolysis activation is usually regarded as one of the mechanisms of ribosome switching on. It has been shown recently that the fertilization or activation of eggs with ammonia leads to specific cortical changes during which a high-molecular-weight component (glycoprotein) is released by the egg, which is capable of suppressing the protein synthesis (Johnson and Epel, 1975). It is assumed that the removal of this substance restores DNA replication, and, possibly, protein synthesis (Mazia *et al.*, 1975). Fertilization also changes the membrane potential and thereby increases the permeability of potassium ions (Tupper, 1973, 1974), which are associated by the author with the stimulation of the protein synthesis.

Finally, a high correlation between the value of intracellular pH and the protein synthesis rate was observed recently in sea urchin eggs (Grainiger *et al.*, 1979). Measurements with the aid of microelectrodes showed that after fertilization intracellular pH increases from 6.9 to 7.2 within the first five minutes, simultaneously with the activation of protein synthesis. Artificial reduction of intracellular pH by only 0.2 reduces the incorporation of amino acids into the protein by 6 times, while an increase by 0.13 augments the incorporation 3- to 4-fold (Grainiger *et al.*, 1979).

A model closely resembling sea urchin eggs are the embryonic cysts of the brine shrimp *Artemia salina.* Its eggs develop to the gastrula stage, then dry out and lapse into anabiosis, which is terminated by water entering the cysts and triggering the synthesis of macromolecules, primarily those of proteins. Although one cannot speak about the protein synthesis activation "at the inception of development" in this case, the problems arising here are the same as in the fertilization of sea urchin eggs. The first studies of *Artemia salina* showed that 15 to 30 min after incysted gastrulae were placed in water, newly formed polysomes were detected, and there was incorporation of precursors into the protein. If the cysts were soaked at 0°C, when the temperature rose the increase in the number of polysomes was detected already after 3 min (Clegg and Golub, 1969). This means that the limiting factor is the rate of water penetration into the cyst, and the protein synthesis is switched on practically at once, although the entire process of increase in synthesis rate lasts more than 2 h (Fig. 54). During this period, not only the quantity of polysome-forming ribosomes increases, but so does the translation efficiency, i.e. the protein synthesis rate per one ribosome (Clegg and Golub, 1969; Finamore and Clegg, 1969; Hultin and Morris, 1968). This increase is mainly due to the mRNA store formed before anabiosis, rather than to the increase in the quantity of mRNA's (although their synthesis begins soon after the development is resumed) (Nilsson and Hultin, 1972).

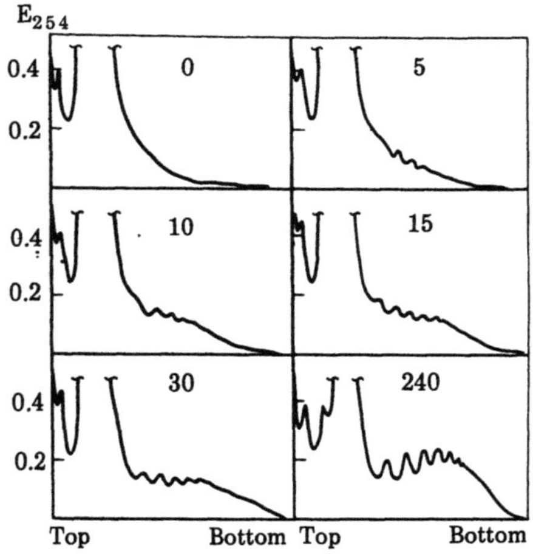

Fig. 54. Increase in the number of polysomes in postdiapause gastrulae of *Artemia salina.* Encysted dry eggs (gastrulae) of the brine shrimp *Artemia salina* were activated by sea water at 30°C. After different time intervals they were homogenized, and the polysomes analyzed by centrifugation in a sucrose gradient. The numerals in the figure show the incubation time in sea water (min). The top of the ribosome peak (left) is not shown. The polysomes (right) appear already after 5 min, and their number subsequently increases (Finamore and Clegg, 1969).

The detailed investigation of this phenomenon at Ochoa's laboratory at first resulted in an erroneous conclusion that the limiting element in the protein synthesis in *Artemia salina* embryos is the quantity of mRNA's (Sierra *et al.*, 1974). But at a later date the same authors showed that the protein synthesis is limited by several initiation factors. If these translation factors, denoted as IF-MP, M3, M2A, and M2B, are washed off the ribosomes of the developing embryo, they may activate translation on ribosomes in nondeveloping eggs *in vitro* (Filipowicz *et al.*, 1975, 1976).

The protein synthesis activation systems in sea urchin eggs and in *Artemia salina* embryos seem to be different, although in both cases the regulation mechanisms operate at the level of translation.

Protein Synthesis Rate

The expression "protein synthesis rate" is not unambiguous. This term may express rather different notions. Defined most accurately, it means the rate of increase in polypeptide chain, i.e. in the number of amino acid residues added to this chain per unit time. This value can be determined if one knows the specific radioactivity of a given labeled aminoacyl-tRNA pool, the frequency of occurrence of this particular amino acid residue in the polypeptide synthesized, and, finally, the quantity of functioning ribosomes contained in the polysome complexes. Determination of all these values is complicated enough even in an *in vitro* system, and the difficulties increase still more for studies *in vivo* (cell).

Direct determination of the specific radioactivity of leucyl-tRNA in unfertilized eggs and gastrulae of the sea urchin showed that the absolute synthesis rate of protein (0.11 and 12.4 μg of protein per 10^6 eggs per hour, respectively) increases 113-fold within the time elapsed between these stages (Regier and Kafatos, 1977).

The specific activity of the amino acid pool may be estimated and taken into account in a simpler way by comparing incorporation of the label into the protein after injection of amino acid solutions with different specific activity into the egg (Shih *et al.*, 1978).

The values of absolute synthesis rate thus obtained are of limited interest and not always relevant if one deals with protein synthesis in the embryo. For the purposes of developmental biology it is, as a rule, sufficient to compare the protein synthesis rates in the embryo at different stages of development, in its different parts, or under different experimental conditions. To this end, the protein synthesis rate can be expressed by the amount of the label incorporated into the protein per unit time per 1 mg of protein or per embryo. Such a calculation yields the relative synthesis rate values, since the ratio of the number of the newly formed protein molecules to the total amount of protein has no meaning *per se*. Radioactivity related to the amount of protein is, however, a reliable index of the mass of the cytoplasm in which the synthesis being measured is taking place. Therefore, if the amount of the embryonic proteins does not increase in the course of development, this index reflects, in principle, the protein synthesis activity per one embryo. But since losses may occur in the course of preparation of the specimen for determining the radioactivity, calculation per 1 mg of protein is more reliable. We have already considered the possible errors due to the differences in permeability to amino acid precursors and in their free pool in the cell. To these factors could be added errors associated with the differences in the amount of inert protein (yolk granules) in different tissues (for instance, in the amphibian ectoderm and endoderm) and at different stages. All this introduces a considerable arbitrariness in the values of the protein synthesis rates and impedes the comparison of data obtained for different species.

For sea urchin eggs, labeled amino acids, to which the former are readily accessible, are used as specific precursors of protein synthesis. For amphibians, one has to inject labeled amino acids into eggs and cut the embryos into parts or dissociate into cells. Fish embryos, prior to incubation with labeled amino acids, are separated from the yolk microsurgically (sturgeons) or by centrifugation on a sucrose cushion (loach; Kostomarova and Neyfakh, 1964). For amphibians and fishes, $^{14}CO_2$ can also be used as protein precursor, since it easily penetrates whole eggs and is incorporated into different amino acids by carboxylation. The advantage of this precursor lies not only in the fact that it enables one to work with whole eggs, but also in that its use practically eliminates the problem of egg permeability, while its incorporation into different amino acids makes considerable fluctuations in the specific activity of the amino acid pool in different tissues and at different stages of development most unlikely. Incidentally, a specially conducted comparison of incorporation of labeled amino acids into isolated loach blastoderms and of $^{14}CO_2$ into whole eggs (Krigsgaber and Neyfakh, 1968) showed that the two methods produce similar results, i.e. similar relative changes in the label incorporation rate in the course of development. The absolute values of label incorporation were, of course, different. In amphibians and fishes, the protein synthesis after fertilization increases hardly at all, but it does increase slowly in the course of cleavages (Smith *et al.*, 1966; Neyfakh *et al.*, 1968; Ecker and Smith, 1971; Shih *et al.*, 1978).

In loach embryos, the synthesis intensity goes up sharply from the late-blastula stage (9th hour at 21°C) and reaches a maximum at the period of early formation of axial organs (18th-20th hour at 21°C). After this the protein synthesis intensity decreases appreciably and rather sharply, attaining minimum values at the first somites stage (22nd-24th hour). This fall-off is succeeded by a new rise. In different egg batches these patterns vary somewhat but the overall trend persists rather unchanged, with fluctuations in the time of changes not exceeding 2 h (10% of total development time), and fluctuations in the amplitude not exceeding 10-20%. From early blastula (6th hour) the protein synthesis intensity increases by 6-8 times (18th-20th hour), then it decreases by a factor of 1.5-2 (22nd-24th hour) and increases several-fold again (Neyfakh *et al.*, 1968).

Despite all the differences in egg development between the sea urchin and the fishes, the shape of their protein synthesis rate curves is rather similar. They differ only in the first hours of development, when the protein synthesis rate in the sea urchin increases manifold from near zero. In embryos of *Strongylocentrotus nudus* it reaches a maximum at early-mid-blastula (6th-7th hour at 25°C), then declines by 25-30% with a minimum at the 11th-12th hour (mesenchyme blastula) and increases several-fold again in the course of gastrulation (Berg, 1965, 1968; Neyfakh and Krigsgaber, 1968; Krigsgaber and Neyfakh, 1972; see Giudice, 1973). The absolute values of protein synthesis rate were determined in sea urchin embryos and it was shown that the embryo synthesizes $7\text{-}10 \times 10^{-4}$ μg of protein in 1 h, which is about 1% of the total protein (Fry and Gross, 1970b; Berg and Mertes, 1970).

It has been impossible, on the whole, to relate the changes in protein synthesis intensity directly to morphogenesis. The acceleration of synthesis from late blastula in the loach is easily explained by the formation of new proteins necessary for gastrulation and subsequent organogenesis. But the synthesis intensity decreases when the formation of the axial organs is completed and before the somites, brain compartments, etc. begin to form. The subsequent increase coincides with segmentation, further differentiation of the head, and other organogeneses. The decrease in protein synthesis could perhaps be somehow related to some reprogramming of organogenesis when one of its stages is replaced by another. But to provide an experimental basis for these speculations, one must find some quantitative criterion for differentiation intensity or at least compare the protein synthesis intensity in development of different organs, which is quite feasible, in principle, but has not been done so far.

The differences in protein synthesis rate between the various parts of the embryo have been estimated autoradiographically at the blastula-early gastrula stage. Like RNA synthesis, the protein synthesis at these stages first increases in the basal layer of the blastoderm, then spreads throughout the embryo (Burakova and Kostomarova, 1975).

In the sea urchin, the protein synthesis attains a maximum at blastula, when many new nuclei, membranes, etc., are formed. But at the next stage, when the hatching enzyme is synthesized, the cilia and the tuft are formed, the hatching takes place, and the migration of the mesenchyme cells into the blastula cavity begins, the protein synthesis weakens and then increases again with the beginning of gastrulation. In this case, too, it would be premature to relate the protein synthesis intensity to some morphogenetic processes or other.

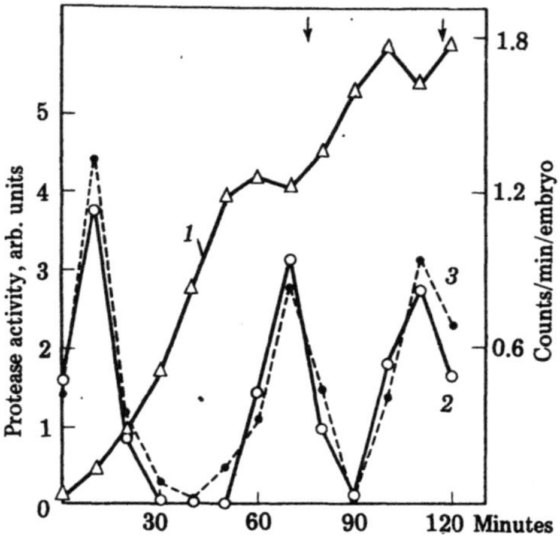

Fig. 55. Cyclic variations in the incorporation of labeled amino acids into the proteins and in protease activity during cleavages in sea urchin eggs. At different times after fertilization, eggs of *Pseudocentrotus depressus* were short-incubated with a mixture of ^{14}C-amino acids and their incorporation into protein was investigated (*1*). The protease activity was determined in parallel samples using endogenic proteins (*2*) or caseine (*3*) as substrates. The arrows indicate the time of the first and second cleavages. The amino-acid incorporation into the protein increases rapidly, but slows down prior to the first and second cleavages. The protease activity increases at the same time and also immediately after fertilization (Mano, 1970).

Special mention should be made of the fluctuations in protein synthesis rate due to cleavages. Such fluctuations have been detected in a sufficiently detailed investigation of pulse-labeling of sea urchin egg proteins during cleavage (Gross and Fry, 1966; Sofer *et al.*, 1966; Mano, 1968, 1969; Timourian and Watchmaker, 1970) (see Fig. 53). These fluctuations are not large, and they are superimposed on the overall increase in synthesis rate occurring at these early stages. These two concomitant processes are termed *basal growth* and *cyclic variation.* Cyclic variations with an amplitude of 10-20% of the total synthesis were thoroughly studied by Mano (1968, 1969, 1970, 1971a, b) (Fig. 55).

Incorporation of labeled amino acids into the protein in each of the mitotic cycles reaches a maximum at the metaphase and a minimum at the anaphase-telophase. Because of the basal growth each subsequent maximum is higher than the preceding one. That cyclic variations are, in fact, fluctuations in protein synthesis rate and not in the rate of amino acid incorporation was shown by Mano in experiments with determination of the ability of labeled amino acids to enter the cells, experiments with radioactive preloading, and finally, experiments with cell-free systems. He found that although fluctuations in permeability to amino acids do exist, they have a different periodicity than that of fluctuations in label incorporation into the protein, and that they damp out at earlier stages than do the cyclic variations in protein synthesis. With the preliminary administration of labeled amino acids (preloading) the cumulative radioactivity does not increase uniformly and, moreover, even shows slight drops, which indicate not only a slowing-down of the synthesis but also the decay of part of the labeled proteins (see below). Finally, no increase in protein synthesis occurs (i.e. there is no basal growth) in the fertilized egg homogenates after the sedimentation of large granules (12,000*g*), but the cyclic increases and decreases in the rate of incorporation persist.

These three versions of experiments would seem to eliminate the problem of determining the specific activity of the amino acid pool and enable one to relate the fluctuations in amino acid incorporation to those in protein synthesis rate. But other researchers deny the existence of such fluctuations. Fry and Gross (Gross, 1967; Fry and Gross, 1970a, b) confirm the fluctuations of amino acid incorporation, but they attribute these entirely to changes in the permeability of the embryo cells to amino acids.

Thus, the existence of cyclic variations cannot be considered proved conclusively until it is confirmed at other laboratories. Besides, it is difficult to explain Mano's data on partial cyclic degradation of the proteins synthesized. Indeed, the cumulative incorporation curve (after preloading) presented by Mano clearly shows reductions in activity, i.e. reductions in the quantity of labeled proteins, rather than in the synthesis rate. The total quantity of protein in the embryo remains unaltered, which means that only newly synthesized protein molecules undergo degradation. Mano ascribes this new fact to cyclic increases in protease activity, which he detects at the same moments.

Although Mano's data cannot yet be considered generally accepted, they are represented by a large number of papers and therefore deserve to be briefly discussed. In addition to whole embryos, Mano also studied cell-free systems. The homogenization of sea urchin eggs at different moments after fertilization immediately arrests any further increase in synthesis rate (basal growth), but the level of synthesis achieved is preserved and cyclic variations continue against its background. The nature of the dependence of these variations on the homogenate concentration (quantity of proteins in the medium) and on the temperature is amazing (and paradoxical). Thus, the fluctuation frequency decreases with increasing protein concentration. On the other hand, a temperature reduction from 21 to 10°C does not change the fluctuation frequency (!), but reduces its amplitude. A supernatant from the homogenate of fertilized eggs added to that from unfertilized ones induces in it the same fluctuations. Mano describes three groups of factors responsible for the fluctuations in protein synthesis rate. One of them, macromolecular but not yet identified, is present both in fertilized and unfertilized eggs and determines the frequency and amplitude of the fluctuations (its dilution increases the frequency). The second macromolecular component are maternal mRNA's. As noted above, prior to fertilization they form according to Mano, relatively large granules precipitated in the presence of Ca ions at 12,000*g*. After fertilization the granules break down, and the mRNA released becomes incorporated into the polysomes. Finally, the third, low-molecular-weight, factor necessary for cyclic variations in protein synthesis are ATP and GTP, whose concentration increases after fertilization. All these data are of definite interest but they must be confirmed. The authors of many reviews treat them cautiously (see, for instance, Giudice, 1973; Stearns, 1974).

Autoradiography shows that in the sea urchin gastrula the incorporation of labeled amino acids into the ectoderm cells is appreciably higher than into the endoderm and mesoderm cells. In exogastrulation, however, when the endodermal anlage (primordial gut) is turned inside out, label incorporation into proteins is identical in ectoderm and endoderm (Karp and Weems, 1975). It can presumably be concluded that the noticeable differences in protein synthesis rate between the various parts of the embryo are an artefact caused by the differences in the rate of delivery of labeled amino acids from the medium to cells lying on the surface (ectoderm) and inside the embryo (endoderm).

The protein synthesis rate during early development has also been investigated in other animals, but the problem of embryo permeability to amino acids has not always received sufficient attention. In the mollusc *Ilyanassa*, the incorporation of amino acids into the protein was determined with simultaneous measurement of label incorporation rate and determination of the amino acid pool. All the three values changed substantially in the course of development, and each of them showed a specific trend in its variation. The protein synthesis rate, as calculated from these values, increases after fertilization, but then remains unaltered up to the beginning of gastrulation. During gastrulation it increases manifold and remains high during organogenesis, then drops to the initial value (Collier and Schwartz, 1969). The quantity of polysomes in the course of the development of this mollusc to the veliger stage increases 4- to 5-fold, and their specific activity, i.e. the synthesis efficiency per one ribosome, increases 4- to 6-fold. As a result, the protein synthesis rate at the veliger stage is 26 times higher than in the unfertilized egg. The average polysome size also slightly increases in the course of development (from 15 to 20 ribosomes per polysome) (Mirkes, 1972).

In the mollusc *Spisula* the share of polysomes increases slowly in the course of development (from 10% in the unfertilized egg to 60% at the trochophore stage) and so do the permeability to amino acids and their rate of incorporation into the protein. Measured by the quantity of polysomes, the protein synthesis rate increases. Synthesis rate calculations carried out with due regard for the changes in permeability to amino acids also confirm this. Protein synthesis on light polysomes at the 16-blastomere stage is more sensitive to actinomycin D than that on heavy ones, i.e. at the early stages large proteins are synthesized predominantly on maternal templates (90%), while the smaller-sized proteins are translated, to a great extent (32%), on the mRNA transcribed in the embryo nuclei (Firtel and Monroy, 1970). In *Ascidia nigra* the protein synthesis increases in the course of early development — from fertilization to the 4-blastomere stage and then increases again during gastrulation. Actinomycin D does not affect the first increase in synthesis but represses it during gastrulation (Smith, 1967).

In early mouse embryos, different concentrations of several amino acids were used and their uptake by the embryo and incorporation into the protein were measured in order to eliminate the effect of permeability to them and of their pool size, and to determine the protein synthesis proper. Incorporation remains at a low level for 2 days, up to the 8- to 16-blastomere stage, after which it increases 3- to 9-fold (depending on the precursors) (Epstein and Smith, 1973; Pikó, 1975). The authors believe that since at the earliest stages the synthesis rate remains constant with an increase in the number of cells, the translation occurs on maternal templates, and then, as the synthesis rate increases more rapidly than does the number of cells, it switches to the new mRNA's synthesized in the embryo's nuclei. This conclusion has some grounds but is difficult to prove by a simple comparison of synthesis rates and of the number of cleavages.

Brinster *et al.* (1976), using early mouse embryos, measured not only incorporation of ^{3}H-leucine into the protein but also the rate of its uptake by the embryo and the size of the leucine pool. This enabled them to calculate accurately the protein synthesis rate and to show that the synthesis rate in unfertilized and fertilized eggs and at the 2-cell stage is low and constant ($7\text{-}8 \times 10^{-13}$ mole of leucine per hour per egg). It becomes much higher at the blastocyst stage (56×10^{-13} mole). The protein decay rate in the blastocyst is also higher than in the egg.

In the rabbit, the protein synthesis also increases considerably but at somewhat later stages (Manes and Daniel, 1969; Karp *et al.*, 1973). Manes (1975) believes that this increase is due to the active synthesis of tRNA at the earliest stages of mammalian development and to the appearance of new ribosomes at blastula.

Summarizing this section, we can evidently conclude that the protein synthesis rate increases in the course of development in all the species. In many of them, the synthesis slows down markedly at the end of oocyte growth and is stimulated at the beginning of maturation and ovulation (fishes, amphibians). In worms and molluscs, in which maturation (germinal vesicle rupture) occurs after insemination, the protein synthesis is also enhanced several-fold at the beginning of maturation-fertilization. Fertilization activates the protein synthesis most noticeably in sea urchin eggs, which are fertilized already after maturation and meiotic divisions.

In the course of early cleavage the protein synthesis increases very slowly in most of the animals (amphibians, fishes, mammals, molluscs, and worms). In echinoderms, however, these stages coincide with protein synthesis activation from a near-zero level. The increase in protein synthesis rate in most of the animals is associated with the beginning of differentiation — late blastula, gastrulation, and organogenesis.

An increase in protein synthesis rate always manifests itself in an increase of the share of ribosomes contained in the polysomes, and is, in some instances, accompanied by enhancement of translation efficiency. The mechanisms underlying the increase in synthesis rate are largely unknown and may be different in different animals. It is believed that at the late stages the synthesis rate increases predominantly owing to the appearance of new mRNA's, while at the early stages, mainly owing to the more effective utilization of the stored templates.

Composition of *de Novo* Synthesized Proteins

In order to study the composition of the synthesized proteins they must be first separated from each other. At present there are a variety of methods in protein chemistry allowing separation of practically all the proteins. These methods are based on the differences in molecular weight (ultracentrifugation, molecular sieves — sephadexes, biogels, etc.), in solubility in saline solutions (salting out in ammonium sulfate), in the electric charge (electrophoresis), in selective binding by ion-exchange resins, in the isoelectric point (isofocusing), in antigenic properties (precipitation in gel, immunophoresis), and so on. The most widely used method for investigating embryo proteins, however, is electrophoresis, which yields up to 50-100 fractions on very scant material. Two-dimensional electrophoresis increases this number to many hundreds or a thousand of fractions. Nevertheless, the data obtained by these methods have so far only helped to find the differences between the stages and establish an increase in protein diversity in the course of development. A more promising method appears to be, for instance, coelectrophoresis of a mixture of proteins from different organs, or from different stages, differently labeled with ^{3}H- and ^{14}C-amino acids. The ratio between the ^{3}H- and ^{14}C-counts in the same fraction indicates the quantitative differences in the content or synthesis of a protein at different stages, in different parts of the embryo, etc. But even at the best electrophoretic resolution, only protein fractions present in the embryo at high concentrations can be detected. They correspond to just a small percentage of active genes, and, apparently, of all translatable polypeptides. These fractions may not include precisely the most specific proteins, and it is no wonder that the appearance in the course of development of new protein species is usually detected with difficulty. Most of such investigations have been carried out on sea urchin eggs, where it is not too hard to obtain a large number of prelabeled proteins at different stages of development.

In the 1950s, when the modern protein separation methods were only in the making, much hope was pinned on immunochemical methods. In those years, data on the appearance of several new antigens in the course of development were obtained (Perlmann, 1953; Ranzi, 1957). Of special interest seemed to be the works where antigens of the paternal type were found in hybrid embryos, which definitely pointed to a synthesis of new proteins under the genetic control of the embryo (Harding *et al.*, 1954). These investigations, however, produced very scant results. Their reproducibility is also low. Thus, more recent studies of sea urchin hybrids yielded no relialbe data on the appearance of either new antigens in the course of development or the paternal antigens (Badman and Brookbank, 1970). Nevertheless, the consistent appearance of new antigens, their localization in the cells and in different types of cell and, finally, the dependence of their appearance on the nuclear activity have been found for sea urchin embryos by Westin and Perlmann (Westin *et al.*, 1967; Westin, 1969, 1972; Westin and Perlmann, 1972).

Beginning with the 1960s, methods of chromatographic and electrophoretic protein separation gained wide recognition. In one of the first investigations of this kind Monroy *et al.* (1961) labeled soluble proteins of the sea urchin with ^{35}S-methionine at different stages and separated them on DEAE cellulose. With the rather low resolution of this method, the protein profile remained unaltered from stage to stage, and only some differences in radioactivity (synthesis rate), which indicated changes in the composition of the proteins synthesized, are of interest.

In a similar investigation carried out later (Spiegel *et al.*, 1965) electrophoresis in polyacrylamide gel was used, which provided a better resolution of the protein mixture. Seven clearly defined peaks of radioactivity were thus observed. The authors found a similar distribution at the 16-blastomere stage and at gastrula. Moreover, in the presence of actinomycin D, the same (within the resolution limits) types of protein are synthesized. The authors noted the limitations of their method. They used only soluble proteins, and the resolution was determined by the thickness of the sections into which the polyacrylamide gel was cut.

In protein separation on the DEAE Sephadex, the number of peaks was greater (Ellis, 1966), but only a few peaks among them were clearly reproducible. This time a gradual increase in the number of radioactivity peaks in the course of development was detected, indicating the synthesis of new protein species which were absent in the egg. The actinomycin D practically did not affect the radioactivity profile at the early stages, but repressed the inception of new protein synthesis at the gastrula stages.

The electrophoretic separation of the basic (acid-soluble) proteins revealed early changes in the protein composition of amphibian and sea urchin eggs after fertilization: the disappearance of some and appearance of other protein fractions during early cleavage and at the gastrula stage (Silver and Comb, 1967; Spiegel *et al.*, 1970).

A higher resolution was obtained when after the electrophoretic separation of labeled proteins the polyacrylamide gel was not cut but placed against an X-ray film and exposed with subsequent autoradiographic recording of activity (Terman and Gross, 1965). A comparison of the total protein content in the gel (determined photometrically after staining) with its radioactivity (determined by photometry of the corresponding film) showed that the synthesis rate of different proteins is not proportional to their content in the embryo: the specific activity, i.e. the rate of renewal, is much higher for some fractions than for others. The radioactivity profile changes in the course of development, particularly at the blastula-gastrula stage. On addition of actinomycin D the pattern changes little at the early stages, but the differences become considerable at gastrula. It is significant that some changes in protein profile occur despite the presence of actinomycin D, which proves the independence of these changes of direct nuclear control.

In a later investigation Terman (1970) used a still more efficient method by comparing the electrophoretic mobility of proteins doubly labeled at two stages: immediately upon fertilization (^{14}C-lysine) and at blastula (^{3}H-lysine). If the two labels are introduced simultaneously, their ratio remains constant in all the fractions, since lysine incorporation into all the proteins naturally does not depend on which isotope was used as label. This ratio was taken as unity. But in electrophoretic separation of a mixture of proteins obtained from embryos labeled at two stages (zygote and blastula), the $^{3}H : {}^{14}C$ ratio varied over the length of the gel, i.e. some proteins were synthesized more intensively at the beginning of development, while the others, at the blastula stage (Terman, 1970). When using the same method to compare protein synthesis in normal and animalized embryos it was shown that the average molecular weight of the proteins synthesized decreased in the course of development, and that animalization, which disrupts considerably the normal course of morphogenesis, has a relatively small effect on the profile of the proteins synthesized (Carroll *et al.*, 1975).

A different approach to estimating the composition of the proteins synthesized was proposed at Spiegelman's laboratory (Slater and Spiegelman, 1968). RNA was isolated at different stages of development in a sea urchin embryo, and fractionated in the sucrose gradient. The template activity of the fractions in an *in vitro* system was then determined. During development from unfertilized egg to gastrula the template activity of the RNA remained practically unaltered (despite the considerable changes in translation rate in the embryo). Though there was no qualitative change in the template activity of the RNA fractions (the RNA in these experiments was separated into seven fractions according to the size), one could notice that the light template synthesis predominated in the eggs, and the heavy template synthesis, in the gastrulae.

Investigations of labeled (synthesized *de novo*) antigens showed that new proteins appear in sea urchin embryos only after hatching at the stage of mesenchyme blastula (Westin, 1969, 1972): one labeled antigen appears anew, and one, which has been present up to this stage, begins to be labeled, i.e. renewed. Actinomycin D blocks these pregastrula changes.

The immunofluorescence method not only reveals antigens but also pinpoints their localization in the cells and parts of the embryo. In the sea urchin embryo, the amount of two such antigens (one in the cytoplasm and the other in the outer membrane) diminishes gradually in the course of development, and they disappear prior to gastrulation. The third antigen is present in the egg in very small amounts but its concentration increases in the course of development, and after hatching it becomes localized in the mesenchymal migrating cells (Westin and Perlmann, 1972).

An indirect, but elegant method was used in order to determine the stage of development of the sea urchin at which new surface proteins appear (McClay *et al.*, 1977). Embryos of two distant species, *Lytechinus* and *Tripneustes*, and their hybrids were dissociated into cells and the surface proteins were detected by the ability of the cells to form aggregations. Cells of the two species failed to aggregate. At early stages, cells of hybrid embryos aggregate only with those of the maternal species, but beginning with gastrula they begin to aggregate with cells of the paternal species as well. New, species specific, antigens appear on the cell surface at the same stages (McClay *et al.*, 1977).

In amphibians, the electrophoretic separation of labeled proteins reveals the appearance of new protein species already at the 64-blastomere stage (Malacinski, 1971), i.e. long before the activation of the RNA synthesis (Fig. 56). By the mid-late blastula stage, about 30% of the proteins synthesized were those absent in the just fertilized egg. The eggs of mutants with abnormal oogenesis have a somewhat different radioactivity profile, since they lack some of the normally synthesized proteins. This is detected already at late blastula, i.e. before the beginning of the functioning of the embryo genes.

In a similar investigation (Spiegel *et al.*, 1970), electrophoretic separation of nonradioactive proteins of crude homogenate revealed in the liver and ovary of an adult frog 28 and 14 bands, respectively, which points to a relatively low resolution of the methods. Nevertheless, when comparing the egg and embryo proteins, at the 8-blastomere stage a new fraction was noticed, which disappeared at early blastula, and at early gastrula there was still another fraction which persisted throughout the subsequent development. In the axolotl, the same method revealed the appearance of three new fractions at the gastrula stage. Since in this instance the change in protein composition, rather than the incorporation of the labeled precursor, was measured, the appearance of the new fraction pointed to the synthesis of appreciable

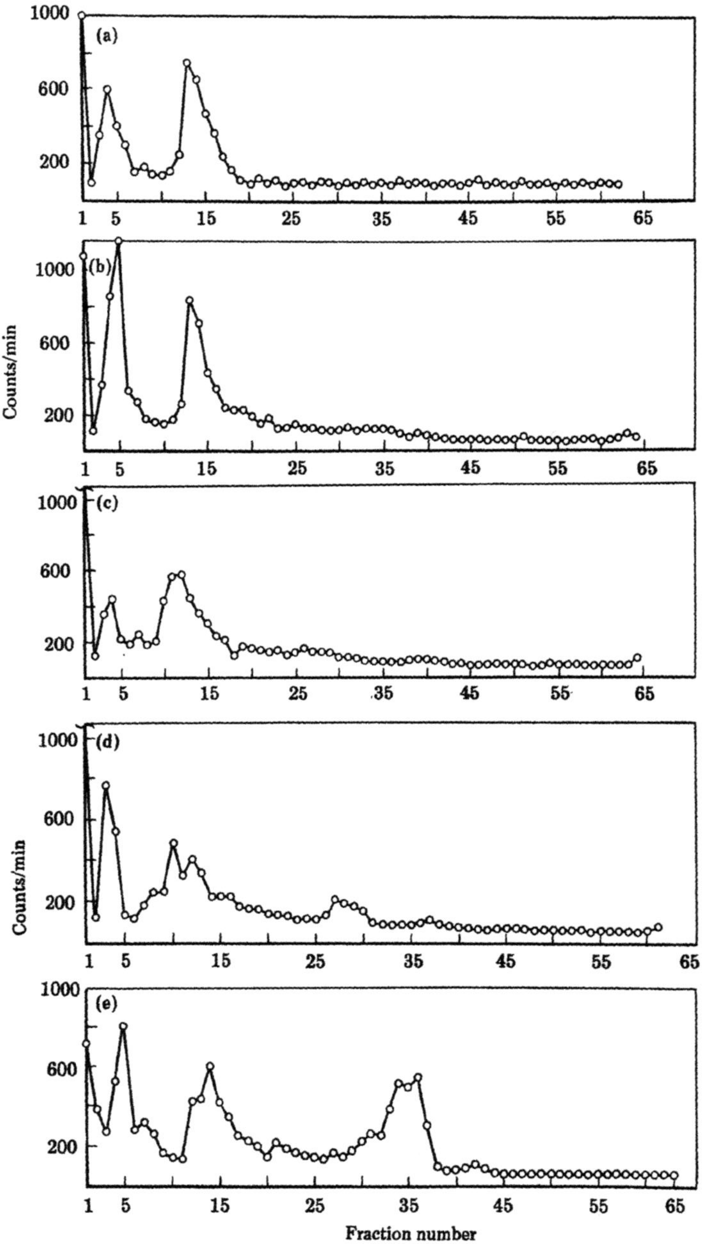
1000
(a)
600
200
1 5 15 25 35 45 55 65
(b)
1000
600
200
Counts/min
(c)
(d)
(e)
Fraction number

amounts of a new protein or to a consiberable accumulation of one of the previously existing proteins.

A further progress in investigating the composition of proteins synthesized in the course of amphibian development was achieved by utilizing a number of techniques, such as injection of high-labeled amino acids into an egg or an embryo, two-dimensional protein separation technique of O'Farrell (isoelectrofocusing in the first direction and SDS-electrophoresis in the second), and fluorography of gels (Brock and Reeves, 1978; Bravo and Knowland, 1979). These methods make it possible, in principle, to study hundreds of proteins, but until now researchers have practically compared only several tens of the most abundant proteins: (1) those synthesized at all the stages of development from the oocyte to the tadpole tissues; (2) stage-specific proteins synthesized exclusively in oocytes, or only during cleavage, or at blastula, or only in a definite tissue; and (3) proteins whose synthesis begins in the course of cleavage. With an accumulation of data on new proteins the protein classification will necessarily become more complicated (Bravo and Knowland, 1979). It is significant that protein synthesis begins during early cleavage, when the genetic apparatus of the embryo is practically not functioning yet. This suggests a protein synthesis control at the level of translation of previously transcribed mRNA's, rather than at the transcription level.

An important problem is to detect local differences in the composition of synthesized proteins as early as possible. Such an investigation was carried out on 16-cell sea urchin embryos. They were divided into micro-, meso-, and macromeres and their proteins were compared using a two-dimensional electrophoresis (Tufaro and Brandhorst, 1979). Despite the high resolution achieved in this study (about 1000 spots of different proteins), no differences in composition between the synthesized proteins were observed. Recall that at this stage differences in the composition of synthesized RNA's are already revealed (see above).

The local qualitative differences in the composition of the proteins synthesized were detected at the early stages of development of the mollusc *Ilyanassa* (Donohoo and Kafatos, 1973). At the 2-blastomere stage (AB and CD) they were separated from one another and grown for several hours in the presence of ^{14}C-and/ or ^{3}H-labeled amino acids. Then the embryos obtained from both halves were homogenized together and their proteins were separated electrophoretically, immediately or after preliminary ammonium sulfate precipitation (Fig. 57). When each of the blastomeres was grown in the presence of a doubly labeled

Fig. 56. Changes in the composition of the synthesized proteins during early development of the axolotl. At different stages of development, the embryo was injected with ^{3}H-leucine, and 2 to 10 h later the labeled proteins were electrophoretically separated. *a*, Fertilized egg; *b*, the first cleavage; *c*, the 16-blastomere stage; *d*, the 64-cell stage; *e*, mid- to late blastula. At the 64-cell stage, a new, previously absent, protein appears (fractions 27 to 32) (Malacinski, 1971).

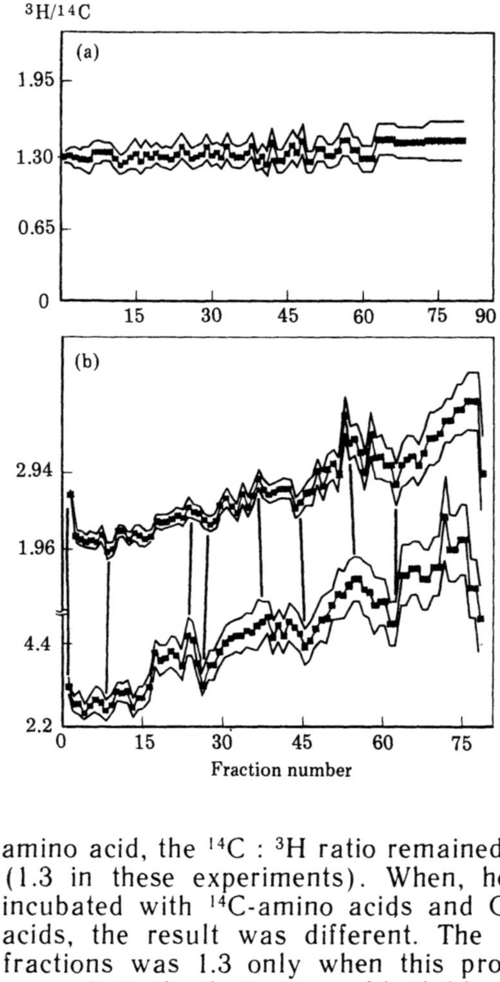

Fig. 57. Differences in the composition of proteins synthesized in embryos grown from AB and CD blastomeres of the mollusc *Ilyanassa.* In the control (*a*) the embryos were grown in a medium with ^{3}H- or ^{14}C-leucine, mixed together, and the proteins were separated electrophoretically. In experiments (*b*), the embryos at the 2-cell stage were separated into AB and CD blastomeres and grown with ^{14}C- and ^{3}H-leucine, respectively; the proteins were subjected to joint electrophoresis and the ^{3}H to ^{14}C radioactivity ratios in the gel fractions were determined. In the control, the ^{3}H to ^{14}C ratio is the same in all the fractions; in experiments, this ratio varies, reflecting the different composition of the proteins synthesized in the embryos grown from AB and CD blastomeres (Donohoo and Kafatos, 1973).

amino acid, the ^{14}C : ^{3}H ratio remained unchanged in all the fractions (1.3 in these experiments). When, however, AB blastomeres were incubated with ^{14}C-amino acids and CD blastomeres with ^{3}H-amino acids, the result was different. The label ratio in each of protein fractions was 1.3 only when this protein was synthesized in equal amounts in the derivatives of both blastomeres. If the blastomeres differed in the rate of synthesis of some proteins or other, the ratio changed as well. It exceeded 1.3 if a given protein synthesis rate was higher in the derivatives of AB blastomere, and was less than 1.3 if it was higher in the derivatives of CD blastomere.

These results show very clearly that in the early cleavage the proteins synthesized in a half-embryo derived from one blastomere differ, at least quantitatively, from the proteins synthesized in a half-embryo derived from the other blastomere. The polyacrylamide gel in which the electrophoresis was carried out was cut into 40 slices, i.e. all soluble proteins were divided into 40 fractions. This means that each fraction could contain several protein species in different quantitative ratios.

It is quite possible that in actual fact the blastomeres differed from each other in the composition of the protein synthesized not only quantitatively, but also qualitatively. For instance, a higher incorporation of ^{14}C-amino acids into a given fraction could mean that there is a more intensive synthesis in AB blastomeres not of corresponding proteins but of quite different proteins with a similar electrophoretic mobility.

Morphogenetic differences between AB and CD blastomeres in mosaically developing *Ilyanassa* eggs are revealed already at the earliest stages, because a so-called polar lobe forms in CD blastomere, which is an extrusion determining the subsequent differentiation, mainly of the mesodermal structures. Removal of the polar lobe leads to the formation of an abnormal embryo which has only the ectodermal cover with cilia and endodermal derivatives. The protein synthesis in such abnormal embryos differs from the normal at the very outset: it is much lower and its increase during gastrulation occurs later and attains lower values (Collier, 1961).

A comparison of the composition of the proteins synthesized in normal embryos and in those deprived of the polar lobe by the double label method shows substantial differences which arise 24 h after fertilization, even before the beginning of organogenesis (Newrock and Raff, 1975). These differences partly persist in the presence of actinomycin D, i.e. the effect of the polar lobe is also manifested at the translation level.

Considerable changes in the electrophoretic profile of proteins take place in the course of differentiation in the culture of eye, limb, and wing anlages of *Drosophila*. The differences in the set of protein fractions between the types of anlages are not so substantial as those between the various stages of differentiation of these anlages (Seybold and Sullivan, 1978).

Similar data were obtained by Arking (1978), who studied the electrophoretic profile of synthesized proteins from different stages and different tissues of *Drosophila*. The differences were wider between distinct stages of differentiation of the same tissue than between different tissues at the same stage of differentiation. On the other hand, fractions were revealed which were unique for each given stage of development or for a given tissue: one or two fractions for larvae, 13 fractions for imaginal discs and the brain. Eight fractions proved to be common to all the stages and tissues studied (Arking, 1978).

The electrophoresis of proteins synthesized during the early development of mammals (mouse and rabbit) carried out in recent years indicated changes in the profile of the proteins synthesized at stages where no visible differentiation occurs as yet (Van Blerkom and Manes, 1974; Epstein and Smith, 1974; Manes, 1975; Tucker and Schultz, 1978). Manes, using a high resolution electrophoresis and contact gel autoradiography, found about 100 protein fractions at all (even the earliest) stages of development of rabbit embryos. Qualitative and quantitative changes in the composition of these proteins were detected already at the 8-blastomere stage (1.5 days of development) and then prior to the formation of the blastocyst (3.5 days of development). The appearance of new bands was attended by partial disappearance of the old ones. The authors note that the detection of 100 bands is not yet sufficient for concluding that each of them is homogeneous and represents a single protein. The changes described persist during development of eggs *in vitro* (Van Blerkom and McGaughey, 1978).

The double-label method has been used on early mouse embryos to compare the different stages. But the cutting of gel into a finite number of slices and determining the ^{14}C and ^{3}H radioactivity in each of them made it possible to compare only 25 fractions (Epstein and Smith, 1974). It turned out that the synthesis rate increased in three fractions and decreased in two. These changes take place between the first and the second 24-hour period of development, at the 2- to 8-blastomere stages. It is significant that these changes occur before the overall increase in protein synthesis rate, which begins after the 8- to 16-cell stages.

A higher resolution (above 400 fractions) was obtained when the proteins of mouse embryos were analyzed in a two-dimensional system by isoelectric focusing in one direction and by electrophoresis (with or without sodium dodecylsulfate) in the other. The composition of synthesized proteins changes substantially at the beginning of maturation of the mouse oocyte, and then remains unaltered at the earliest stages of development (up to the first cleavage) (Schultz *et al.*, 1979a, b). Certain changes in protein composition take place in the course of subsequent early development (Martin *et al.*, 1978; Failly-Crepin and Martin, 1979). The same method showed that about 40% of the high-molecular-weight proteins disappear on the 9th day of development and a lot of new proteins appear between the 9th and 14th day. In the course of further development, the number of tissue-specific proteins gradually increases in various organs (Klose and Wallenberg-Pachaly, 1976).

Thus, it has been found that for mammals, too, many changes in the composition of the proteins translated are not determined by morphological processes but rather precede them and are controlled by a "clockwork" of their own.

Homogenate separation helps one to investigate protein synthesis in various cell fractions of the embryo. A particularly high specific activity in the nuclei at the early stages of development has been shown in sea urchin (Kedes *et al.*, 1969a) and loach (Krigsgaber *et al.*, 1971) embryos (the synthesis of nuclear proteins is considered in the next section). The specific activity of mitochondria and hyaloplasm (cytosol) differs but slightly from that of the total protein.

Mitochondria. The number of mitochondria hardly increases at all in the course of embryonic development of fishes (Abramova and Vasilyeva, 1973; Neyfakh and Abramova, 1974), amphibians (Chase and Dawid, 1972), and echinoderms (Gustafson and Lenicque, 1952, 1955; Shaver, 1956; Gustafson, 1965). In the frog *Xenopus laevis* the amount of mitochondrial protein begins to increase after hatching (stage 38) and is doubled by the feeding tadpole stage (stage 45) (Chase and Dawid, 1972). The cytochromeoxidase activity precisely corresponds to the amount of mitochondrial protein.

All these data make it possible to view separately the two processes of renewal of mitochondrial proteins and of increase in their amount. The synthesis of mitochondrial proteins presumably proceeds from the early stages of development, and this occurs partly or completely in the mitochondria themselves, which is indicated by the relatively high content and early synthesis of rRNA in them. In sea urchin embryos the transcription of new mRNA's in the mitochondria is detected already at the stages of early cleavage (8-cell embryo). These mRNA's are found in mitochondrial polysomes and, hence, they are involved in the synthesis of mitochondrial proteins (Innis and Craig, 1978). The data on the mitochondrial enzymes in hybrids show that the formation of new enzyme molecules (controlled by the paternal genotype) begins earlier than does the increase in the number of mitochondia — already at stage 19 (heartbeat) in amphibians (Wright and Moyer, 1966; Wright and Subtelny, 1971), and at the 14-hour stage in the loach (Abramova *et al.*, 1978).

One can evidently conclude that throughout the embryonic development there occurs a synthesis of mitochondrial proteins, which is counterbalanced by their decay. The fact that the synthesis of mitochondrial DNA also begins relatively early suggests that the replacement not only of individual protein molecules, but also of whole mitochondria, may occur in embryogenesis. At the late stages of development, the synthesis level exceeds the decay, and the number of mitochondria begins to increase. In amphibians this coincides with the beginning of appreciable yolk utilization, i.e. an increase in the mass of active cytoplasm. Thus, a relatively constant mitochondria concentration is maintained in the embryo (Neyfakh and Abramova, 1974).

Ribosomes. The number of ribosomes in amphibians, fishes, and echinoderms also remains constant during most of embryogenesis and begins to increase slowly toward its end. In the loach, the quantity of RNA's (ribosomes) at the larval stage is only 1.5 times greater than at the beginning of development (Timofeeva and Kafiani, 1964).

But the relatively early synthesis of rRNA (see above) shows that all the ribosomes are replaced already in the first half of embryogenesis. The ribosomal proteins are evidently also replaced then. According to the data obtained in Brown's laboratory (Hallberg and Brown, 1969), the synthesis of ribosomal proteins is coordinated with that of rRNA and is absent in anucleolate mutants (see Hallberg, 1974; Hallberg and Smith, 1975). If this is so, the synthesis of these proteins must begin in fishes and amphibians at the end of gastrulation, and in the sea urchin, slightly earlier. But there is an ever-increasing amount of data on the functional metabolism of some of the ribosomal proteins, and if this is so, then their synthesis may precede the *de novo* formation of ribosomes (Kaulenas and Unsworth, 1974). In the loach, a slight incorporation of the proteins into the ribosomal fraction is detected already at the blastula stage, but a definite synthesis of ribosomal proteins is revealed only from the organogenesis stage (Krigsgaber *et al.*, 1976).

According to more recent data, the synthesis of ribosomal proteins continues after selective suppression of the rRNA synthesis, i.e. their correlation may not be rigid (Graig and Perry, 1971; Kaulenas and Unsworth, 1974).

The synthesis of 12 ribosomal proteins was studied in detail during oocyte maturation and at early stages of embryonic development in the mouse (La Marca and Wasserman, 1979). In oocytes, it constitutes 1.1 to 1.5% of the total protein synthesis but increases to 8% in the 8-cell embryo (4 ng/h per embryo). This happens because while the total protein synthesis increases 1.5-fold from the oocyte to the 8-cell stage, the synthesis of ribosomal proteins increases 11-fold. Although these 12 proteins are represented in the ribosomes in equimolar amounts, their synthesis rate may differ by an order of magnitude (La Marca and Wasserman, 1979).

The quantity of membrane proteins increases in the course of development. It is easy to calculate that the surface area of cell membranes increases during cleavages as the cube root of the number of cells. This means that the quantity of these proteins must increase 10-fold by early-mid-blastula in amphibians and fishes and by gastrula in the sea urchin (about 1000 cells). It is possible that part of the cell membrane proteins have been stored in some form or other during oogenesis, but it is more probable that the synthesis of these structural proteins begins at the earliest stages of development (see, for example, Chuguyev *et al.*, 1976). Thus, the increase in incorporation of labeled choline (the precursor of the membrane lipoproteins) and in choline phosphorylation in the course of cleavages of sea urchin eggs is approximately simultaneous with the increase of the cell membrane surface area (Pasternak, 1973). However, no appreciable increase in the activity of the cell membrane enzymes in the course of cleavages of sea urchin eggs has been detected (Trams *et al.*, 1974).

Cell differentiation is associated with the formation of membrane proteins specific to a given cell type. For instance, the appearance of two such antigens has been shown for mouse spermatogenesis. These antigens are absent in the liver and kidney cells, in the brain, spleen, and whole ovaries (Bechtol *et al.*, 1979). It has been demonstrated that retina differentiation is due to quantitative and qualitative changes in the cell membrane glycoprotein composition. On mutations which lead to cell degeneration a number of glycoprotein fractions disappear consistently (Wallenfels, 1979).

The same kind of consistent changes in glycoprotein composition has been described for leukopoietic cells. They are absent in leukemic cells (Gahmbery *et al.*, 1979).

In the course of cytodifferentiation the area of intracellular membranes (the endoplasmic reticulum, the Golgi apparatus, etc.) increases considerably. If all of them are formed during development, the specific activity of these proteins at the early stages must be much higher than that of the total protein. Some changes in the proteins of the endoplasmic reticulum in the course of ontogenesis have actually been detected (Black and Bresnick, 1972).

Spermatogenesis is a good model for investigation into a gradually changing protein composition and the synthesis of proteins specific to a given differentiation. At present, the morphology, the timing of the cell cycle, and the biochemistry of spermatogenesis have been studied thoroughly in the mouse. It has been shown, in particular, that the synthesis of C-subunit of LDH begins at the late prophase of meiosis, decreases in the course of spermatogenesis, and is practically absent in the mature spermatozoa (Meistrich *et al.*, 1977). In contrast, the synthesis of protamine takes place at the late stages after the condensation of the nucleus, when transcription has ceased completely (Monesi, 1964; Kierszenbaum and Tres, 1975). In addition to these proteins, the synthesis time course for other, SDS-soluble and SDS-insoluble proteins has been studied (O'Brien and Bellvé, 1980). Soluble proteins are mainly synthesized at the early stages of meiosis. Among the insoluble proteins, four fractions are distinguished, each of which has a synthesis pattern specific to spermatogenesis. Two protein fractions (acid and moderately basic) are synthesized as protamine at the end of spermatogenesis in the absence of transcription (O'Brien and Bellvé, 1980).

Thus, early development is attended by an intensive protein synthesis, but only some of these proteins are absent in the egg and appear for the first time in the course of development. The share of these new proteins is not large at the early stages as regards their amount, but their diversity must be substantial. The methods used until now have revealed only a few types of such proteins, but their number will be increasing with improvement in techniques or with increasing embryological application of the existing separation methods.

The number of proteins under synthesis can be estimated by the number of messenger RNA species found in polysomes. Such a measurement has been carried out at Davidson and Britten's laboratory (Galau *et al.*, 1974) in sea urchin gastrulae. The excess of mRNA's extracted from polysomes was hybridized with single-copy DNA sequences. Proceeding from the quantity of bound RNA and from the average sizes of mRNA's per one protein species, the authors calculated that 1.4×10^4 different proteins are synthesized at the sea urchin gastrula stage. This number seems excessive. Nevertheless, it is obvious that the number of proteins synthesized in the embryo runs into many thousands, whereas the presently available methods reveal only tens or hundreds of such proteins.

Nuclear Proteins

During cleavages the number of nuclei increases in geometrical progression, and after 10 synchronous divisions (amphibians, fishes) it reaches a thousand. In the sea urchin the number of synchronous divisions is smaller, but in all the species the number of nuclei rapidly increases in the course of early development and, hence, the total mass of nuclear proteins increases as well. The amount of cytoplasm remains practically unaltered, therefore the share of nuclear proteins increases by hundreds and thousands of times. The accurate calculation of these values is impeded by the fact that at the early stages development is attended by a decrease in the size of nuclei and a change in protein composition, whereas at later stages the amount of cytoplasm may slightly increase (at the expense of the yolk).

It has already been mentioned that labeled germinal vesicle proteins synthesized during maturation are detected in the blastomere nuclei after the beginning of development (Ecker and Smith, 1971). In a mature egg, these labeled proteins are in the cytoplasm but rapidly move into the nuclei formed in the course of cleavage or transplanted into the egg (Merriam, 1969; Gurdon, 1970; see Gurdon, 1974).

At the same time, during incubation of early sea urchin embryos with labeled amino acids the specific activity of the nuclear proteins is much higher than that of the total protein. According to the data from Gross' laboratory, part of the nuclear proteins in early sea urchin embryos are translated on RNA's synthesized in the nuclei of a 2- to 8-cell embryo (Kedes *et al.*, 1969a, b; Shutt and Kedes, 1974). Calculations show that the mRNA of early polysomes consists, by at least 85%, of RNA which preexisted in the egg (Humphreys, 1971).

The number of nuclear proteins increases during early development due to two sources: proteins synthesized in oogenesis and stored in the egg cytoplasm, and proteins synthesized at the early stages of embryo development. The latter can, in turn, be subdivided into proteins which are translated on templates transcribed in oogenesis, i.e. on the DNA's of the maternal genome, and proteins which are synthesized on newly formed RNA's transcribed in the nuclei of the embryo (Gross *et al.*, 1973a, b; Ruderman and Gross, 1974).

In the nuclei, one distinguishes chromatin proteins, relatively tightly bound with the DNA, and the nucleoplasm proteins. With the traditional methods for isolation of nuclei, part of the nucleoplasm proteins (up to 30% of all the nuclear proteins) may be lost. The chromatin proteins consist of basic, acid-soluble proteins or histones and acidic nonhistone proteins. The histones, which have been thoroughly investigated in the past decade, are by now well systematized and characterized. The present-day classification distinguishes five main and two additional fractions of histones: lysine-rich histones — H1, H2a, and H2b (formerly f1, f2a, and f2b) with a lysine : arginine ratio of 22, 2.5, and 2.0, respectively, and arginine-rich histones — H3 and H4 (formely f3 and f2a1) with a lysine : arginine ratio of 0.8 and 0.7. In electrophoresis, each of these fractions may produce several bands which reflect the modification of histones — their acetylation or phosphorylation. Some histone fractions (for instance, H1) exist in the form of several subfractions differing in primary sequence, i.e. encoded by different genes.

But at the earliest embryonic stages, when isolation of a sufficient number of high-purity nuclei presents a complicated technical problem, the traditional methods for separation of nuclear proteins yielded different results at different laboratories. Thus, identification of acid-soluble proteins with histones is not always considered justified, and even electrophoretic identification seems insufficient (cf. Hnilica and Johnson, 1970; Johnson and Hnilica, 1971; Evans and Ozaki, 1973).

After incubation of embryos with labeled amino acids, the overall specific activity of the nuclear proteins in early sea urchin embryos is 4 times that of the cytoplasmic proteins and of the nonchromatin nuclear proteins (Silver and Comb, 1967; Thaler *et al.*, 1970). This means that the chromatin proteins are synthesized more than four times faster than the other proteins of the nucleus. Despite the fact that the volume of the nuclei at the early stages is many times smaller than that of the cytoplasm, the synthesis of nuclear proteins in the sea urchin makes up at least 40-60% of the total protein synthesis, 25% being accounted for by histones (Kedes *et al.*, 1969a; Nemer and Lindsay, 1969) (Fig. 58).

In loach embryos, the specific activity of the nuclear proteins at the blastula-early gastrula stage is 3-3.5 times higher than that of the total protein (Krigsgaber *et al.*, 1971). The size of the nuclei and the content of dry matter in them decrease gradually (Marshak and Neyfakh, 1973). Knowing the number of nuclei at the different stages of development (Rott and Sheveleva, 1967) and the protein content in them, it is not difficult to calculate the increase in the amount of nuclear proteins and their share in the total synthesis,

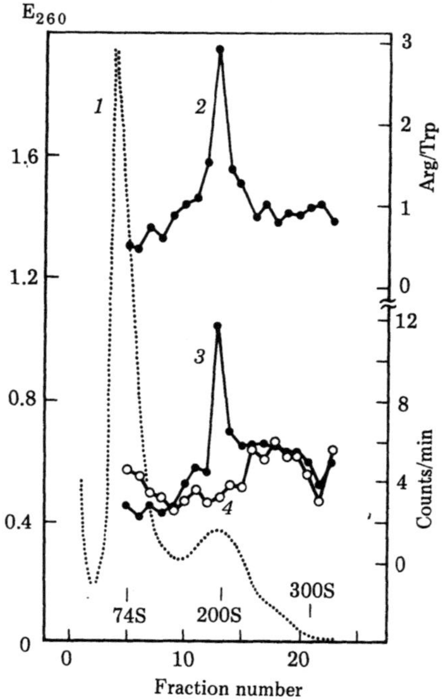

Fig. 58. Histone synthesis on light polysomes of the sea urchin. Embryos of *Strongylocentrotus purpuratus* at the morula stage were incubated for 5 min with a mixture of ^{14}C-arginine and ^{3}H-tryptophan. The homogenate was studied by centrifugation in a sucrose gradient, and the radioactivity of ^{3}H and ^{14}C was determined in the resulting fractions. *1*, Optical density (E_{260} nm); *2*, arginine to tryptophan incorporation ratio; *3*, arginine incorporation; *4*, tryptophan incorporation. The values of the sedimentation constants for ribosomes (74S) and for the heavy and light polysomes are shown. The higher incorporation of arginine as compared to tryptophan in the light-polysome region (200S) reflects the synthesis of a considerable portion of histones (Nemer and Lindsay, 1969).

which is 6-15% in the loach at the blastula-gastrula stage (Marshak and Neyfakh, 1973).

The question as to the site of the nuclear protein synthesis in embryogenesis is still discussed sometimes, although the prevailing viewpoint is that nuclear proteins are synthesized only in the cytoplasm, and then migrate into the nuclei (Sevaljević, 1974; Feldherr and Richmond, 1976). However, some investigations, including those of isolated sea urchin nuclei, seem to demonstrate incorporation of labeled amino acids directly into the nuclear proteins (Løvtrup-Rein, 1972). It is possible that in such cases one deals with the synthesis of protein on ribosomes located on the outer surface of the nuclear membrane.

A very thorough study carried out recently showed that in the sea urchin nuclei at blastula there is probably no protein synthesis, or if there is, it does not contribute more than 0.2% to all the new proteins (Allen and Wilt, 1975).

At most of the laboratories, histone proteins are detected in sea urchin nuclei at all the stages of development, including the nuclei of the unfertilized egg (Thaler *et al.*, 1970; Evans and Ozaki, 1973; and others). Hnilica and Johnson (1970) are, however, of a different opinion and deny, in a number of reports, the appearance of histones before the blastula stage, or at least until the 32- to 64-blastomere stage. Later Johnson and Hnilica (1971) found that one of the histones — H3— is synthesized already at the 32- to 64-blastomere stage but becomes incorporated into the chromatin only after several divisions. In later reports Johnson *et al.* (1973a, b) no longer insist on the absence of histones in the sea urchin egg nuclei during cleavage. At the 32- to 64-blastomere stage, they found "nuclear basic proteins" which move electrophoretically together with the histone fractions. At these stages there occurs acetylation of the basic proteins, mostly those moving together with the arginine-rich histones H3 and H4, and, to a lesser extent, of the lysine-rich H1, H2b, and H2a proteins. The rate of synthesis of histones H4, H2b, and H2a at the subsequent stages of development is twice as high as that of DNA, which means that part of the histones can be renewed without simultaneous DNA replication. It can evidently be concluded that, according to the data of the Johnson-Hnilica team, histones must also appear in the sea urchin embryo nuclei at sufficiently early stages.

Other authors isolated, already from the nuclei of unfertilized sea urchin eggs (pronuclei), histones differing in composition from those of later stages (Thaler *et al.*, 1970).

Changes in histone composition in the course of sea urchin development were revealed in a number of studies (Fig. 59). The synthesis of different histones proceeds at a different rate, and their nuclear control dependence is also different. Thus, H2b and H3 histones incorporate labeled valine at gastrula 50-70 times more rapidly than H1 and H2a. In the presence of actinomycin D, the label incorporation ceases almost completely for H1, whereas it increases for H2a,

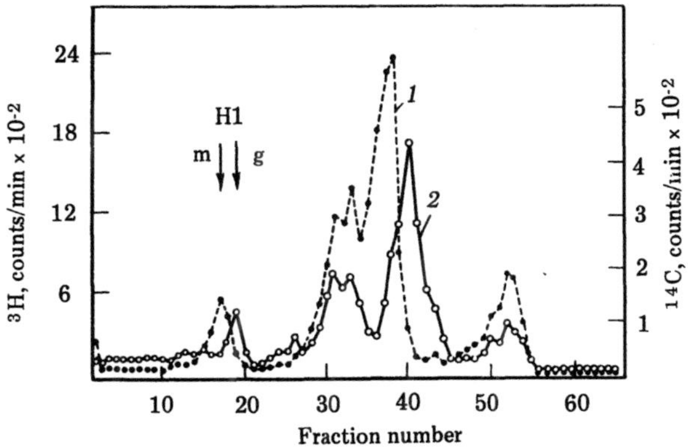

Fig. 59. Changes in histone synthesis in the course of sea urchin development. Embryos of *Lytechinus pictus* were incubated with ^{3}H-leucine at the morula stage (*1*) or with ^{14}C-leucine at the pluteus stage (*2*). The histones were isolated and separated electrophoretically. The arrows indicate subfractions H1m and H1g. The histone composition changes somewhat in the course of development, in particular fraction H1m is replaced by H1g (Ruderman and Gross, 1974).

i.e. the decrease in translation on short-lived RNA's (for H1) is offset by its acceleration on long-lived templates (for H2a) (Thaler *et al.*, 1970).

Benttinen and Comb (1971) detected all the histone fractions at the 32-blastomere stage though their ratio changed in the course of development due to an increase in the share of arginine-rich fractions. Since they did not find any histones in the embryonic cytoplasm, they concluded that the increase in their total amount and change in their composition are completely due to their synthesis.

Changes in histone composition during early sea urchin development have been studied repeatedly, and at first, when some or other histone fractions could not be detected, contradictory results were published. Nevertheless, the first papers showed replacements of histone Hl subfractions at gastrula (Seale and Aronson, 1973b), which were later denoted as H1m (morula) and H1g (gastrula) (Poccia and Hinegardner, 1975). Replacements of subfractions at early stages of development have also been shown for histones H2a and H2b (Cohen *et al.*, 1975).

Modern conceptions of replacement of histone subfractions were obtained with the use of an improved technique of separation, analysis of the histone RNA composition (see above), etc. It has been demonstrated for histone Hl that at the cleavage stages there are two subfractions which are denoted as H1m and H1x (Senger *et al.*, 1978) and are replaced by Hlg since the gastrula stage. According to another classification, histones or histone-like proteins detected in early-cleavage nuclei are designated as "cs" (cleavage stages), i.e. $H1_{cs}$, $H2a_{cs}$, $H2b_{cs}$, and the subfractions appearing later, by α, β, γ, and so on (H1α, H1β, etc.) (Newrock *et al.*, 1977; Childs *et al.*, 1979).

Histones H1m and H1x are present in the nuclei from the very beginning of development and are maternal histones. No histone-like proteins carried in with the spermatozoa are found in the embryonic nuclei (Carroll and Ozaki, 1979), although early histone RNA's are equally synthesized in both genomes (Maxson and Egrie, 1980).

The H1m/H1x ratio is not constant (Senger *et al.*, 1978). Subfraction "m" is much larger than "x" up to the 8-cell stage, after which there is a rapid but transient increase in subfraction "x," which becomes larger than subfraction "m" at the 8- to 16-cell stage. However, at the 16- to 32-cell stage the previous, "maternal," ratio of these subfractions is restored. At the 16-cell stage one can detect some relative increase in subfraction "x" in micromeres as compared with the meso- and macromeres. These differences are probably due to some asynchronism of the fourh division (and also to asynchronism of subfraction "x" degradation after the 16-cell stage). The biological significance of this earliest replacement of the histone H1 subfractions is not clear, but it is apparently regulated at the transcription level (see above, the section on histone mRNA's) (Senger *et al.*, 1978).

A detailed picture of replacements of histone RNA's and histone subfractions was presented by Newrock *et al.* (1977) and by the Kedes group (Childs *et al.*, 1979) (Fig. 60). According to their scheme the early subfraction of histone H1cs is replaced by histone H1α by the 16-cell stage, and by histones H1β and H1γ in the period between mesenchyme blastula and gastrula. H2a cs and H2bcs disappear by the 16-cell stage where they are replaced by $H2a\alpha_1$ and $H2b\alpha_1$ (here one also distinguishes minor subfractions $H2a\alpha_2$ and $H2b\alpha_2$). Beginning with mesenchyme blastula, subfractions "x" start being replaced by several simultaneously appearing subfractions (H2aß, γ, δ, ε and H2bγ and δ). Finally, histones H3 and H4 also undergo replacement of early embryonic subfractions by later ones (Childs *et al.*, 1979). The diversity of

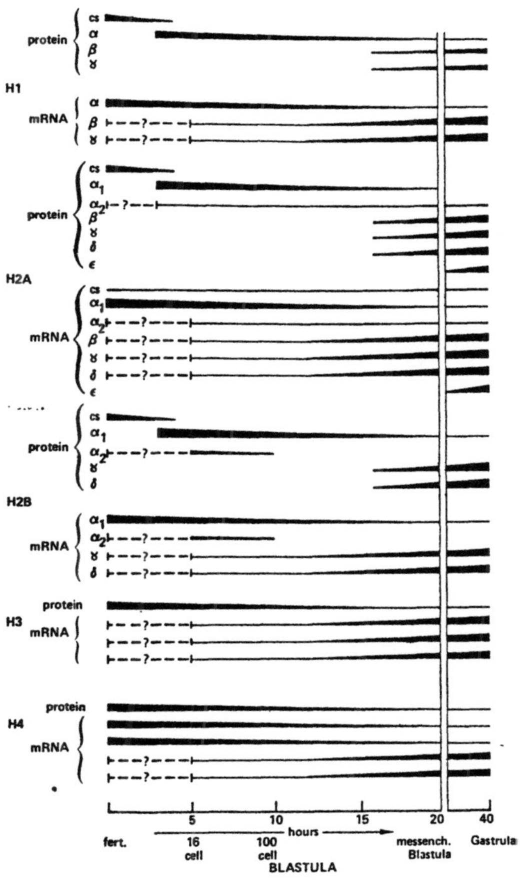

Fig. 60. Developmental expression of sea urchin histone genes. The figure is a chronicle of gene expression of histone subtypes during early development of *S. purpuratus* modified from Newrock *et al.* (1977) with additional data. The thick line indicates a relatively abundant component and the thin line represents a relatively minor component. Broken lines indicate uncertainty in the synthesis of the component. Vertical lines indicate the first point in development when the component has been positively identified. Data for each component in the figure are based on *in vivo* protein synthesis, production of a protein in the wheat germ cell-free system, or identification of a pulse-labeled radioactive mRNA species at a particular developmental stage. Histone proteins are stable and remain associated with chromatin even after its synthesis is no longer detected. Fert. = time of fertilization; mesench. blastula = mesenchyme blastula. Histone genes represent a complex multigene family that is developmentally modulated (Childs *et al.*, 1979).

histones may be partly due to posttranslation modifications, in particular acetylation (Treigyte and Gineitis, 1979).

According to Asao (1969, 1970) the total content of histones in amphibian embryos as related to nuclear DNA increases from 0.4-0.6 at the blastula stage to 1.0 at later embryonic stages. In distinction to the sea urchin, this increase is mainly due to the lysine-rich histones. The change in histone content was investigated in different anlages of amphibian embryos, and clearly defined local differences were detected

The histone : DNA ratio begins to increase, primarily in the ectoderm cells, already at late blastula and reaches 0.8 by late gastrula, the share of lysine-rich histones being 60-65%. At the neurula stage these histones make up 80-83%, while the share of arginine-rich histone H3 declines accordingly. In the nuclei of mesodermal cells this process is much slower: at late gastrula the histone : DNA ratio is still very low (0.3), the share of histone H3 being 67%. At the neurula stage these values are 0.7 and 24%, respectively. The histone composition in the endoderm changes still more slowly (Asao, 1969, 1970).

The histone synthesis in *Xenopus* is detected already during cleavage (Byrd and Kasinsky, 1973) and constitutes at these stages about 8% of all the proteins synthesized (Hallberg and Brown, 1969). At blastula, a newt embryo showed an increasing synthesis of all the main histone fractions, with the synthesis of four of them (H3, H2b, H2a, and H4) proceeding simultaneously with that of DNA. The synthesis of histone H1 rises sharply up to the gastrula stage, then remains at a constant level, lagging behind the synthesis of DNA. Newt embryos show local differences both in synthesis rate and in relative content of histone H1 (Imoh and Kawakami, 1973). At the same time Kasinsky and coworkers did not detect any noticeable differences in the composition of the histones synthesized during cleavage and in the swimming tadpole (Kasinsky *et al.,* 1973). The investigation by Adamson and Woodland appears to be the most thorough. They studied the histone synthesis in the oocyte and at the early stages of *Xenopus* development. An important advantage of this work is the very rigorous identification of histones with respect to a number of parameters (incorporation of lysine and arginine, the absence of tryptophan, two-dimensional coelectrophoresis with pure histone preparations and, finally, peptide analysis for histone H4) (Adamson and Woodland, 1974). They showed that the synthesis of four histone species proceeds at all the stages of development, with the exception of histone H1 whose synthesis begins at late blastula. The synthesis of histones and their storage occur in oogenesis, when there is no DNA synthesis, and continue after fertilization, when the total DNA synthesis is very small. According to Adamson and Woodland, the egg contains a histone stockpile equivalent to the histone content in 500 diploid nuclei. By stage 7 (blastula), when the embryo contains approximately 300 nuclei, the histone stockpile becomes equivalent to 3.5×10^3 nuclei, but by stage 9 (1.5×10^4 nuclei) it is entirely exhausted, and the further synthesis of DNA and histones presumably proceeds in a coordinated manner. The histone synthesis in oocytes and early embryos of amphibians has also been investigated by Malacinski (1974).

During the subsequent years (1977-1980) the problem of providing histones for the chromatin of newly formed nuclei at the early stages of amphibian development has been studied by many research groups, but a clear picture is still lacking, since the data obtained at different laboratories and at different stages of development do not quite coincide. These difficulties stem from the unreliability of methods of isolation and purification of histones, both stored and entering the chromatin, which sometimes constitute *in toto* a millionth fraction of the total protein. Nevertheless, the investigations of those years confirmed the main feature of early synthesis of histones: their noncoordination with DNA synthesis (in distinction to somatic and late-embryonic cells, where this coordination is very strict) and, frequently, the lack of coordination between different histone species synthesis.

Oocytes and early embryos of *Xenopus* were thoroughly investigated by Woodland, Adamson and their coworkers (Adamson and Woodland, 1974; Woodland and Adamson, 1977; Ruderman *et al.*, 1979; Woodland, 1979; Flynn and Woodland, 1980; Woodland and Wilt, 1980a, b). A considerable store of histones is built up already in oogenesis. But for *Xenopus* one should distinguish between the processes of histone H1 and other histone species (H2b, H2a, H3, and H4) accumulation. The four latter histones form the core of the nucleosomes and are denoted as core histones. The store of the core histones in the oocyte is built up gradually, but their synthesis rate rises particularly (by a factor of 20 to 50 for different histones) during oocyte maturation (Adamson and Woodland, 1974), although the overall protein synthesis in this period speeds up only 2- to 3-fold. In the course of cleavage the synthesis of the core histones is accelerated by another factor of three and is maintained at this high level throughout the early embryogenesis (Flynn and Woodland, 1980). Thus, the egg already contains a store of core histones equivalent to 2×10^4 nuclei, which constitutes an appreciable portion of the amount required for the nuclei of late blastula (3×10^4 cells). With a progressive increase in the number of nuclei the histone reserve is exhausted despite their rapid synthesis and, beginning with gastrula, when the rate of nuclear divisions lowers considerably, the histone and DNA synthesis becomes coordinated.

Part of the stored *Xenopus* histones are in the germinal vesicle of the oocyte and part in the cytoplasm. As shown by Woodland, the histones stored in the cytoplasm, those in the germinal vesicle, and those contained in chromatin of the oocyte may undergo different modifications. Thus, histone H4 contained in chromatin is usually only slightly acetylated. In the germinal vesicle, however, which contains half of the stored histones, H4 is in the diacetylated form, whereas in the oocyte cytoplasm it is predominantly phosphorylated (Woodland, 1979). It is assumed that, as in somatic cells, histone H4 is phosphorylated immediately after synthesis in the cytoplasm (and is actually stored there in this form); on entering the nucleus it is acetylated (and is stored in this form in the germinal vesicle), and, upon incorporation into chromatin, it is diacetylated again. Unlike H4, the stored histone H3 is always acetylated and phosphorylated, and its modifications take place after its incorporation into chromatin (Woodland, 1979).

According to Woodland and his coworkers, histone H1 behaves differently in *Xenopus* oocytes and embryos (Flynn and Woodland, 1980). Its synthesis rate in the course of oogenesis probably does not differ from that of the other histone species. But it does not speed up during oocyte maturation, while during cleavage the histone H1 synthesis proceeds much more slowly than that of the core histones. A detectable synthesis of H1 begins only from late blastula and increases rapidly up to neurula, when it apparently achieves the same rate as the synthesis of the other histones (Flynn and Woodland, 1980). On the other hand, the mRNA for histone H1 reaches appreciable quantities in the *Xenopus* eggs (Ruderman *et al.*, 1979), which shows that its synthesis is controlled at the posttranscriptional level. The involvement of the genome in the histone H1 synthesis, at any rate from neurula, has been shown on the hybrids of two *Xenopus* species in which H1 can be distinguished (Cassidy and Blackler, 1978). The question as to the content of histone H1 in chromatin of early *Xenopus* embryos remains open, since a low level of the H1 synthesis would have led to its considerable deficiency. However, direct methods for determining HI in chromatin of early embryo nuclei are still not sensitive enough.

The paradoxical situation with histone H1 also seems to be confirmed by the data of Imoh (1977, 1978) obtained on the newt, in which histone H1 is detected in relatively smaller amounts than the other histones beginning with blastula. Koster *et al.* (1979), however, using *Xenopus* again, obtained a different result. Employing various methods of extraction and electrophoretic separation, the authors detected a discernible synthesis of histone H1 also before gastrulation (they incubated the embryos with an isotope from the 8-cell stage to late blastula). The authors attribute the difference of their data from those of Woodland to a possible loss of H1 in his experiments. In their later investigation, however, Flynn and Woodland (1980) carried out a number of experiments confirming the adequacy of the methods used. In particular, in experiments where histone mRNA of the sea urchin was introduced in *Xenopus* oocytes the synthesis of the sea urchin H1 was detected readily (Woodland and Wilt, 1980a, b).

The situation did not become less complicated following the work of Shih *et al.* (1980) investigating the early development of another frog species, *Rana pipiens*. They demonstrated that, as in *Xenopus*, the histone synthesis in the eggs of *R. pipiens* is 10 times as intensive as in oocytes. By blastula it accelerates by another factor of three. The fundamental difference observed in *R. pipiens* is that at the cleavage stages this species showed a high H1 synthesis rate but almost no synthesis of H4 (Shih *et al.*, 1980). However, it is difficult at present to depict a clear picture of fluctuations of the amount, synthesis rate, and the content in chromatin of different histones.

Late stages of differentiation (in particular, terminal differentiations, such as erythropoiesis) are also associated with some changes in histone composition. This refers primarily to the subfractions of histone H1, for which the species and tissue specificity have been demonstrated, although it is not yet known whether the H1 subfractions are determined by different genes or by polypeptide modifications. During erythrocyte differentiations in birds the nuclei experience heterochromatinization and a drop in RNA synthesis. This is accompanied by the synthesis of histones and nonhistone proteins (Sadgopal and Kabat, 1969), and also of the erythrocyte-specific histone H5 (f2c) whose share increases from 13% at early erythropoiesis to 21% in mature red blood cells (Moss *et al.*, 1973). The role of this histone in repressing the RNA synthesis was shown by Gasaryan and Andreeva (1972) in experiments with saline extraction of histones from chromatin: removal of H5 fraction restored the initial template activity.

A special case of differentiation which is apparently attended by a complete replacement of the nuclear proteins is spermatogenesis. Following the histochemical investigations of this process (see, for instance, Das *et al.*, 1964), the replacement of the nuclear histones by specific protamine-like histones or protamines has been shown biochemically in a number of species. It occurs at the last stages of spermatogenesis, against the background of a previously arrested RNA synthesis (Monesi, 1964; see Ling and Dixon, 1970; Brasiello, 1971; Vaughn and Thomson, 1972; Kierszenbaum and Tres, 1975; McMaster-Kaye and Kaye, 1976).

The composition of nonhistone nuclear proteins in differentiating systems has been investigated much less, for one thing because this group includes many different proteins for which hardly any identification methods have been worked out yet. Nevertheless, it has been shown on sea urchin embryos that this heterogeneous group of proteins contains 3 main and 10 minor fractions (Gineitis *et al.*, 1976). These proteins change quantitatively and qualitatively in the course of development (Seale and Aronson, 1973). The rate of synthesis of these proteins changes only slightly at all the stages investigated, from blastula to pluteus, but does not parallel the DNA synthesis. The highest synthesis rate is observed at blastula (20 h). According to Seale and Aronson, nonhistone proteins are synthesized in the cytoplasm and form a pool, from which they slowly penetrate the newly formed nuclei and are incorporated into the chromatin. A short-term incubation with labeled precursors followed by tracing the proteins synthesized (the chase method) showed that the proteins formed as far back as the cleavage continue being incorporated into the chromatin up to gastrulation.

The total quantity of nonhistone proteins in the embryo increases throughout the development, but up to the mesenchyme blastula stage a considerable part of them (about half) are in the cytoplasmic pool. They become incorporated into the chromatin beginning with gastrulation (Seale and Aronson, 1973).

Somewhat earlier, the differences between the electrophoretic profiles of nonhistone proteins at the 16-blastomere, blastula, and pluteus stages had been shown in sea urchin embryos (Cognetti *et al.*, 1977).

A similar investigation on *Xenopus* embryos also showed that the size distribution of nonhistone nuclear proteins during the sodium dodecylsulfate (SDS) electrophoresis remains constant up to stage 18, but begins to change with the beginning of embryo mobility (stage 26), and these changes increase by the time of larva hatching (Theriault and Landesman, 1974) (Fig. 61).

Changes in the synthesis rate of nonhistone proteins have also been noted in the embryogenesis of the bug *Oncopeltus fasciatus* (Teng, 1974). The synthesis rate achieves a maximum by the 45th hour of development, and by the 70th hour the embryo shows a maximum ratio of acidic nuclear proteins to DNA, which then decreases to a constant ratio by the 120th hour. Apparently, in this case, too, the synthesis of nonhistone proteins is not coordinated with that of DNA, and the proteins stored at the early stages are used in subsequent development. The composition of nuclear proteins becomes more diversified in the course of development (Teng, 1974).

It may be concluded that, because of technical difficulties,

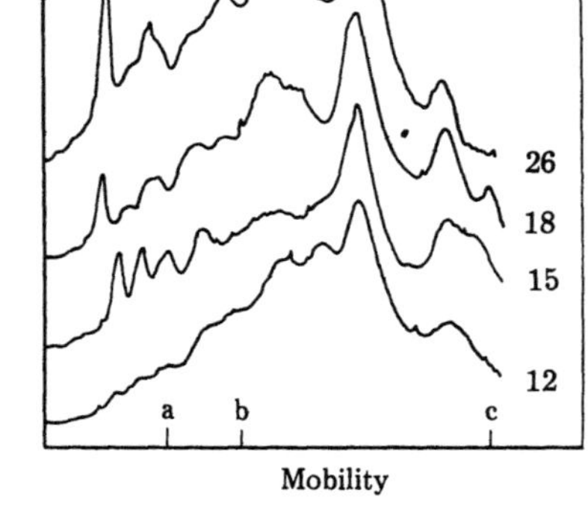

Fig. 61. Changes in the composition of acid nuclear proteins during *Xenopus* development. Acid proteins were isolated from nuclei at different stages (numerals at the right of the curves), separated electrophoretically with SDS, and their amount was measured by stained gel densitometry. The fraction numbers are given at the top; *a*, *b*, and *c* indicate positions of the tracer proteins: serum albumin, ovalbumin, and cytochrome *c*. The spectrum of the nonhistone proteins becomes enriched in the course of development due to the higher-molecular-weight fractions (fractions 1 to 5) (Theriault and Landesman, 1974).

investigation into the synthesis, and particularly into the content, of nuclear proteins at the early stages of development is only beginning to yield relatively reliable data. It seems to have been shown convincingly that histones and nonhistone proteins accumulate in the oocytes and cells of early embryos and are gradually incorporated into the chromatin as the number of nuclei increases. But the forms of histone storage in the cytoplasm are unknown. The data on the changes in the histone : DNA ratio and the content of the various histones in the chromatin at the early stages of development must also be checked and refined. These data do not agree with the notion of a stable ratio between the histones in the chromatin nucleosomes, and it will probably be established in the near future whether the structure of the chromatin and, in particular, of the nucleosomes at earlier stages of development differs from that at later stages and of the adult organism, or whether the data on DNA : histone ratio in the chromatin of early embryos are an artefact.

SYNTHESIS OF TISSUE-SPECIFIC PROTEINS

In this section we deal with several proteins which are almost or completely absent in the egg and appear only in definite tissues of the embryo in the course of its development. Thus, these proteins can be regarded as typical differentiation proteins, although they should be viewed as a manifestation of differentiation, rather than its cause. At the same time many of the proteins under consideration, such as tubulin, collagen, myosin, or lens crystallin, are structural proteins and determine, in a large measure, the shape of the cells and their arrangement, i.e. morphogenesis. These proteins are similar in that they are isolated and determined as individual chemical substances. This distinguishes them from the enzymes (see the next section), whose quantity is, as a rule, estimated indirectly, by their activity.

Collagen

Collagen, as a protein associated with relatively late stages of differentiation — mainly the differentiation of the connective tissue and cartilage — is detected by the routine methods rather late in development (Edds, 1958; Deuchar, 1964).

However, determination of radioactive oxyproline (an amino acid which is found only in collagen and elastin) reveals the collagen synthesis at earlier stages. Labeled proline was injected to *Xenopus* embryos at different stages of development, and 3 h later the proteins

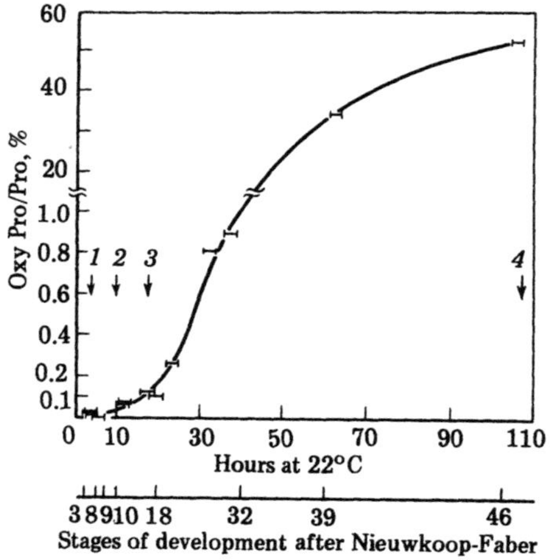

Fig. 62. Collagen synthesis in *Xenopus* embryogenesis. Embryos were injected with ^{3}H-proline at several stages, and the content of ^{3}H-proline and ^{3}H-oxyproline in the protein was determined 3 h later. The collagen synthesis rate was expressed as percentage of proline converted to hydroxyproline (ordinate). The arrows indicate the stages of development: cleavage (*1*), gastrula (*2*), neurula (*3*), free feeding (*4*). Collagen synthesis is first detected already at gastrula and increases appreciably from then on (Green *et al.*, 1968).

were hydrolyzed and the oxyproline-proline radioactivity ratio was determined in them; this ratio reflected the share of newly synthesized collagen among the other proteins (Green *et al.*, 1968) (Fig. 62). Collagen is not formed until the blastula stage, but it is clearly distinguishable already at early gastrula, after which its synthesis accelerates rapidly up to the feeding tadpole stage. Oxyproline forms as a result of the action of the specific enzyme collagenhydroxylase on definite amino acid residues of proline, which are contained in the just formed polypeptide chain, protocollagen. It turned out that this enzyme, which is necessary for the formation of collagen, is already present in the unfertilized egg but does not function up to the beginning of gastrulation, when the collagen precursor, protocollagen, begins to be synthesized and is transformed into procollagen by hydroxylation of a part of the proline residues.

Protocollagen mRNA was isolated from the chick embryo bone anlage. It sediments at 30S, which approximately corresponds to a chain of 4.9×10^{3} nucleotides, and can encode a polypeptide 1.6×10^{3} amino acids in length. This greatly exceeds the actual size of the protocollagen chain, 1.25×10^{3} amino acid residues (Boedtker *et al.*, 1974). It may be assumed that either the collagen mRNA has such a configuration that it sediments faster than, for instance, rRNA, or that the actual size of the polypeptide immediately after the synthesis is still larger than has been assumed until now.

Apart from oxyprolin, oxylysine is formed in the collagen molecule with the aid of the corresponding hydroxylase; sugar residues are then added to the oxylysine, thus making this protein a glycopeptide. The collagen molecule is formed of identical or different (α1, α2, etc.) subunits. Their combinations determine the formation of four principal types of collagen. Thus, collagen synthesis is controlled by several (at least five) genes which determine the synthesis of the polypeptides themselves and of their modification enzymes.

The formation of collagen fibers includes one more rather important stage: the procollagen subunits have a molecular weight exceeding by 6.5×10^4 daltons that of the collagen subunits. Immediately after the emergence of procollagen from the cell, a specific proteolysis takes place, which cuts off additional sequences at both ends of the polypeptide chain, thus ensuring the formation of the final quaternary structure of collagen (Fessler and Fessler, 1978; Bruns *et al.*, 1979).

So, the formation of collagen is controlled by one more structural gene which encodes this specific protease.

This is true for collagens of types I to III. Type IV is secreted and accumulates in the form of procollagen, it is unstriated and nonfibrillar (Minor *et al.*, 1976; Dehm and Kefalides, 1978). It is believed that collagens determine the architectonics of organs and tissues, and in this sense morphogenesis is caused by expression of various collagen-genes. Such investigations began to develop in recent years on the basis of specific immunological tests. In this respect studies on the synthesis of collagen IV in the development of the mouse are of interest (Adamson and Ayers,. 1979). This type of collagen is contained in basement membranes in such structures as the lens capsule, kidney glomerules and tubules. In the course of early development of the mouse (5th day of pregnancy) collagen IV is synthesized in the primordial endoderm, then it accumulates on the inner surface of the trophoblast. At later stages it is detected between the cell layers of the embryonic membrane separating the visceral mesoderm from the endoderm, reaching Reichert's membrane and underlying the parietal endoderm. Then collagen IV is detected both in the amnion and the chorion. Thus, the collagen IV-containing basement membranes begin to separate the mesoderm from the endoderm and ectoderm rather early in development. In addition, the basement membranes apparently serve as a barrier to penetration of immunoglobulins to the embryo (Adamson and Ayers, 1979).

Determination of collagen synthesis in various embryonic tissues has shown that they are subjected to expression of various collagen genes. The endoderm of the visceral yolk sack synthesizes type I collagens, and the mesoderm layer, types I and IV. Since the visceral endoderm of the yolk sack forms from the visceral endoderm which synthesizes collagen IV, it may be assumed that there is a switching of expression from some collagen genes to others in the course of differentiation of embryonic tissues (Adamson and Ayers, 1979).

Contractile Proteins

This class contains three main proteins: myosin, actin, and tropomyosin, and also several myosin-like minor proteins present in small amounts in muscle fibers (troponin and α- and β-actinines). The myosin molecule consists of two identical large subunits (molecular weight 2.2×10^5 daltons) and two small ones (molecular weight $1.5\text{-}3 \times 10^4$ daltons). Actin is not divided into subunits (molecular weight $6\text{-}7 \times 10^4$ daltons). Tropomyosin is formed by two subunits with a molecular weight of $3\text{-}3.5 \times 10^4$ daltons. In accordance with the size of these molecules the synthesis of myosin, actin, and tropomyosin occurs on large polysomes consisting of 50-70 (Fig. 63), 15-25, and 5-9 ribosomes, respectively (see Young, 1970). These three types of contractile protein are contained in muscle fibers in stoichiometric amounts, but they naturally differ widely in weight.

Two more muscle proteins have been detected recently: titin, which is present only in striated muscles, and filamin, which is characteristic only of smooth muscles. These are probably the largest proteins, since the molecular weight of titin is $0.5\text{-}1 \times 10^6$ daltons and that of filamin, about 0.4×10^6 daltons (Wang *et al.*, 1979). The muscle proteins also contain other, recently detected, structural proteins (e.g. desmin), whose diversity increases owing to their posttranslation modifications (O'Connor *et al.*, 1979).

Certain, sometimes considerable, amounts of myosin and actin have been found in many types of cell (fibroblasts, myoblasts, eggs) (Mabuchi, 1976). But these contractile proteins differ from the similar muscle proteins in their molecular weight, antigenic properties, half-life, and relative content (Chi *et al.*, 1975; Rubinstein *et al.*, 1976). Contractile proteins are even detected in nonhistone chromatin proteins (Douvas *et al.*, 1975). During muscle differentiation the nonspecific myosin and actin of myoblasts are replaced by specific contractile muscle proteins. It is presumed that different myosins are controlled

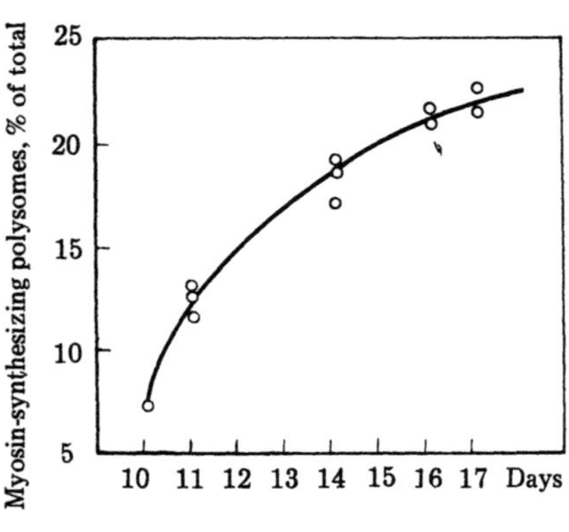

Fig. 63. Increase in the fraction of myosin-synthesizing polysomes during development of the chick embryo. Myosin-synthesizing polysomes containing 50 to 70 ribosomes each were isolated as the heaviest among the overall polysome profile (Herrmann *et al.*, 1970).

by different genes (Chi *et al.*, 1975; Rubinstein *et al.*, 1976; Storti and Rich, 1976).

The synthesis of contractile muscle proteins begins rather early. In the chick embryo this occurs primarily in the heart primordia, which begins to contract already at the end of the second day of incubation. Myosin synthesis is detected at the earliest stages of myogenesis, usually immediately after the fusion of myoblasts into myotubes. In recent years this process has been investigated successfully, not in the embryo but in the differentiating tissue culture of the myogenic cells (Coleman and Coleman, 1968; Schubert *et al.*, 1973; Delain and Wahrmann, 1975; Rubinstein *et al.*, 1976, and others).

As regards the contractile protein synthesis pattern the data found in the literature have been somewhat contradictory. Some authors note that actin is the first to be synthesized, followed by myosin, and finally by tropomyosin (see Young, 1970). EM observations also first detect the short actin filaments (50-60 μm) in the myoblasts, and then the longer myosin filaments (140-160 μm) (Fischmann, 1970).

A more sensitive method for determining myosin is its detection by binding with fluorescent antimyosin antibodies (Herrmann *et al.*, 1970). In this way myosin is detected when about 10^5 molecules on this protein are accumulated in the cell. This method reveals myosin even in the period of myoblast proliferation in the chick's heart muscle anlage already on the 50th-55th hour of development, and in the limb anlage on the 4th day, i.e. much earlier than by other methods.

According to these data the synthesis of contractile proteins of all the three types begins simultaneously and at once in equimolar amounts, i.e. in a coordinated manner. The coordination of the synthesis of contractile proteins has been shown reliably in cultures of myogenic and nonmyogenic cells (Rubinstein *et al.*, 1976). The synchrony of the synthesis of myosin, actin, tropomyosin, and α-actinin has been shown accurately by means of isotopic dilution (Allen *et al.*, 1979).

In the course of differentiation in the culture, the myosin synthesis rate increases from 1% of the total protein synthesis in dividing myoblasts to 7% in myotubes (Schubert *et al.*, 1973). In the course of embryogenesis the myosin concentration increases from 4% of the total protein in the muscles of an 11-day-old chick to 12% in the muscles of an adult chicken (Young, 1970). This relatively small concentration is ascribed to the fact that myosin itself constitutes about half of all the contractile proteins and that a considerable part of the muscle proteins are enzymes, whose amount rapidly increases during myogenesis. An experimental analysis of the role the myoblast fusion plays in the induction of the contractile proteins and specific muscle enzymes synthesis showed that artificial restriction of the cell fusion with the aid of calcium-binding substances does not arrest the synthesis of specific muscle macromolecules (Delain and Wahrmann, 1975; Emerson and Beckner, 1975; Moss and Strohman, 1976; Vertel and Fischmann, 1976; Turner *et al.*, 1976).

Transition of dividing myoblasts from proliferation to fusion into myotubes is stimulated by the peptide hormones insulin and somatotropin (Haba *et al.*, 1966). In bird embryos these hormones are located in the egg and stimulate myogenesis long before they are secreted by their own glands. Another stimulating agent of myogenesis is the energetic metabolite creatine which is characteristic of the muscles (Ingwall *et al.*, 1974). Its addition activates the synthesis of myosin but not of the total protein.

Simultaneously with contractile proteins, a multitude of specific muscle enzymes (creatinekinase, myokinase, glycogen phosphorylase, aldolase A and others) (Schubert *et al.*, 1973; Delain and Wahrmann, 1975; Turner *et al.*, 1976) and also cell membrane proteins (adenylate cyclase, acetylcholine receptor and acetylcholinesterase) (Wahrmann *et al.*, 1973; Prives and Paterson, 1974) are synthesized in the differentiating muscles.

The synthesis of rat myosin continues after the hybridization of the rat myoblasts and mouse fibroblasts, whereas no synthesis of mouse myosin is detected (Buckingham *et al.*, 1974b).

Quantitative measurements of the synthesis rate of heavy myosin chains before and after the myoblasts fusion were based on incorporation of labeled leucine and valine into them with simultaneous measurement of the specific pool radioactivity. From the moment of fusion the rate of myosin synthesis increases by a factor of 50-100. The same increase is observed in the unfused myoblasts (in the presence of EDTA or a lowered concentration of the serum and the embryonic extract) (Emerson and Beckner, 1975). The synthesis time of one myosin molecule is 5-10 min, and 5 to 15 myosin molecules per minute are translated on one mRNA molecule. The myosin half-life being 6 days, 3×10^4 myosin molecules per nucleus must be synthesized in one minute (Emerson and Beckner, 1975) (Fig. 64).

The formation of protein subunits of the contractile apparatus alone, including the minor components, involves 7 to 10 structural genes. Their activation begins at the earliest stages of myogenesis, but increases considerably at the final stages of muscle differentiation.

There is an ever-increasing body of data accumulating on another protein, actin. Many different forms of actin are distinguished now. Some of them are involved in the formation of the contractile muscle cell complex, while others are present in all the other types of cell (Vandekerkhove and Weber, 1978). Besides the direct contractile function they perform as part of actomyosin in the muscles, actins form microfilaments and are therefore detected in nearly all the cells, where their content reaches 5-10% of the total amount of proteins. Their content in nonmuscle cells particularly increases in the period of differentiation of nerve cells, skin, liver, and other

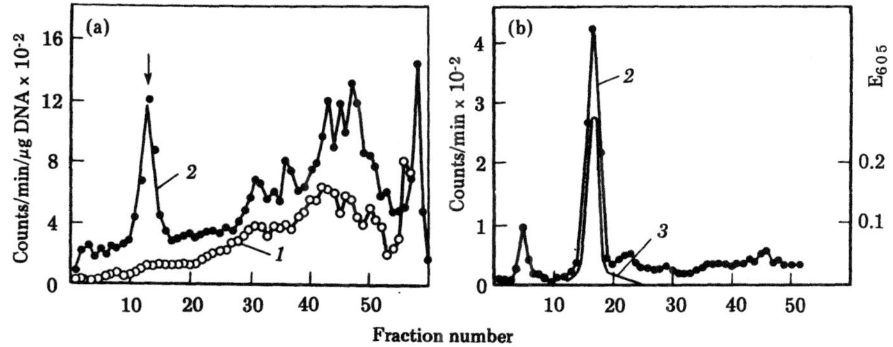

Fig. 64. Stimulation of myosin synthesis following myoblast fusion. Dividing (2-day culturing) and fused (6-day culturing) myoblasts were incubated with ^{3}H-leucine. The total protein (*a*) and the purified myosin (*b*) were separated electrophoretically with SDS together with the tracer heavy myosin chains. *1*, Proteins from dividing myoblasts; *2*, proteins from the myoblasts fused into myotubes; *3*, tracer myosin. The synthesis of myosin (fractions 12 to 16) is intense in the myotubes and very low in the myoblasts (Emerson and Beckner, 1975).

tissues (Santerre and Rich, 1976). It would seem likely that on entering the microfilaments actin determines the cell movements in morphogenesis and also egg cleavages (Schmidt *et al.*, 1980). But a relatively large fraction of actin (3-5%) remains in adult tissues, where, as a component of microfilaments, it takes part in mitosis (Blose, 1979) and apparently performs some other function.

The current availability of rather pure preparations of myoblasts and myotubes, the method of two-dimensional electrophoresis and computer analysis provide quantitative data on the change in the protein composition during muscle differentiation (Garrels, 1979). The synthesis rate of such contractile proteins as tropomyosins and light myosin chains increases more than 1000-fold. Among the 278 thoroughly investigated proteins, over 40% speed up or slow down their synthesis by more than two times. The difference in the composition and synthesis of proteins between two consecutive stages of myogenesis proved to be no less than between such distinct types of cell as myoblasts and the nerve cells (Garrels, 1979). This example clearly indicates that differentiation affects expression of a large number of different genes.

Crystallins

The crystallins probably belong to the group of the best-studied structural proteins. One distinguishes α-, β-, and γ-crystallins according to their solubility, mobility in the electric field, molecular weight, and antigenic properties. In birds and reptilians, δ-crystallins, or

FISC (First Important Soluble Crystallin), are also found. Some of these proteins consist of several subunits. Thus, the formation of the lens structure involves at least 20-30 different polypeptide chains, which are evidently encoded by the corresponding number of genes.

The lens differentiation begins when the ectodermal epithelium above the eye anlage (eye cap) thickens and is drawn into the eye cavity, forming the lens placode. Then this drawing-in forms a closed epithelial vesicle which separates from the ectodermal epithelium. The cells of the rear, thickened wall of the vesicle elongate still more and fill its entire cavity, forming elongated transparent cells of lens fiber, while its front wall forms the lens epithelium. In the equatorial region one can see gradual conversion of the epithelium cells into lens fibers. The lens growth and the formation of new fibers occur at the expense of the epithelium: its cells in the equatorial region divide and gradually move to the rear, where they become elongated and turn into lens fiber cells.

Thus, even after the formation of the eye, at the early stages of embryonic development the lens undergoes successive stages of differentiation from slightly specialized epithelial cells and up to irreversibly differentiated and highly specialized lens fibers.

The first crystallins in chick embryos appear in the cells of the lens placode already after 56-hour incubation. These crystallins belong to the FISC class (δ). On the 80th hour of incubation there are many of them both in the epithelium and in the lens fibers. An electrophoretic analysis shows that after 96 h the composition of the δ-crystallins in the epithelium and in the fibers is different: the former contains subunits 1 and 5, and the latter, subunits 3 and 4. By the 15th day of development the δ-crystallins disappear from the chick epithelium but remain in the adult chicken in the fibrous part of the lens, constituting a considerable part of its proteins (Zwaan, 1968; Clayton, 1970).

Alpha-crystallins first appear in the fibers of the early lens (80 h) and then in the epithelium. By the 18th day of development a new subunit is added to the existing two. Traces of β-crystallins appear at the placode stage, and then in all the cells of the lens. The number of subunits of this type increases in the course of development.

The electrophoretic analysis of the composition and synthesis of lens proteins, carried out by Piatigorsky *et al.* (1972), shows that in the lens of a 6-day-old chick embryo the predominating protein is that with a molecular weight of 4.5×10^4 daltons (Fig. 65). It is, however, less abundant in the lens epithelium than in the lens fibers. To offset this, the epithelium is endowed more lavishly with proteins having different molecular weights, including those of 3×10^4 and 6.5×10^4 daltons, which are practically lacking in the fibrous part of the lens. Protein synthesis data confirm this distribution: 80% of protein with a molecular weight of 4.5×10^4 daltons, which is δ-crystallin (FISC),

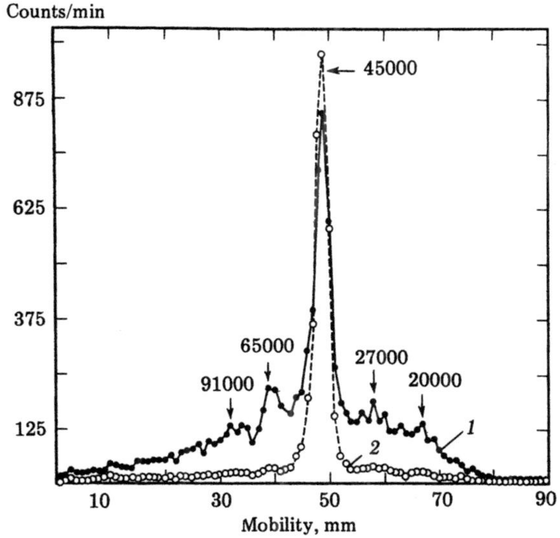

Fig. 65. Crystallins of the epithelium and fibers of the chick embryo lens. Lenses from 6-day-old embryos were labeled for 4 h *in vitro* with ^{3}H- or ^{14}C-valine and divided into epithelium and fibers. The ^{3}H-labeled epithelium proteins (*1*) underwent electrophoresis in SDS jointly with ^{14}C-labeled fiber proteins (*2*). The numerals above the peaks denote the molecular weights. The lens epithelium contains many proteins of different size but δ-crystallin with a molecular weight of 4.5×10^4 predominates (37%). In the fibers, it constitutes 80% of all proteins (Piatigorsky *et al.*, 1972).

is synthesized within this period in the cells of the lens fibers. In the epithelium, its synthesis constitutes only 35% of the total protein synthesis, but a number of other proteins with molecular weights from 2×10^4 to 2×10^5 daltons (peaks in the regions of 2×10^4, 2.7×10^4, 6.5×10^4, and 9.1×10^4 daltons) are also synthesized here. When pieces of lens epithelium are placed in a medium containing bovine serum, its cells rapidly elongate; this process resembles the formation of lens fibers. Indeed, the spectrum of synthesized proteins changes in the course of 24 h from the one characteristic of lens epithelium to the one typical of lens fibers, although the synthesis of δ-crystallins (molecular weight 4.5×10^4 daltons) reaches only 50% and not 80%, as in the fibers of a normal lens (Piatigorsky *et al.*, 1972). This investigation shows that the growing and dividing cells of the lens epithelium at the beginning of specialization synthesize many proteins, although here, too, they largely consist of the specific lens protein. As the cells move through the equatorial region to that of lens fibers, their protein composition changes: proteins with a molecular weight of 3×10^4 and 6.5×10^4 daltons disappear, and the synthesis of all the proteins, with the exception of δ-crystallin, practically

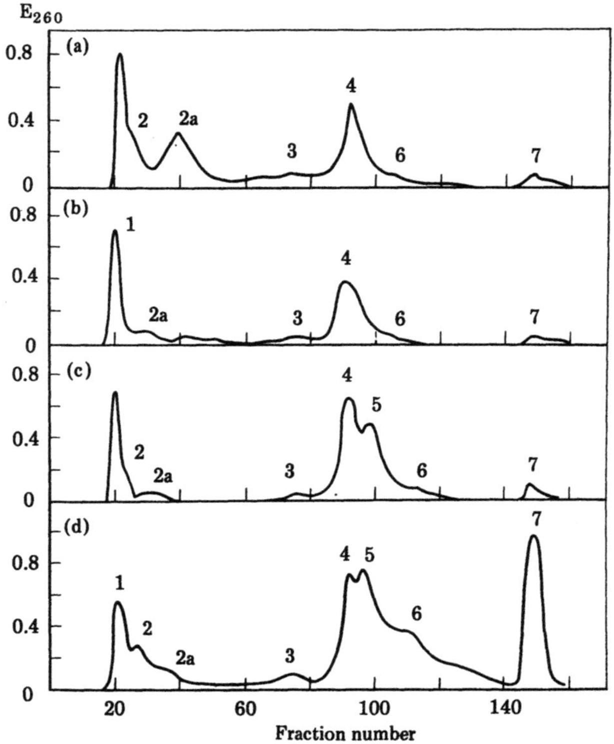

Fig. 66. Changes in the α-crystallin subunit composition during the calf development. α-Crystallins were isolated from the lenses of a 4-month (*a*) and a 4.5-month (*b*) embryo, a calf (*c*), and the lens cortical fibers of an adult bull (*d*), and then separated into subunits by chromatography on DEAE-cellulose. The numerals above the peaks denote the subunits. Fraction 2a decreases with development; there appears fraction 5, and fractions 6 and 7 increase (Palmer and Papaconstantinou, 1968).

ceases (see Fig. 65). Similar processes, although to a lesser extent, occur when the lens epithelium is cultivated under conditions promoting its terminal differentiation. The synthesis of lens proteins and the elongation of the lens epithelium cells may also proceed for some time in the presence of actinomycin D, i.e. when the gene activity is suppressed, due to previously transcribed templates (Craig and Piatigorsky, 1973). Simultaneously with the change in the synthesis of δ-crystallin subunits one observes an increase in the Na^+/K^+ ratio and an overall increase in the Na^+- and K^+-dependent ATPase activity. During *in vitro* incubation of the lens of 15-day-old embryos the removal of the vitreous body or the addition of ouabain inhibits these processes.

The observed changes in the composition of the synthesized proteins and concentration of electrolytes are stage-specific, since the crystallin synthesis and the Na^+/K^+ ratio during the lens differentiation in the retina cell culture from 6-day-old embryos do not change after addition of ouabain or removal of the vitreous body (Shinohara and Piatigorsky, 1979; Araki *et al.*, 1979).

In amphibians, the order of appearance of the different crystallins has been described for normal development and for the process of lens regeneration from the iris (Yamada, 1967). The γ-crystallins characteristic of amphibians appear in fiber-forming cells and persist at later developmental stages. In addition to crystallins, the lens epithelium cells synthesize the collagen forming the lens capsule.

In the bull, changes in the subunit composition of α-crystallins of the lens take place in an ordered manner during the development from a 4-month embryo. The relative content of subunit 1 (eluted as the earliest fraction from DEAE-cellulose) is halved, the content of subunit 2 is almost doubled, and the content of the minor subunit 3 increases several-fold (Palmer and Papaconstantinou, 1968) (Fig. 66). When 14S RNA obtained from the polysome fraction of the bovine eye lens is injected into amphibian oocytes, the latter undergo a synthesis of α-crystallins which thus does not require any specific translation factor (Berns *et al.*, 1972).

The crystallins are probably not the only proteins determining the development of the lens. They first appear at the lens placode stage, i.e. when the lens is already determined and its morphogenesis is being effected. But crystallin synthesis is a good model for studying the general pattern of synthesis, accumulation, and replacement of tissue-specific proteins.

Hemoglobins and Other Blood Proteins

Hemoglobin is produced in blood-forming cells which undergo a whole series of transformations in the course of their differentiation; morphologically, this transformation can be divided into 6 or 7 stages: stem cell—proerythroblast—basophilic erythroblast—polychromophilic erythroblast—normoblast—reticulocyte—erythrocyte.

It has already been mentioned that the synthesis of mRNA and its transfer into the cytoplasm ceases at the stage of polychromophilic erythroblasts, while the hemoglobin synthesis begins and proceeds most intensively in basophilic erythroblasts, decreases 4-fold in polychromophilic erythroblasts, but still continues (20 times slower than in the basophils) in reticulocytes. Blood formation in an embryo usually occurs in other organs than it does in an adult organism: in the chick embryo, these are blood islands located outside the embryo itself, in the tadpole this is the primordial kidney, in fishes these are blood islands located over the yolk sack but not outside the embryo. Finally, the embryonic development of mammals is attended by a complicated replacement of blood-forming organs. At first, red blood cells appear in blood islands of the yolk sack. In the mouse, these primitive red cells containing nuclei synthesize on the 8th day three embryonic hemoglobins: EI (x_2y_2), EII (α_2y_2), and EIII (α_2z_2). Then, soon after the 10th day and up to birth, cells are formed in the liver, which are precursors of the erythroid series and which differentiate inside the liver parenchyma into mature anucleate red blood cells synthesizing exclusively adult hemoglobin ($\alpha_2\beta_2$). In the same period, on the 10th to 14th day the primitive nuclear red blood cells also begin to accumulate the adult type of hemoglobin (Brotherton *et al.*, 1979). At the end of embryonic development and after birth the hemoglobin synthesis takes place in the spleen and the bone marrow. Embryonic erythropoiesis always occurs in cells of mesodermal origin, but usually needs a contact with the endoderm for its determination (Wilt, 1967).

At later stages of mammalian development the embryonic hemoglobin (F), which consists of two α- and two γ-chains, is fully replaced by the hemoglobin of the adult organism (A) with two α- and two β-chains. Hemoglobin F differs from hemoglobin A by a higher affinity for oxygen, which allows the oxygen to be transported from the maternal to the embryonic blood in the placenta.

In amphibians undergoing metamorphosis, there is a similar transition from oxygen diffusion from the water through the gills to air respiration through the lungs. In this period the tadpole experiences a replacement of hemoglobins, the total intensity of their synthesis being reduced. The formation of erythrocytes in the kidneys is terminated at this time, and the formation of "adult" erythrocytes, containing "adult" hemoglobin A, begins in the liver, spleen, and the kidneys (see review by Maclean and Jurd, 1972; Sullivan, 1974; Hentschel *et al.*, 1979).

In the chick, the biological significance of hemoglobin replacement is also associated with the transition from oxygen uptake by the vessels of the yolk sack surrounded by a liquid white coat to oxygen uptake by the air-contacting vessels of the chorioallantois, which lines the shell from the inside (7th day of incubation), and thence to pulmonary respiration (17th-18th day of incubation).

The replacement of hemoglobins in chick embryos occurs in a relatively early period of embryonic development when the yolk sack is the main erythropoietic organ. An electrophoretic analysis reveals at least two types of hemoglobin in the period of embryonic development and at least three in the adult state. Each type consists of two α-chains and two chains of other classes. Hemoglobin A is already detected in a 5-day-old embryo. On the 7th day embryonic and adult hemoglobin types are present in equal amounts, and by the 11th day of development the type replacement is completed (Wilt, 1967).

In the human embryo, hemoglobins E ($\alpha_2\varepsilon_2$, $\zeta_2\gamma_2$, and $\zeta_2\varepsilon_2$), in which the ζ-chains are similar to α-chains and the ε-chains to β- and γ-chains, function up to eight weeks of development. The erythropoiesis site in this period is the yolk sack. After the eight weeks, HbF ($\alpha_2\gamma_2$) becomes the main hemoglobin type. There is a certain heterogeneity by the γ-chains of Gγ and Aγ with glycine and alanine in the 136th position, respectively. In newborn children the Gγ/Aγ ratio in HbF is 3/1.

Activation of the synthesis of the α- and γ-chains occurs in the period of liver erythropoiesis.

Synthesis of adult HbA ($\alpha_2\beta_2$) is activated at the early stages of embryo development in the liver, spleen, and bone marrow, and at first (up to 36 weeks) the β-chains constitute only 10% of all the non-α-chains, but then the chain synthesis sharply increases with simultaneous reduction in the γ-chain synthesis. At birth and in the adult organism HbA makes up 97% of the total hemoglobin and there are 2-3% of HbA_2 ($\alpha_2\delta_2$) and 1% of HbF. The main sites of hemoglobin synthesis in adult man are the spleen and the bone marrow (see review by Weatherall and Clegg, 1979). Switching of the synthesis from one hemoglobin type to another is strictly determined only by the time of development, is not subjected to any endocrine control and arises simultaneously, even in different sources of erythropoietic cells: liver, spleen, and bone marrow (see review by Weatherall and Clegg, 1979).

Two mechanisms of hemoglobin replacement are suggested: the repression of γ-genes and switching on of β-genes in the same cells, and the replacement of cells synthesizing hemoglobin F with cells in which β-genes are active and only hemoglobin A is synthesized.

The existence of the former mechanism is supported by detection of both types of hemoglobin in some erythrocytes using the spectrophotometric and cytochemical methods (Matioli *et al.*, 1962; Maclean and Jurd, 1972; Chapman and Tobin, 1979). But the latter mechanism actually prevails, since the erythropoiesis site changes in embryogenesis (yolk sack→liver→spleen→bone marrow) and, besides, the time of activity of globin genes during erythropoiesis in man is appreciably shorter than the period of hemoglobin replacement, and erythrocytes containing hemoglobin F predominate in the embryo, while in the postnatal period the share of erythrocytes containing hemoglobin A gradually increases.

Thus, the formation of hemoglobin as a tissue-specific protein involves several nonallelic structural genes responsible for the synthesis of α-, β-, and γ-chains. Besides, myoglobin, which has a monomer structure and differs widely from all globins in its amino acid composition, forms in muscle cells. Nevertheless, all these proteins are obviously of common phylogenetic origin and have formed by duplication of genes (see Ohno, 1973) and their divergent evolution. It is significant, however, that the evolution of the structural genes themselves was accompanied by the evolution of the regulatory genes determining the switching-on of structural genes in different types of cell and at different times.

The appearance of other blood proteins, immunoglobulins, in embryonic development has been investigated in *Xenopus laevis.* In *Xenopus*, like in other amphibians, the blood antibodies belong to group IgM. This protein sediments at 19S and consists of heavy chains (7.65×10^4 daltons), light chains (2.65×10^4 daltons), and low-molecular-weight J-chains (Hadje-Azimi and Michea-Hamzehpour, 1976). There are also small amounts of 7S Ig-like protein and a number of other immunoglobulins (Hadje-Azimi, 1971).

Synthesis of IgM begins at stages 34-36 during hatching (Moyer *et al.,* 1977).

Tubulin

In some cells, tubulins constitute tens percent of the soluble fraction proteins (Bamburg *et al.*, 1973), although their amount in early embryos is much smaller. The specific method for isolation of tubulins is based on their ability to bind with colchicine and to be precipitated by vinblastine. Tubulins form, in cells, microtubules, which make up such important structures as spindle filaments, cilia, and flagellae and, together with other structural proteins, apparently play an essential role in determining the shape of the cells. Because of this last property they can be regarded as "morphogenesis proteins," although the data on this function are only beginning to accumulate.

Some difficulties in understanding their function are due to their low species and tissue specificity, which is in discord with the fact that they participate in such different processes as chromosome segregation, the formation of intricately organized structures of the cell, such as cilia, and the determination of the shape of different types of cell. It might be more expedient to regard them, for the time being, as low-specificity proteins forming bearing structures whose specific arrangement is determined by some additional factors.

The tubulin molecule consists of two nonidentical subunits with a molecular weight of 5.7×10^4 and 5.2×10^4 daltons, respectively (Raff and Kaumeyer, 1973). A comparison of these subunits in tubulin of such different origin as the nerve cells and flagellae of Chlamydomonas showed that some of the subunits in these cells are identical, while others are distinct (Olmstead *et al.*, 1971).

Tubulin is accumulated in oogenesis and makes up an appreciable portion of the total amount of synthesized proteins. In sea urchin oocytes tubulin synthesis constitutes 4% of the total protein synthesis (Cognetti *et al.*, 1977). In axolotl its accumulation is nearly proportional to the protein content in the oocyte (Raff, 1977). The tubulin content in the eggs of different animals varies within one order of magnitude: 0.3% of the soluble proteins in the ascarid, 0.4% in *Drosophila*, 0.8% in the worm *Urechis*, 0.4 to 1.7% in the sea urchin, and 3.3% in the mollusc *Spisula* (Friedman *et al.*, 1980; Raff, 1977). In axolotl, the egg tubulin constitutes less than 0.3% of the total protein but about 10% of non-yolk proteins (Raff, 1977).

In embryogenesis, tubulin synthesis begins immediately after the fertilization of the sea urchin egg, and the synthesis rate changes cyclically, attaining a maximum at metaphase and a minimum at telophase (Meeker and Iverson, 1971) (Fig. 67). The authors have shown that the specific activity of tubulin at the early stages is higher than that of the other proteins, and its synthesis constitutes a substantial part of the total protein synthesis. It may be assumed that tubulin, which is synthesized at definite phases of the first cleavages, is used for the formation of the mitotic apparatus. But its synthesis also occurs at subsequent stages of sea urchin development and increases 4-fold from the 16-blastomere to the gastrula stage (Hynes *et al.*, 1972a). This tubulin is probably used in morphogenesis processes, in particular in the formation of cilia.

Raff *et al.* (1971, 1972) have shown in a series of investigations that tubulin synthesis in sea urchin eggs begins at the earliest stages and increases gradually until late pluteus.

Tubulin synthesis was also investigated in oogenesis and early embryogenesis of mammals (Schultz *et al.*, 1979a, b). In the mouse oocyte, 0.61 ng of tubulin per hour is synthesized. During maturation this value falls off to 0.36 ng, then increases to 0.60 ng immediately after fertilization and to 0.66 ng at the 8-cell stage. These changes are proportional to the total protein synthesis and comprise about 1.3% of it. Although α and β subunits of tubulin are represented in equimolar amounts in the microtubules, the β subunit is synthesized in a double amount as compared with α (Schultz *et al.*, 1979a, b).

Tubulin degradation rate is presumably proportional to its synthesis rate, since in the course of embryogenesis many animals (ascarid, *Drosophila*, sea urchin, *Spisula*, axolotl, etc.) have a relatively constant tubulin pool (Friedman *et al.*, 1980).

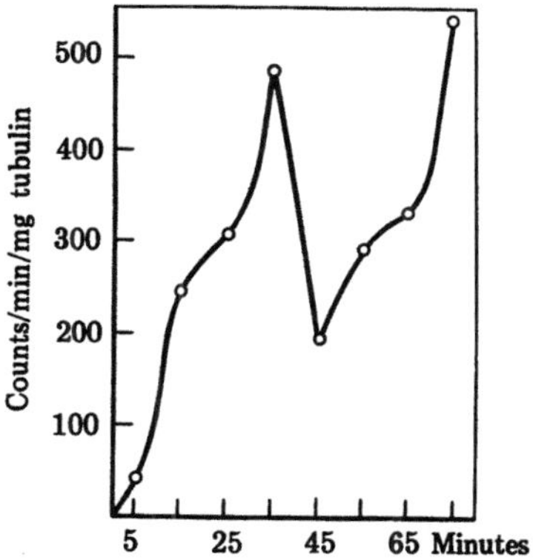

Fig. 67. Tubulin synthesis at the early stages of the sea urchin development. Eggs of *Lytechinus variegatus* at different times after fertilization were incubated for 5 min with a mixture of ^{14}C-amino acids, then the tubulin was precipitated using vinblastine and its amount determined by binding with ^{3}H-colchicine. Tubulin synthesis begins after fertilization and varies cyclically during cleavage (Meeker and Iverson, 1971).

In the embryonic development of the ascarid the tubulin binding of colchicine increases 4- to 5-fold, which would seem to indicate an increase in tubulin. However, an analysis of the binding and dissociation constants of the tubulin-colchicine complex in ascarid embryos has shown that these changes are due to an increase in the affinity of tubulin for colchicine rather than to an increase in tubulin amount (Friedman *et al.*, 1980). Some changes in tubulin properties have also been noted in other cases. This suggests the existence of certain factors affecting the assembly of microtubules and changing in the course of development (see Friedman *et al.*, 1980).

Both tubulin subunits are involved in the formation of cilia, one of the earliest manifestations of differentiation in sea urchin embryos. Besides tubulin, the cilia contain other proteins: dinein A which possesses an ATPase activity, structural dinein B, and also minor ("architectural") proteins nexin and ribon. Their synthesis and their pool have been studied in experiments on cilia regeneration in sea urchin embryos after artificial deciliation (Stephens, 1977). The experiments have shown that although there is a continuous synthesis of tubulins in early development, one of them (B) has a smaller pool and a more intense synthesis than does the other (A). Dinein A accumulates already in oogenesis, whereas dinein B is not synthesized until after fertilization. By the beginning of cilia formation (ciliogenesis) these main components are present in the cells in a sufficient amount. The synthesis of minor (architectural) proteins begins only with the beginning of ciliogenesis, their amount is limited and apparently determines the rate of cilia formation. During cilia regeneration in the presence of labeled amino acids this is reflected in the different specific activity of the cilia-forming proteins (Stephens, 1977).

Egg Chorion Proteins

In conclusion we consider the specific synthesis of proteins in the follicular cells of the silk moth — a system found and investigated by Kafatos' group (Paul *et al.*, 1972; Blau and Kafatos, 1978; Kafatos *et al.*, 1978; Sim *et al.*, 1979; Thireos and Kafatos, 1980). The female moth has dozens of oocytes simultaneously at different stages of development all lying in the ovary in a single row in a strict age sequence. Each oocyte is surrounded by a layer of follicular cells. The follicular cells perform different functions depending on the oocyte age, passing through the successive stages of differentiation together with the growth and maturation of the oocyte. The function of the follicular cells of young oocytes consists in the transfusion of yolk, and then water, from lymph to oocyte. At the next stage of development the same follicular cells form the vitellin membrane and, still later, they also synthesize several proteins of the outer egg membrane, the chorion.

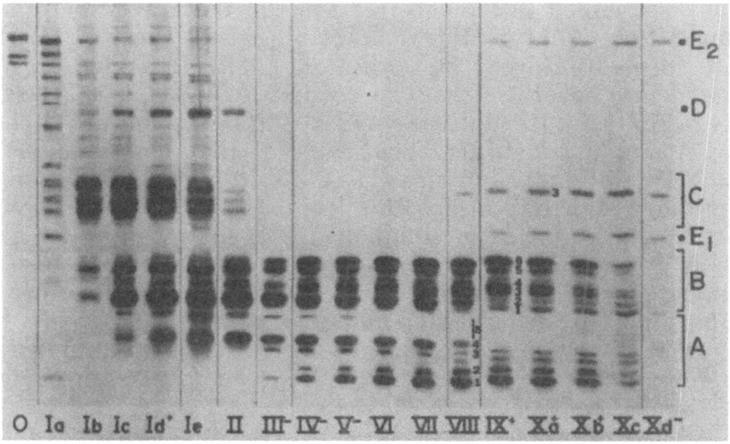

Fig. 68. Changing patterns of chorion protein synthesis during development of the silk moth. Protein-synthetic profiles are presented, corresponding to all 17 stages of choriogenesis, plus the last prechoriogenic stage (0). A single ovariole was labeled for 30 min in culture with ^{3}H-leucine (60 Ci/mmole). Sequential follicles were then deyolked, dissolved in sample buffer, and analyzed by SDS-polyacrylamide slab gel electrophoresis. The fluorogram shows the newly synthesized proteins at each developmental stage. Stage identifications are consistent with those previously based on gels of lower resolution (Paul and Kafatos, 1975); (+) and (—) represent late and early phases, respectively, for each stage. The A, B, C, and D classes of chorion proteins have been defined previously (Paul and Kafatos, 1975). E1 and E2 are a newly recognized class of chorion proteins (Regier, Mazur, and Kafatos, 1980). Numbers identify specific bands within the A, B, and C classes (Sim *et al.*, 1979).

As shown by electrophoresis, the number of these proteins is about 100 and each of them is synthesized at a definite period of oogenesis (Fig. 68). As the follicular cells are differentiated, the synthesis of all the other proteins in them ceases, and chorion proteins, differing in electrophoretic mobility, molecular weight, and amino acid composition, begin to be synthesized one after another. The follicular cells are thus an excellent model of successive synthesis of several proteins. As demonstrated by experiments with actinomycin D, the regulation of this time-ordered synthesis takes place at the transcription level, i.e. consists in sequential switching-on of different genes.

On release from follicular cells the chorion proteins form a complex system of fibers and matrix on the oocyte surface: a multilayer egg membrane (chorion). Once this structure has been formed, the solubility of chorion proteins in water and SDS is gradually lost and they are cemented owing to disulfide bonds (Blau and Kafatos, 1979).

During cell-free translation of mRNA from follicular cells the chorion proteins are synthesized in the form of precursors carrying signal peptides at their ends. When microsomal membranes (from a dog's pancreas) are added to the cell-free system, the precursor processing (splitting off of signal peptides) takes place and specific chorion proteins are formed (Thireos and Kafatos, 1980). The chorion proteins are divided into several classes, according to their sizes: class A (18 proteins of $0.7\text{-}1.1 \times 10^4$ daltons), class B (25 proteins of $1.1\text{-}1.5 \times 10^4$ daltons; A and B constitute about 90% of the chorion mass), class C ($1.5\text{-}2 \times 10^4$ daltons), and class D ($2\text{-}3 \times 10^4$ daltons). A comparison of the primary structure of proteins from classes A and B shows their close similarity and, hence, common phylogenetic origin of all these classes. A genetic analysis showed that genes encoding chorion proteins form "multigene families," which are clustered together on the second chromosome (in *Bombyx mori*) (Kafatos *et al.*, 1978). Further investigation of the genes of chorion proteins is being carried out by gene cloning techniques (Sim *et al.*, 1979). The results are considered below.

Similar investigations have been started on a genetically more suitable model: the protein synthesis of the *Drosophila* egg chorion. It was shown initially that the bulk of the chorion consists of six proteins (Petri *et al.*, 1976), but high-resolution pH gradient electrophoresis has revealed 20 different protein fractions. Their synthesis in the follicular epithelium clearly correlates in time and amount with the synthesis of the corresponding mRNA's at stages 11 to 14 with a maximum at stages 12 and 13 (Waring and Mahowald, 1979).

A group as numerous (about 70 components) of related structural proteins is contained in the egg membrane of the silkworm *Bombyx mori*. They can also be identified by their electrophoretic mobility, synthesis kinetics, and the relative content in the chorion. Each of the proteins has its own synthesis kinetics during chorion formation. On the other hand, at definite stages the synthesis of a whole group of such proteins is switched on (Nadel *et al.*, 1980; Nadel and Kafatos, 1980). The genes controlling the synthesis of these proteins are localized in the second chromosome and form three clusters in the genome of *Bombyx mori* (Goldsmith and Basehoar, 1978; Goldsmith and Clermont-Rattner, 1979).

Several more examples of synthesis in differentiated cells of specific proteins which have no enzyme activity can be adduced: that of keratin-like proteins in the epidermis (Balmain *et al.*, 1979), fibroin in the silk gland, casein in the mammary gland, mucin in the salivary gland, etc. As has been shown for the synthesis of a simple bearing protein, collagen, formation of each of them involves several structural genes responsible for the synthesis of the corresponding polypeptides, and also a number of genes controlling the synthesis of modifying enzymes, hydrolases, specific proteases, etc. Such a seemingly simple differentiation system as the formation of the lens, where crystallins perform a mechanical, or rather a physical, function, i.e. the building up of a dense body with definite optical properties, involves dozens of proteins and, consequently, dozens of genes. Here, the proteins present in minor amounts and determining the shape of the lens and the sequence of the processes of its formation, i.e. the morphogenesis proteins proper, remain undetected.

There is not a shadow of doubt that the synthesis of each of the polypeptides participating in a single differentiation is determined by a specific nucleotide sequence of DNA's, and these genes are often located in different chromosomes (for instance, globin genes). Nevertheless, a strict mutual correspondence of the synthesis is maintained in the cells being differentiated, thus ensuring the concentration constancy of each of the protein species or a consistent and gradual replacement of one species by another. This is manifested in the equimolarity of the synthesis of subunits, such as α- and β-globins, or the large and the small myosin subunits. Coordination is also manifested in the synthesis of different proteins which later form the common functional complex — for instance, myosin, actin, and tropomyosin. It is obvious that there are regulation mechanisms at the translation level, which can hardly be envisaged without a feedback, i.e. without the inhibiting effect of the excess end product. Such an excess accumulates, for instance, if the α- and β-chains of globin are formed in unequal amounts and one of the polypeptides remains unused in the cell after the formation of the quaternary structure. Similarly, contractile proteins hardly exist at all in the cell in the free state, but form myofibrils at once. In this case, even small differences in protein synthesis rate will sharply increase the concentration of one of them in the cell. Therefore there must be a mechanism ensuring the cessation or retardation of the synthesis of the protein which is accumulated in the free state in the cell.

ENZYME ACTIVITY

This section considers the activity of enzymes irrespective of their role in the chain of biochemical transformations but just as a specific indication of the presence of a definite type of protein molecules, which enables one to estimate their quantity in the embryo in the course of development.

Speaking of enzymes, we can consider three quantitative parameters separately: (1) the enzymatic activity determined *in vitro* under conditions we have found optimal or, at any rate, standard; (2) the quantity of enzymes in weight or molar units, which can be determined only in some cases; (3) the actual activity of enzymes in the cell, i.e. not the maximum possible but the actually manifested one. These three parameters are interrelated, but the relation is not direct, and the proportionality factors depend on many conditions and are, as a rule, unknown. It is extremely difficult to determine the actual reaction rate in the cell, and this has been done in just a few cases. Nevertheless, a considerable increase in the activity of an enzyme, as determined *in vitro*, suggests that the rate of the corresponding reaction in the cell has also increased. In some instances this is obvious. Thus, an increase in acetylcholinesterase activity during development coincides in time with the beginning of the activity of neuromuscular synapses, and hence reflects the actual function of the enzyme, i.e. the hydrolysis of the acetylcholine released on transmission of the nervous impulse. An increase in the activity of the hatching enzyme reflects, as a rule, the beginning of its function of digestion of the embryonic membrane. But all this is true only to a first approximation. In actuality, the enzyme must form before it begins to act, and, for instance, an increase in the activity of the pancreas enzymes in mouse embryos on the 12th to 16th day of pregnancy clearly precedes their actual function, which is effected only postnatally.

While a change in enzymatic activity points, although indirectly, to its functioning, the absence of such changes means nothing at all. Thus, in the course of embryonic development of the loach the respiration rate of the embryo (a directly measurable parameter) increases by 7-8 times (Neyfakh, 1960). At the same time the activity of mitochondrial enzymes which sustain this process changes little or not at all (Abramova and Neyfakh, 1965; Abramova and Vasilyeva, 1973). Thus, there is an increase in the respiration rate with the enzyme activity remaining constant.

The wide differences between the maximum activity and the actual function are ascribed to the fact that the enzymes usually have a different localization in the cell, because they are bound with the intracellular structures. This ensures the spatial isolation (compartmentalization) of the intracellular processes and restricts the contact between enzymes and substrates. The substrate concentration in the cell is usually much lower than in an artificial optimal medium. Finally, the activity of an enzyme in the cell is controlled by various metabolites whose composition and concentration in artificial media are different, as a rule.

All these problems constitute important aspects of biochemistry, including developmental biochemistry. But they have little bearing on the changes in the qualitative and quantitative composition of enzymes in the course of development. In this sense, of greater importance is the question of relationship between the enzymatic activity, as determined in artificial optimal conditions, and the content of the enzymes in the embryo cells. This question could be resolved using the immunochemical method: determination of the enzyme quantity by its titration with a specific antiserum (Murison, 1969).

It should be assumed, in principle, that, all things being equal, the activity of an enzyme is proportional to its content. In a number of works, some of which are mentioned below, this relationship is not questioned or discussed. But precisely because the linear proportionality is usually taken for granted, mention should be made of some, perhaps not too rare, exceptions.

When determining the enzymatic activity in the course of development it is not always possible to evaluate the importance of various factors. In comparing the different stages of development or even the different parts of the embryo, one cannot be sure that enzyme molecules have the same localization or the same concentration of nonprotein metabolites affecting their activity. We do not consider obvious artefacts which are caused, for instance, by differences in degree of homogenization between early embryos (consisting of a few easily destructible large cells) and a late larva whose small cells may remain intact. The activity of an enzyme, determined *in vitro*, may depend, to a great extent, on how it is bound with the intracellular structures and how accessible it is to the substrates. Thus, for instance, in a commonly used method for determining the cytochrome oxidase activity (Smith, 1955), the oxidation of reduced cytochrome *c* is measured spectrophotometrically. In comparing this value at different stages of loach development one can detect an increase in oxidation rate, which would seem to indicate an increase in activity, and hence in the amount, of enzyme in the course of development (Neyfakh and Dontsova, 1962). But if the mitochondria in which the enzyme is localized are first destroyed with deoxycholate, the oxidation rate will rise considerably but will remain constant at all the stages of development (Abramova and Vasilyeva, 1973). Evidently, the quantity of enzyme remains in fact unaltered throughout development. In mitochondria isolated at different stages, the accessibility to a substrate added from outside (reduced cytochrome) increases and an apparent "activity increase" is observed.

A similar example is the change in activity of β-1,3-glucanase (E.C.3.2.1.39) in just fertilized sea urchin eggs (Fig. 69). This enzyme splits off glucose from glycoproteins. Two minutes after fertilization its activity in the homogenate increases by a factor of 2.5. Using differential centrifugation it has been found that the activity of this enzyme slightly increases in large granules (3000*g*, 7 min), remains unaltered in particles precipitated after centrifugation at 28,000*g* (15 min), but increases by about 3.5 times in the supernatant (Epel *et al.*, 1969a). In an unfertilized egg this enzyme is bound with cortical granules and manifests itself only partially after routine homogenization. But after a detergent laurylsulfate is added to the medium, the granules are destroyed, the homogenate activity of the unfertilized eggs increases more than 5-fold and that of the fertilized ones, almost 2.5-fold, so the two become similar. Hence, the total amount of

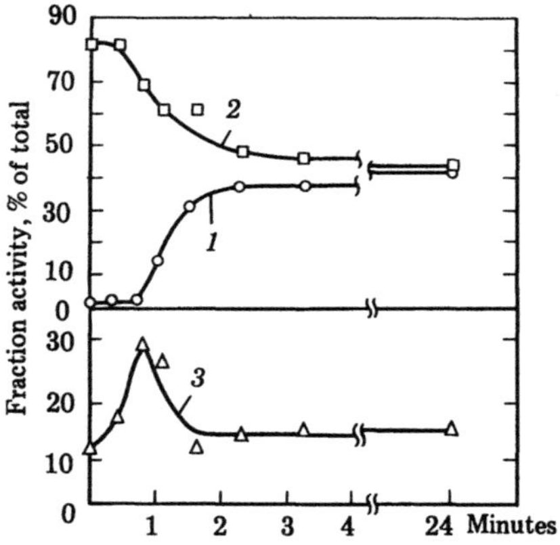

Fig. 69. Conversion of glucanase to the soluble form and its release by sea urchin eggs upon fertilization. Each 30 sec after egg fertilization the activity of β-1,3-glucanase was determined in the medium (sea water) (*1*), in the homogenate pellet after centrifugation at 28 000*g* (*2*), and in the supernatant (*3*). Within the first 1 to 2 min, half of the enzyme amount from the cortical granules becomes solubilized and then passes into the medium (Epel *et al.*, 1969a).

the enzyme remains unaltered during fertilization.

The rate of incorporation of ^{3}H-choline into the phospholipids of whole sea urchin eggs increases appreciably after fertilization. This could point to the activation of the choline kinase, but further investigation has shown that it is the permeability of the egg to choline, rather than the enzyme activity, that increases (Byrd, 1975).

The enzyme in the embryo may be in an active and in an inactive form, and the ratio between these forms may vary in the course of development. Thus, the phosphorylase in loach eggs is only partially in the active state: a considerable portion of it, localized predominantly in the yolk, is inactive under usual conditions of determination and can be detected only in the presence of high concentrations of ammonium sulfate, though it can also be partly activated by ATP and by magnesium ions (Milman and Yurovitsky, 1973).

In conclusion we can quote one more example where the mechanism of changes in enzyme activity has not been established, though it is obviously not associated with the decay and synthesis of the enzyme, i.e. with the change in its amount. In sea urchin embryos, the aspartate aminotransferase activity changes periodically at the beginning of development, and the nature of these changes is in discord with the cleavage rhythm. Puromycin in concentrations completely suppressing the protein synthesis and blocking the cleavage does not affect these activity changes, and they occur synchronously with the changes in normally developing eggs from the same batch (Botvinnik and Neyfakh, 1969).

The amplitude of these changes slightly decreases in the presence of coenzyme pyridoxal phosphate or, on the contrary, when the proteins are washed off from the low-molecular-weight components by gel filtration (Abramova and Neyfakh, 1971). It may be assumed that the aspartate aminotransferase activity is determined by some firmly bound component, but the mechanism and biological significance of this phenomenon remain unclear. It certainly cannot be attributed to the synthesis and decay of the enzyme, since no synthesis would be possible in the presence of puromycin.

The above examples demonstrate the diversity of actual situations which may face an enzymologist working with embryological objects.

The enzyme activity can be calculated per one embryo or per 1 mg of protein, nitrogen, or dry weight. Either method has a meaning of its own, as in calculation of RNA synthesis (see Fig. 14). Thus, for echinoderms, amphibians, and fishes calculation of activity per whole embryo or per 1 mg protein is almost equivalent, because the amount of protein remains practically unchanged in the course of development. However, in many fishes the embryo can be separated from the yolk, and then the amount of protein in it slowly increases during development.

If the investigation is concerned with the embryonic development of birds or mammals, calculation per embryo loses its meaning to a great extent: the embryo grows rapidly and the amount of all its proteins increases. In this case, it is expedient to determine only the specific activity of the enzyme (per 1 mg protein) and thus establish whether the synthesis of this protein proceeds more rapidly or more slowly than that of the other proteins.

Calculation per cell (or per quantity of DNA) is justified only in some cases, since if the cell size changes in the course of development, so does the volume of the cytoplasm in which the enzyme is localized. But if we deal exclusively with newly appearing molecules, it will be reasonable to express the changes in enzymatic activity per one cell, one nucleus, or amount of DNA, since this quantity will then indicate the activity of the corresponding genes. This type of calculation, which is rarely applied, is used, for instance, in the works of Rutter *et al.* (1968) on pancreas anlages.

Enzyme Activity during Fertilization

Changes in enzymatic activity during fertilization have been investigated mainly in sea urchin eggs not only because this object is most convenient. These investigations were associated with attempts to find the mechanisms of postfertilization activation of protein synthesis in sea urchins. The studies of Epel's group (Epel, 1967; Epel *et al.*, 1969b) were most comprehensive in this respect (Fig. 70). At a developmental temperature of 17°C, which is relatively low for the sea urchin, the first events are revealed already after 45 sec. They consist in a change of the membrane potential, a breakdown of cortical granules, and an increase in protease activity. After another 3 sec (48 sec postfertilization) the NAD-kinase activity increases (Epel, 1964a). Immediately after, glucose-6-phosphate dehydrogenase is also activated, and there is a sharp elevation of respiration rate which increases by approximately 5 times within 15-25 sec (Epel,

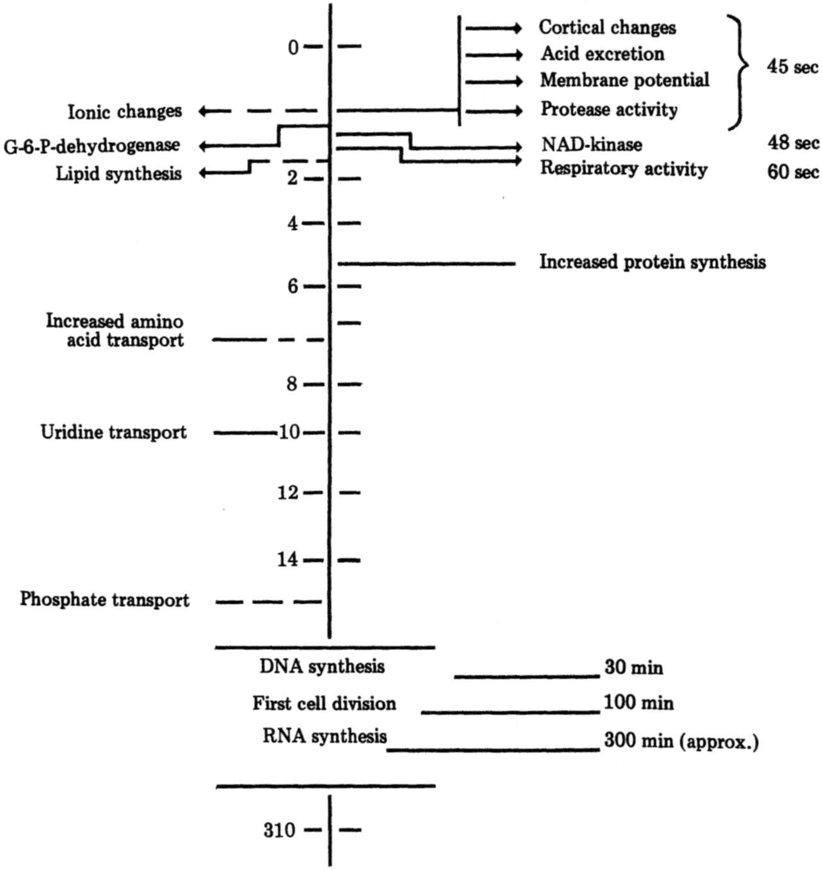

Fig. 70. Diagram of the first postfertilization events occurring in the sea urchin egg (top to bottom: time in min) (Epel *et al.*, 1969b).

1964b). It is to be noted that the concentration of known respiration metabolites — adenosinephosphates (AMP, ADP, and ATP) — remains unaltered. After 2 min, or earlier, comes a considerable increase in activity of β-1,3-glucanase, which was mentioned above. This enzyme is released from the cortical granules into the perivitelline space where it presumably splits the glucose off from the glycoproteins of the fertilization membrane, thus increasing its strength (Epel *et al.*, 1969a). A similar role is evidently played by the partial release of the proteases into the perivitelline space (Vacquier *et al.*, 1973).

Only 7-10 min after fertilization the membrane permeability to amino acids begins to increase (Epel, 1972), with a subsequent sharp increase in protein synthesis rate (Epel, 1967; Epel *et al.*, 1969b). The change in the membrane permeability to definite amino acids is also probably due to the activity of the corresponding membrane transferases and ATPases.

The increase in the activity of glucose-6-phosphate dehydrogenase, and also of 6-phosphogluconate dehydrogenase in sea urchin eggs had been described, even before Epel, by Bäckström (1959, 1963), who found that the activity increase continues, with some fluctuations, up to the hatching of the swimming blastula from the membranes.

Thus, an increase in activity of a number of enzymes begins, or is even completed, before the protein synthesis is stimulated. All this gives sufficient ground to assert that fertilization is attended by activation of the preexisting, rather than by synthesis of new, enzyme molecules. Judging by the phosphorylation of ^{3}H-thymidine in intact sea urchin eggs, the thymidine kinase activity sharply increases after fertilization. But the enzymatic activity in the homogenates of fertilized and unfertilized eggs is found to be equal. The authors conclude that homogenization is accompanied by the same activation as is fertilization. This activation is due to the release of thymidine kinase from large cytoplasmic granules (Nonaka and Terayama, 1975). Similar findings have been demonstrated for glycogen phosphorylase (Bergami *et al.*, 1968) and some hydrolases (Grossmann *et al.*, 1973a, b).

No such mechanism has been established for other proteases and dehydrogenases. The protease activity continues to change periodically during cleavages, while the dehydrogenase activity changes periodically for a more prolonged stretch of time. Therefore their activation mechanism may be essentially different. In the initial period, it may be associated only with the activation or change in localization of the existing protein molecules. It is possible that this mechanism operates only during the first minutes after fertilization and the subsequent growth in activity of glucose-6-phosphate or 6-phosphogluconate dehydrogenases, which continues for many hours, is determined by their actual synthesis. The protease activity fluctuations coincide with the cleavage period and, in Mano's opinion (Mano, 1970, 1971b), this is associated, in one way or another, with the mechanism of periodic fluctuations in protein synthesis, the predominant decay of newly synthesized proteins, etc.

Thus, the onset of activity of a number of enzymes in the first seconds and minutes of fertilization of sea urchin eggs has no bearing on the *de novo* formation of specific proteins, but is due to their activation. Nevertheless, the proteins synthesized at the earliest stages probably include enzymes. No such data have been obtained so far, and only structural proteins — histones and tubulin — have been identified among the first synthesized proteins.

Enzyme Activity in Early Development of the Sea Urchin

In the early studies of sea urchin eggs two groups of enzymes were found: those whose activity remains practically unaltered, and those whose activity begins to increase from the gastrula stage (see Gustafson, 1954).

The former group included the enzymes of the cytoplasm (aldolase, phenylsulfatase, catalase, and others) and the latter encompassed mainly mitochondrial enzymes (succinate and malate dehydrogenases, glutaminases, and others).

The sea urchin egg cytoplasm also contains isozymes of aspartate and alanine aminotransferase whose activity fluctuates widely but does not show any trend toward growth or fall-off (Abramova and Neyfakh, 1971). It has already been mentioned that these activity fluctuations are not associated with the enzyme synthesis.

Glucose-6-phosphate and 6-phosphogluconate dehydrogenases behave in a different way: their activity begins to rise after fertilization, continues to increase until the beginning of gastrulation, and then diminishes gradually but considerably (Bäckström, 1959, 1963). We know neither the significance, nor the mechanism of these changes, whether they are due to degradation and synthesis of the enzyme or to inactivation and reactivation of its molecules.

A consistent decrease in the activity of deoxycytidylate aminohydrolase, one of the enzymes of nucleotide metabolism, has been detected in the course of sea urchin development. This activity decreases 4-fold from the moment of fertilization to pluteus (Scarano and Maggio, 1959). The reduction in the activity of this enzyme is a genetically controlled process, because it can be arrested by actinomycin D (Scarano *et al.*, 1964, 1968).

In sea urchin embryos, appreciable fluctuation is observed in the activity of ornithine and S-adenosylmethionine decarboxylases, which are responsible for the synthesis of polyamines: putrescine, spermidine, and spermine (Manen and Russell, 1973). The activity of these enzymes changes synchronously with the cycle of first cleavages (the maxima of ornithine decarboxylase activity fall, for instance, on the S phase of the first and second divisions). The concentration of putrescine and spermidine, which are products of reactions catalyzed by the enzymes of this chain, changes correspondingly within the first hours of development. The interest in these enzyme systems is due to the fact that although the precise function of these polyamines is not clear, they probably play an important role in the synthesis of RNA and protein, and their concentration, as a rule, correlates with the synthesis rate of these macromolecules. The wide activity fluctuations of these enzymes may be attributed to their extreme instability: the half-life of ornithine decarboxylase in the rat liver does not exceed 10-20 min (Russell and Snyder, 1969).

The understanding of the regulation mechanisms of DNA and RNA synthesis requires information on the activity of the respective polymerases. During the development of sea urchin embryos the total activity of DNA polymerase increases slowly, whereas its intracellular localization changes quite considerably and by the hatching stage it is transferred almost completely from cytoplasm into nuclei (Loeb *et al.*, 1969; Fansler and Loeb, 1969; Loeb and Fansler, 1970). The ability of the nuclei to bind DNA polymerase apparently has some limit determined by the rate of DNA synthesis or simply by the amount of chromatin. The DNA polymerase stored in oogenesis and partly synthesized during development is localized in the cytoplasm and passes into the nuclear fraction as the number of nuclei in the embryo increases sufficiently to bind the whole enzyme.

The deoxyribonuclease function is not quite clear, but it may be assumed that this enzyme makes part of the DNA repair system. In sea urchin eggs DNase is localized predominantly in the nuclei and mitochondria, where its specific activity is 4 to 5 times higher than in the whole embryo (De Petrocellis and Parisi, 1973a). The DNase activity changes simultaneously with the rate of DNA synthesis in the course of development: it is constant up to the blastula stage and lowers at the beginning of gastrulation (De Petrocellis and Parisi, 1972), which may point to a correlation between this enzyme and the DNA synthesis. The change in DNase activity depends only slightly on the nuclear function, because it shows little sensitivity to actinomycin D (De Petrocellis and Parisi, 1973a, b).

The localization of RNA polymerase changes similarly to that of DNA polymerase (see Roeder and Rutter, 1970). The activity of RNA polymerase, as calculated per one cell, decreases in the course of development: at blastula (400 cells) it is no more than 1.5-2 times higher than in an unfertilized egg. At the same time, in the egg it is almost completely localized in the cytoplasm and transferred to the nuclei at blastula (Maroun, 1973). This transfer may be due either to the activation of RNA synthesis or to the fact that in an unfertilized egg the oogenesis-stored enzyme can be bound only partly within the only haploid nucleus.

Little is known about the tissue-specific enzymes in sea urchin embryos. At early pluteus, when the gut undergoes differentiation in the larva, there appear polysaccharide splitting enzymes: β-1,3-glucanohydrolase (or glucanase) splitting off glucose from lamarine (the alga polysaccharide), β-amylase, and maltase, which were completely lacking before (Vacquier, 1971; Vacquier *et al.*, 1971). Their comparison with similar enzymes of adult animals shows that the pluteus gut isozymes differ from the adult ones in a number of characteristics.

The transcriptional control of the enzyme synthesis in sea urchin embryonic development was assessed by sensitivity to actinomycin D or (in a few experiments on interspecies hybrids) by the appearance of the paternal enzyme. It has been shown for the β-1,3-glucanohydrolase just mentioned that actinomycin D restricts the increase in the activity of this enzyme, though with a rather prolonged latent period. Thus, inactivation of nuclei at the beginning of activity rise (19 h of development) arrests increase in glucanase activity at the level of 28-30 h, while similar inactivation at the 28-hour stage does not impede the activity rise to the level of 35 h (Vacquier, 1971). This means that the biochemical differentiation of the pluteus gut depends on the activity of the genes of the embryo itself, while the already transcribed RNA templates can ensure the subsequent synthesis of the enzyme for 7-10 h.

The arylsulfatase activity in the embryos of *Strongylocentrotus purpuratus* begins to rise from the mesenchyme blastula stage and increases by a factor of 30 by the pluteus stage (Fig. 71). In another sea urchin species, *Allocentrotus fragilis*, this rise begins later, and the activity increases only 10-fold. Hybrids between these species show curves of change in arylsulfatase activity differing from those in parental species. Reciprocal combinations also show some differences. Calculations indicate that an enzyme synthesized under the control of the genes of both parents is added to that located in the egg. The amount of the newly synthesized enzyme in the hybrids is consistent with the assumption of independent functioning of the two allelic genes (Fedecka-Bruner *et al.*, 1971).

A new technique for investigating phosphatases in the eggs of sea urchin embryos was used by Westin (1975). He detected various phosphatases in a gel histochemically after two-dimensional immunophoresis. This enabled him to relate the activity of enzymes to their antigenic nature. It was found, for instance, that nucleoside phosphatases localized in the granular fraction and in the cytosol are

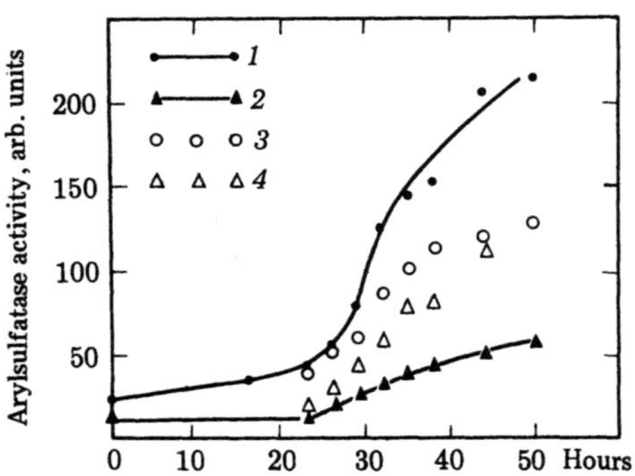

Fig. 71. Activity of arylsulfatase in two sea urchin species and their hybrids. Arylsulfatase activity was measured at different stages of embryonic development of *Strongylocentrotus purpuratus* (*1*), *Allocentrotus fragilis* (*2*), and their hybrids: *S. p.* (♀) × *A. f.* (♂) (*3*) and *A. f.* (♀) × *S. p.* (♂) (*4*). The effect of the paternal genomes begins to show at 22 h of development, and the differences in reciprocal hybrids (maternal effect) disappear by 45 h (Fedecka-Bruner *et al.*, 1971).

not two isozymes since they are represented by the same antigen in both fractions. At the same time, acid phosphatases in the egg are represented by 10 antigens whose number decreases to 5 in the course of development. Westin also discovered that an unfertilized egg contains antigens of alkaline phosphatase, although the activity of this enzyme is manifested only at the subsequent stages (Westin, 1975). The activity of alkaline phosphatase, when determined in the embryonic homogenate, increases gradually by 10 times from the stage of the unfertilized egg to pluteus. However, after pretreatment of the homogenate with butanol the enzymatic activity increases several-fold, and its rise during development is not so large. This indicates that part of the increase in activity in the course of development is due to the transition of the enzyme from the bound inactive state to the free state (Pfohl, 1965). At the same time, the increase in activity of this enzyme is also controlled genetically. Thus, in hybrids between the species *Strongylocentrotus purpuratus* and *Dendraster excutricus*, which differ widely in the kinetics of alkaline phosphatase activity rise, an intermediate position of the activity level is observed at later stages (Flickinger, 1977). It was shown later that two isozymes of alkaline phosphatase — the early and the late — appear successively in the development of the embryo (Pfohl, 1975).

In a similar hybrid combination, though with a different enzyme, the result proved to be different: l-malate dehydrogenase in sea urchin embryos is concentrated by 2/3 in mitochondria and by 1/3 in the hyaloplasm. During development the activity of this enzyme increases uniformly. In the parental species of *Strongylocentrotus* and *Dendraster* this enzyme differs electrophoretically. Crosses between these species live only up to gastrula. Nevertheless, the malate dehydrogenase activity in the hybrids increases, though completely at the expense of the maternal-type enzyme, while the paternal genotype genes do not manifest themselves in any way (Ozaki and Whiteley, 1970). This result may be ascribed to the fact that the increase in malate dehydrogenase activity up to the gastrula stage is not directly controlled genetically and occurs, for instance, on RNA templates synthesized in oogenesis. It may also happen, however, that in a given hybrid combination the paternal malate dehydrogenase allele remains inactive or that paternal chromosomes are eliminated during cleavage.

The hatching enzyme, which appears for a short time before blastula hatching, must also be independent of the genetic apparatus. In two closely related species *Strongylocentrotus purpuratus* and *Strongylocentrotus franciscanus* the hatching enzyme differs in sensitivity to manganese ions. It has turned out that reciprocal hybrids between them inherit the maternal type of the enzyme (Barrett and Angelo, 1969). This means that the synthesis of the hatching enzyme occurs on the maternal mRNA's stored in oogenesis.

Enzyme Activity in Development of Amphibians

In the embryos of the frog *Rana pipiens*, Wallace (1961) investigated the change in activity of 12 enzymes. The activity of most of the enzymes calculated per embryo remains constant until hatching and then begins to increase. This refers to aldolase, lactate dehydrogenase (LDH) and malate dehydrogenase, aspartate and adenine aminotransferases, NADH-reductase, and pyruvate kinase. The increase in their activity coincides with the beginning of active utilization of yolk and of the corresponding increase in the mass of cell proteins of the embryo. Since the yolk in amphibian embryos has intracellular localization, Wallace's data may be interpreted in different ways. Because the activity of enzymes per embryo increases, one can speak definitely of the enzyme synthesis. But their activity may not show any increase if calculated per unit mass of the cytoplasm proper. According to more recent data, the activity of LDH in three investigated amphibian species remains constant only up to the beginning of muscle contractions (stage 26), and then slowly increases (Faulhabe and Lyra, 1975).

In distinction to the activity of other enzymes, that of NAD-cytochrome *c* reductase remains constant until late larva. This indicates that the ratio of the activities of energy metabolism enzymes may change in the course of development. This is also indicated by the increase from the neurula stage (previously detected by Wallace) in the activity of four dehydrogenases: glucose-6-phosphate, 6-phosphogluconate, β-glycerophosphate, and isocytrate dehydrogenases.

Changes in total activity of an enzyme in the whole embryo homogenate may not reflect the actual picture of the syntheses complicated by the replacement of isozymes in differentiating organs. For instance, in the course of frog development, the set of its LDH isozymes is substantially enriched (Claycomb and Villee, 1971; Faulhabe and Lyra, 1975). Only four electrophoretically separable isozymes are found in the frog unfertilized egg and early embryo. During muscle differentiation, a fifth, and then a sixth such isozyme appears, followed by another two at the onset of heartbeat and, finally, by a ninth isozyme on transition to the free feeding tadpole. The formation of such a spectrum must be controlled by at least three genes. Studies of the activity of mitochondrial enzymes localized in the matrix and in the membrane showed that in the course of embryonic and larval development changes in activity occur mainly in the matrix enzymes (for instance, in glutamate dehydrogenase), and are less prominent in the membrane enzymes (glycerophosphate and hydroxybutirate dehydrogenases). These changes proceed differently in the larval liver and in the tail muscles (Kistler and Weber, 1974) (Fig. 72). This means that the development is attended by stage-

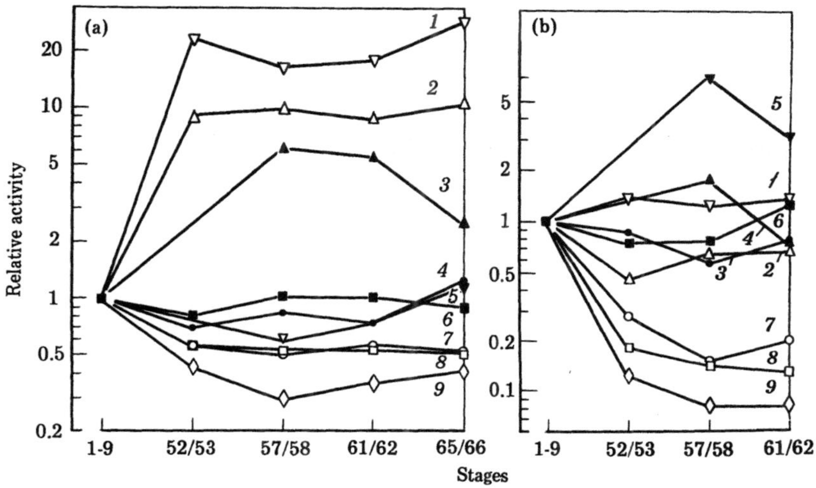

Fig. 72. Changes in activity of mitochondrial enzymes during the larval development of *Xenopus*. The activity of mitochondrial enzymes was determined in the liver (*a*) and in the tail muscles (*b*) and expressed as a ratio to their activity in the early embryos. *1*, Glutamate dehydrogenase; *2*, glycerophosphate dehydrogenase; *3*, hydroxybutyrate dehydrogenase; *4*, cytochrome a_3; *5*, glucosephosphate dehydrogenase; *6*, succinate dehydrogenase; *7*, malate dehydrogenase; *8*, aspartate aminotransferase; *9*, 3-hydroxyacyl-KoA dehydrogenase. The solid symbols denote the membrane enzymes and the open ones, the matrix enzymes (Kistler and Weber, 1974).

and tissue-specific changes in the ratio of enzyme activities in the mitochondria, i.e. by their differentiation.

The activity of polyamine-synthesizing enzymes changes rapidly in the early development of amphibians, but it does so differently than in sea urchin eggs (Russell, 1971). Ornithine decarboxylase (4.1.1.17) transforms ornithine to putrescine. Its activity increases rapidly and uniformly by 10 times from the 2-blastomere stage to early blastula and by another 10 times from early blastula to the tadpole. The next enzyme in this metabolic pathway, S-adenosylmethionine decarboxylase, modifies putrescine by S-adenosylmethionine to form spermidine. The activity of this enzyme is low and does not increase before the gastrula stage, but then grows 40-fold. The concentration of the products of these reactions, putrescine and spermidine, also increases in the course of amphibian development, although not so sharply as does the enzyme activity. Their function can be deduced indirectly by comparing the activity of enzymes and the concentration of products in anucleolate mutants (Russell, 1971). In distinction to dehydrogenases, there is no activity of S-adenosylmethionine decarboxylase in anucleolate mutants at stage 24 (before this stage lethal anucleolate mutants cannot be distinguished from heterozygous embryos with normal phenotype). The ornithine decarboxylase activity in mutants is also strongly repressed, and the putrescine and spermidine concentration is much below the control. The author assumes that these polyamines are related to the synthesis of ribosomal RNA. This is also indicated by the fact that newly synthesized ^{3}H-putrescine first concentrates in the nucleolus and then, like RNA, moves into the cytoplasm. The activity of xanthine dehydrogenase and uricase increases sharply, by a factor of 1000 and more, at stage 18-20, which apparently coincides with liver and kidney differentiation (Barrett, 1964) (Fig. 73). The activity of RNA polymerase begins to increase already from the cleavage stage, increasing from the egg to the morula stage by a factor of 2, and altogether by 3 to 4 times toward the hatching stage (Lievens and Brachet, 1970). The activity of RNA polymerase increases monotonically and does not correlate in any way with the change in RNA synthesis rate. At the blastula-gastrula stage, when the RNA synthesis increases by tens of times, the RNA polymerase activity increases by 25% at best. On the other hand, it would perhaps be more correct to calculate the activity of this enzyme with respect to nuclei in which it functions rather than per embryo. A simple calculation shows that the activity of RNA polymerase per nucleus decreases continuously as the number of nuclei increases. This decrease is especially prominent at the early stages (by about 500-1000 times up to the blastula stage), but continues later as well. While the number of cells increases 100- to 200-fold (from blastula to hatching), the enzyme activity does not increase more than 2-fold per embryo, i.e. per nucleus it decreases by 50 to 100 times.

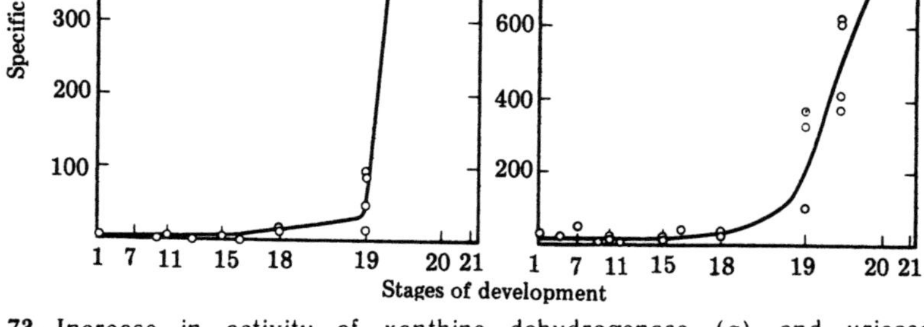

Fig. 73. Increase in activity of xanthine dehydrogenase (*a*) and uricase (*b*) during frog development. The sharp increase in enzyme activity coincides in time with the beginning of cardiac activity (Barrett, 1964).

RNA polymerase, like DNA polymerase in sea urchin embryos, is evidently transferred from cytoplasm to nuclei in ever-growing amounts as the number of cells increases, and it is possible that in actuality the amount of the enzyme in the nuclei is maintained at a constant level or changes depending on the transcription rate. Amphibians and fishes are more convenient for studying this aspect, since in them the RNA synthesis is sharply activated at a comparatively late stage.

In some instances the enzyme synthesis clearly precedes the stage at which its functions can be performed. It has already been mentioned that the collagen synthesis begins only from the gastrula stage, whereas proline hydroxylase, the enzyme necessary for protocollagen modification, has already been stored in the egg (Green *et al.*, 1968).

The tyrosinase activity is revealed in embryos of *Rana pipiens* shortly before hatching. The inactive zymogen of tyrosinase, however, can be detected already from early gastrula by radioimmunoprecipitation. On mild trypsin treatment of the homogenate, the tyrosinase activity can be activated (Benson and Triplett, 1974) (Fig. 74). Here, too, the enzyme is evidently accumulated beforehand in the inactive form.

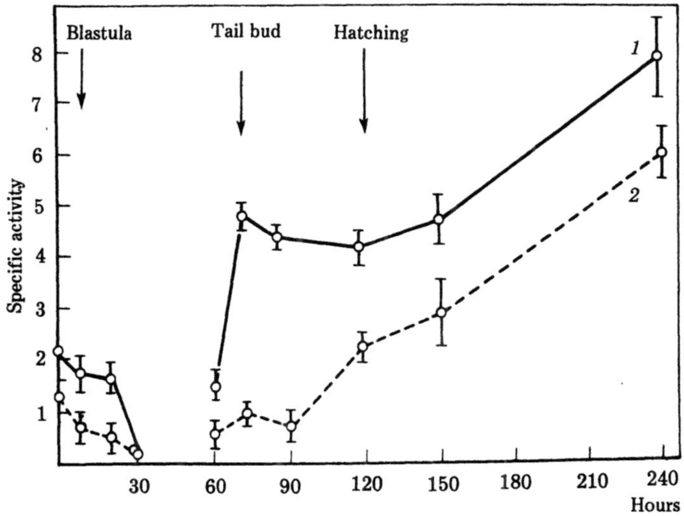

Fig. 74. Tyrosinase activity during frog development. The activity of tyrosinase (DOPA oxidase) at different stages of development was determined in the embryo homogenate pretreated with trypsin (*1*) and without treatment (*2*). The inactive form of the enzyme, which is activated by partial proteolysis using trypsin, is synthesized and accumulates prior to the beginning of tyrosinase function (Benson and Triplett, 1974).

Considerable changes of enzyme activity are induced by the thyroid hormone in tadpoles during their metamorphosis. These organ-specific changes have been studied most thoroughly for the liver and for the tail muscles (Cohen, 1970; Tata, 1970, 1971). Their detailed consideration is beyond the scope of this book.

Enzyme Activity in Development of Fishes and Other Animals

In the loach oocytes and early embryos, the enzymes of glycolysis and of the pentosophosphate pathway of carbohydrate oxidation have been investigated in the most detailed and, what is particularly important, comprehensive manner (Milman and Yurovitsky, 1973). The activity of most of these enzymes (phosphorylase, phosphoglucomutase, glucosephosphate isomerase, phosphofructokinase, aldolase, triosophosphate dehydrogenase and isomerase, phosphoglycerate kinase, enolase, pyruvate kinase, lactate dehydrogenase, 6-phosphogluconate dehydrogenase, phosphoribose isomerase, transketolase, and transaldolase) remains entirely unaltered for the first 20 h of development (at 21°C), i.e. up to the beginning of the axial organs formation. It is to be emphasized that the rate of anaerobic glycolysis and respiration increases about 3-fold at the same stages (aerobic glycolysis is practically nil in the embryos).

None of the enzymes associated with glycogen assimilation shows any activity increase at these stages (up to 20 h). On the other hand, the activity of the others falls off appreciably within this time. This refers to glycogen synthetase (2.3-fold decrease), phosphoenolpyruvate carboxykinase (about 2-fold decrease), fructosodiphosphatase (almost 40% decrease), and glucose-6-phosphate dehydrogenase (35% decrease). The activity of cytochrome oxidase, as determined in whole mitochondria after Smith, increases in the course of development (Neyfakh and Dontsova, 1962). But, as has already been noted, no such increase is observed in mitochondria destroyed by deoxycholate (Abramova and Vasilyeva, 1973). It is presumably due to the increase in permeability of the mitochondria to the reduced cytochrome *c* substrate in the course of development. The protease activity increases during development of fishes, but this occurs mainly in the yolk, where no protein synthesis takes place (Neyfakh and Davydov, 1964; Konovalov and Mestechkina, 1975).

Changes in the glycolytic and respiratory enzymes activity and in the number of mitochondria have also been investigated in the process of loach oogenesis. This became possible after Ozernyuk (1972) proposed a method for separating oocytes according to sizes by gradual dissociation of the ovary tissue with trypsin. In this way one can obtain several groups of oocytes differing in average sizes.

Oocytes with an average diameter of 200, 400, and 800 μm were compared; their volume ratio was 1 : 8 : 64, and their protein content was 1, 4, and 180 mg, respectively. (The differences between the volume ratio and the protein amount are probably due to the different concentrations of lipids and other components.)

To compare the rate of increase in activity of glycolysis enzymes in the course of oogenesis, the activity of small oocytes (200 μm) was taken as unity. Then the activity of the medium-sized (400 μm) and large-sized (800 μm) oocytes increased as follows. The activity of triosophosphate dehydrogenase, isomerase, and aldolase increased by 4-6 times in medium-sized oocytes and 70-100 times in large ones, i.e. almost proportionally to the protein content and the oocyte volume. At the same stages of oogenesis the activity of phosphoglycerate kinase increased 4- and 44-fold, respectively, i.e. already slightly more slowly than did the oocyte size. The activity of the other enzymes (enolase, fructosodiphosphatase, phosphoenolpyruvate carboxykinase, glucose-6-phosphate dehydrogenase, and 6-phosphogluconate dehydrogenase, transketolase, and transaldolase) increased by 3-6 and 9-35 times, respectively, i.e. much more slowly than did the oocyte size (Milman and Yurovitsky, 1973). Thus, not only is oogenesis attended by an increase in the quantity of enzymes, but also the ratio of their activities changes considerably. The ratios typical of early oocytes were replaced by those characteristic of the egg and embryo. For instance, the activity ratio of aldolase and glucose-6-phosphate dehydrogenase increased 12-fold during the oocyte growth.

Part of the enzymes associated with glycolysis are localized in the yolk portion of the egg. Most of the enzymes of the glycolysis and of hexosomonophosphate pathway are characterized by a ratio of 2 : 1 to 4 : 1 between blastoderm and yolk, while the enzyme concentration in the cells is about 20 to 40 times higher than in the yolk portion of the egg. But most of the phosphorylase and fructosodiphosphatase activity is concentrated in the yolk (90 and 70%, respectively), and their concentrations in the cells and the yolk are approximately equal. Besides, the yolk also contains an appreciable amount of latent phosphorylase.

Nonuniform distribution of enzymes between blastoderm and yolk may be regarded as a manifestation of ooplasmic segregation. For instance, preferential localization (up to 80%) of lactate dehydrogenase (LDH) in the blastoderm is achieved within the first 5 to 6 h of development by its migration from yolk into blastoderm (Klyachko and Neyfakh, 1980a, b). In a similar way, other glycolytic enzymes (e. g. aldolase and glucose-6-phosphate dehydrogenase) are accumulated in the blastoderm. Experiments with injection of enzymes into an egg and their subsequent distribution between blastoderm and yolk show that migration of LDH in the blastoderm cannot be the consequence of simple diffusion, because the yolk volume is 5-7 times larger than the blastoderm volume. Finally, it was shown directly that LDH forms an enzyme-containing complex of low specificity and with a low binding constant ($10^7 M^{-1}$) in the blastoderm (Klyachko and Neyfakh, 1980a, b).

In the course of embryonic development of fishes, the set of esterases becomes much more complicated (Holmes and Whitt, 1970; Ivanenkov, 1976; Ivanenkov and Korobtsov, 1976; Botvinnik and Neyfakh, 1977). New esterase species appear at the gastrula stage, at the beginning of organogenesis, and during the hatching of larvae.

The activity and localization of RNA polymerase in early loach embryos have also been investigated. Kafiani and coworkers (Kafiani *et al.*, 1973; Ekizashvili *et al.*, 1974) showed that the activity of RNA polymerase II (activated by Mn^{2+} and repressed by α-amanitine) in a whole embryo increases only slightly in the course of development, including the stages when the RNA synthesis is activated (8 h of development at 21.5°C). At the same time, RNA polymerase II, which is initially localized in the cytoplasm, gradually concentrates in the nuclei during early development. This transition also occurs at the subsequent stages.

At the same early stages, RNA polymerase I (activated by Mg^{2+}) is also transferred from the cytoplasm to the nuclei, although a perceptible rRNA synthesis begins later. It is assumed that the migration of RNA polymerases from cytoplasm to nuclei is not simply the result of an increase in the number of nuclei, but is one of the conditions ensuring an increase in transcription rate. However, the fact that RNA polymerase I also migrates into the nucleus, although it does not participate in transcription at the early stages, indicates the insufficiency of this condition and the need for other regulatory mechanisms.

A large chart showing the changes of enzyme activity in the embryo of the fish chubsucker (*Erimyzon sucetta*) was recently presented at Whitt's laboratory (Shaklee *et al.*, 1974) (Fig. 75).

The data on the enzyme activity of other animals are few and far between. Ascidian embryos have been investigated somewhat better than the others (Durante, 1961). In another object with mosaic development, the mollusc *Ilyanassa*, the development and the spectrum of proteins synthesized change substantially if the polar lobe is removed at an early stage. This is also reflected in the kinetics of *de novo* enzymes formation (Freeman, 1971). The electrophoretic pattern of alkaline phosphatase and esterase isozymes changes considerably in the course of development of this mollusc. Thus, in the first days of development alkaline phosphatase is represented by a single fraction, whose amount then increases, and two new fractions are added successively to it. In embryos deprived of the polar lobe, no new fractions appear, and the amount of the first remains small. In the period of early development esterases are also represented by a single isozyme, to which four more are successively

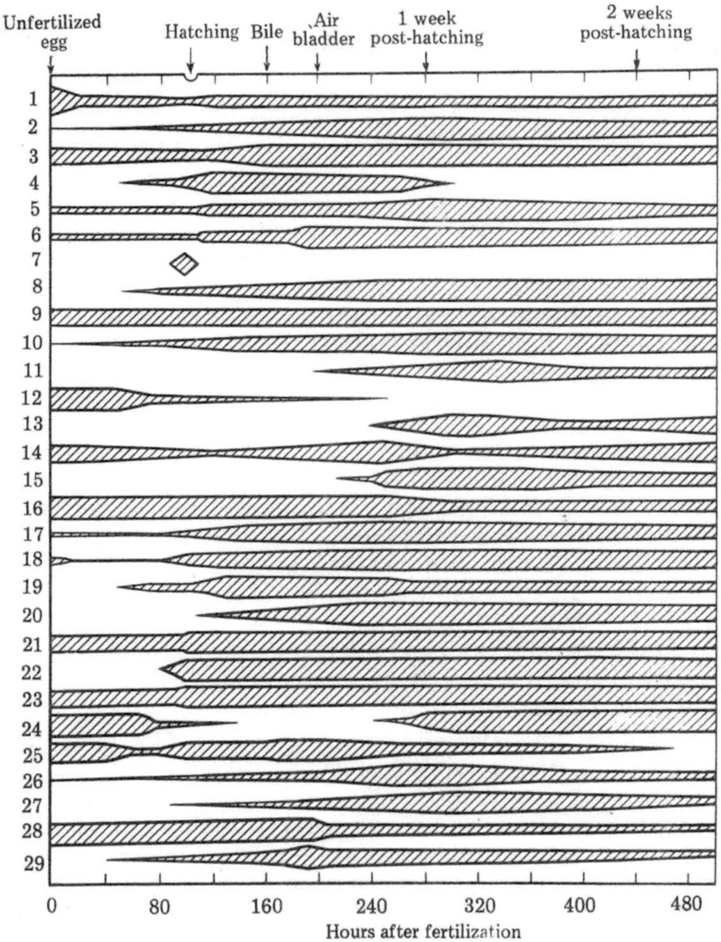

Fig. 75. Changes in enzyme activity during the embryonic and larval development of the lake chubsucker (*Erimyzon sucetta*). Changes in the isozymic spectra of a number of enzymes were investigated electrophoretically and estimated quantitatively in embryos and early larvae of the lake chubsucker. The relative changes in activity during development are shown by the strip width. Enzymes and isozymes (top to bottom): *1*, phosphogluconate dehydrogenase; *2*, *3*, phosphoglucomutases 3 and 1; *4*, hexosodiphosphatase; *5*, *6*, *7*, malate dehydrogenases 3, 2, and 1; *8*, *9*, aspartate aminotransferases 4 and 2; *10*, alkaline phosphatase; *11*, glutamate dehydrogenase; *12*, *13*, *14*, esterases 9, 6-7, 8; *15*, *16*, *17*, *18*, *19*, creatine kinases 2b-3b, 2a-3a, 9, 1, and 6; *20*, mannosophosphate isomerase; *21*, *22*, *23*, glucosephosphate isomerases C, B, and A; *24*, amylase; *25*, glucose-6-phosphate dehydrogenase; *26*, isocitrate isomerase; *27*, *28*, and *29*, lactate dehydrogenases C, B, and A (Shaklee *et al.*, 1974).

added later, and at the late stages the amount of all the five fractions increases. Without the polar lobe this schedule breaks down: one of the new fractions does not appear, while the amount of the other remains unaltered. Thus, a disturbance of morphogenesis caused by the removal of a portion of the cytoplasm affects the normal sequence of enzyme activation.

In the period of embryonic development of *Drosophila* the activity of many enzymes remains unchanged (Wright and Shaw, 1970). Mention should be made of two series of investigations in which the enzyme activity studies of *Drosophila* embryos were accompanied by the genetic analysis of their regulation. It was found that the activity of xanthine dehydrogenase sharply increases in the course of early embryogenesis (3-6 h) of the *Drosophila* wild line. This increase is strongly repressed in heterozygotes for the *rosy* gene ($ry^2/+$) and is absent altogether if the maternal organism was a mutant for the *mal* gene. The *rosy* gene is the structural gene of xanthine dehydrogenase, and its manifestation already 3 h after fertilization, before gastrulation, is probably the earliest gene manifestation in *Drosophila* development. The maternal effect of the *mal* gene is unclear and possibly consists in the formation of a cofactor or specifically affects the synthesis of xanthine dehydrogenase (Sayles *et al.*, 1973).

In the study of the esterase spectrum in Korochkin's laboratory it was shown that one of them, esterase-6, is synthesized in the spermaduct of *Drosophila* males under the control of a number of genes, including both the structural genes of the enzyme and the genes controlling its translation (see Kuzin *et al.*, 1975; Korochkin, 1977).

When investigating enzymes in early mammalian embryogenesis, the main attention, after elaborating the relevant micromethods, was given to the early manifestations of genetic control of the synthesis of various enzymes. Such an early control was detected on the 3rd day of development for glucosephosphate isomerase (Brinster, 1973), on the 4th day for hypoxanthine-guanine phosphoribosyltransferase (Epstein, 1972), and on the 5th day for glucosephosphate isomerase I (Chapman *et al.*, 1971) (Fig. 76).

Variation in activity of glycolytic enzymes and glycolysis-related metabolytes was investigated in early mouse embryos (Barbehenn *et al.*, 1978; Hsieh *et al.*, 1979). These authors showed that up to the morula-blastocyst stage the embryo is incapable of taking up glycogen because of the very low activity of glycogen phosphorylase (the activity of phosphoglucomutase, the second enzyme of glycogenolysis, is high enough). This activity increases at morula, and the embryo acquires the ability for taking up glycogen (Hsieh *et al.*, 1979).

In connection with the early activation of RNA synthesis in mammals, the data on the increase of RNA polymerase activity during cleavage

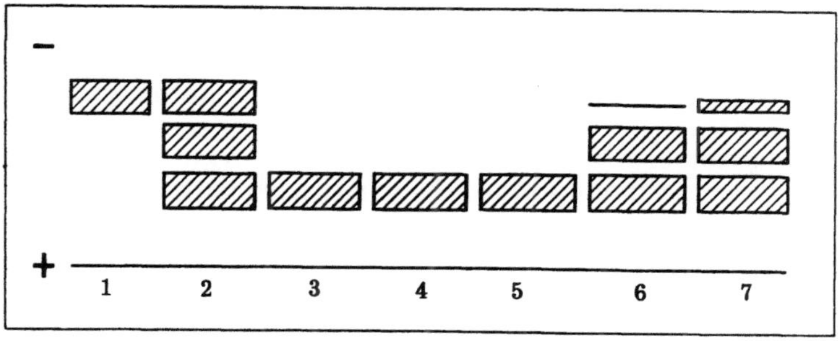

Fig. 76. Expression of the gene which controls glucosephosphate isomerase synthesis in the early embryogenesis of the mouse. Glucosephosphate isomerase from the heart of adult mice of lines GPI-1A (♀) (*3*) and GPI-1B (♂) (*1*), the hybrids GPI-1AB (*2*), and also from 3- (*4*), 4- (*5*), 5- (*6*), and 6- (*7*) day hybrid embryos were analyzed electrophoretically. The first manifestation of the paternal gene of glucosephosphate isomerase was revealed at the 5th day of development (Chapman *et al.*, 1971).

of mouse eggs (Siracusa, 1973) are of interest. The activity increases several-fold even during the first cleavage, and this growth continues at the subsequent stages, increasing 30- to 60-fold by the blastocyst stage. The recorded increase is much higher than the one observed in amphibians and the sea urchin. The reason is probably that the enzyme stockpile in the mammalian egg cannot be too large. The RNA polymerase activity calculated per nucleus decreases, and, like in other objects, there is no visible correlation between the RNA polymerase activity and the RNA synthesis. When the latter is relatively low (up to the 8-cell stage), the polymerase activity increases by 5-20 times, and when it speeds up considerably, the enzyme activity increases by 2-5 times at the most (Siracusa, 1973).

Enzyme Activity in Differentiating Tissues

When studying changes in the enzyme composition of differentiating tissues and organs, one deals with a relatively homogeneous material and can often relate the appearance of an enzyme to the functional specialization of the cells, i.e. to their terminal differentiation.

Many enzymes are specific to specialized organs, and the beginning of their synthesis in the cell is already a manifestation of differentiation no less important than the morphological features. Whenever one deals with the "housekeeping" enzymes, such as aldolase or lactate dehydrogenase, the differentiation of many organs is characterized by the appearance of specific isozymes and by their characteristic ratios, typical only of a given organ. In this section we confine ourselves to enzymes, and not to specialized proteins in general. This division is sometimes quite arbitrary. For instance, speaking of muscles, we consider not myosin itself but its ATPase activity. As mentioned above, determination of the enzymatic activity by no means necessarily indicates the amount of protein. But precisely for tissue-specific proteins the correspondence between the enzyme activity and the amount of the corresponding protein is more evident, and in some cases activity measurements enable one to calculate directly the amount of enzyme molecules per cell (Rutter *et al.*, 1968).

The formation of a specific set of enzymes in specialized cells raises a number of questions. The most important one concerns probably the place of enzyme synthesis in the series of events constituting organogenesis: determination of the organ, its morphogenesis, cytodifferentiation, and, finally, the beginning of its function. Below, when discussing the development of the better-investigated pancreas, we shall see that the first manifestation of a slight enzyme activity either coincides with or immediately follows the earliest stages of organogenesis: the determination and early morphogenesis. Accumulation of organ-specific enzymes occurs at the next stage of differentiation (Rutter *et al.*, 1968).

We do not know the role of all the changes in enzyme activity even in specialized tissues. But the "housekeeping" enzymes, which determine the nonspecific steps of metabolism, also have peculiarities of their own in different organs, and the kinetics of changes in their activity in the course of development is specific to each of these organs. This is illustrated, for instance, by Wang (1968), who compared the changes in activity of eight enzymes of the carbohydrate and amino acid metabolism in various organs of the chick embryo. According to him, the activity of lactate dehydrogenase (per mg protein) in the liver decreases in the course of development but remains practically unaltered in the heart, kidneys, and brain. Glucose-6-phosphatase is characterized by a sharp (7- to 8-fold) increase of activity in the kidneys, with only slight changes in the other three organs. The activity of many dehydrogenases and transaminases increases in the liver, heart, and brain tissues, but changes less pronouncedly in the liver. On the other hand, the activity of glucose-6-phosphate dehydrogenase falls off slowly in all the four organs. In the final analysis, the liver is the most "quiescent" organ, while the kidneys are the most "active" ones, although their function does not begin until hatching.

Morphogenesis of the nervous system begins at an early stage in the development of vertebrates; the establishment of the nervous system functions and the final differentiation continue for a long time in postembryonic development. The specific enzymes of the nervous system and the neuromuscular apparatus include acetylcholinesterase which appears in the tissues simultaneously with the first muscle movements (see Herrmann and Tootle, 1964).

The development of nerve cells is also attended by considerable changes in the activity of glutamate aminotransferase and glutamine synthetase, the enzymes associated with the energy and amino acid metabolism. During the embryonic development of the nervous tissue the activity of glucose-6-phosphate dehydrogenase decreases, but this decrease is particularly pronounced (6- to 7-fold within three days) at the early stages of spinal cord differentiation (Burt, 1965). A characteristic property of a highly differentiated nervous tissue — the embryonic retina — is its high ability for induction of glutamate synthetase synthesis (Moscona *et al.*, 1968). Cortisol is one of the possible inducers. In this particular case, as well as for the steroid hormones in general, its direct effect at the transcription level has been demonstrated. Using this model, it was possible to trace the activation of synthesis of RNA, its transfer into the cytoplasm and entry into precisely those polysomes (12-14 ribosomes) where glutamate synthetase is synthesized (Sarkar and Moscona, 1971b). An active incorporation of amino acids occurs in the same polysomal fraction.

The differentiation of the muscular tissue is attended by a number of changes in enzyme activities and by the appearance of new muscle-specific enzymes. Many data on these changes have been obtained in studying the differentiation of the muscular tissue cultures. At the early stages of myogenesis the myoblast proliferation and their fusion into myotubes are accompanied by the formation of myosin, which possesses myosin-ATPase activity, and by a synthesis of enzymes characteristic of the muscles: myokinase and creatine kinase (Schubert *et al.*, 1973). The isozymic composition of creatine kinase changes in the course of myogenesis. The molecule of this enzyme consists of two subunits. In the early chick embryogenesis (during the first 6 days of development) all the creatine kinases are represented exclusively by isozyme I (BB) typical of the nervous tissue. In the course of muscle differentiation the isozyme I is gradually (from day 7 to 15 of incubation) replaced by the muscle isozyme III (MM), but during the transitional period the mixed form (BM) is also formed (Eppenberger *et al.*, 1964). Since protein subunits do not migrate between cells, this means that the repression of the activity of the creatine kinase B-gene and switching-on of the M-gene occur in the same cells in the course of their differentiation. In the differentiating myoblast culture the process develops more rapidly: on the 2nd day of cultivation the myoblasts fuse together, and on the 4th the activity of creatine kinase increases sharply and isozyme I (BB) is replaced by isozyme III (MM) (Morris *et al.*, 1972).

Muscle differentiation is also attended by the formation of the muscle form of LDH at the expense of the M-subunit. In the *musculus pectorialis* of the chick embryo the activity of the M-subunit begins to rise from the 6th day and increases 1000-fold during 40 days of embryonic and postembryonic development. A similar increase of the M-subunit concentration occurs in the culture of differentiating myoblasts (Cahn, 1964). Along with the increase in activity of the muscle enzymes and the myosin synthesis, receptors of acetylcholine form in the myoblast culture (Merlie *et al.*, 1975).

Muscle differentiation is characterized by changes in activity of the enzymes not associated directly with the muscle function, but apparently important to the differentiation process itself. Thus, the fusion of myoblasts into myotubes is followed by a substantial reduction in activity of adenylate cyclase, an enzyme localized on the cellular membrane and responsible for the synthesis of cyclic AMP when induced by extracellular factors (hormones). If the myoblast fusion is inhibited by a slight temperature decrease, the adenylate cyclase activity does not decrease (Wahrmann *et al.*, 1973). These results are one of the few examples when enzyme activity can be related to morphogenesis. This statement would be more convincing, however, if we could exclude the possibility that the activity of the adenylate cyclase localized on the membrane may fall off because of the reduction in the surface area of the myoblasts fused together.

At the next stage of muscle differentiation, on the 11th day of chick development, one detects a rapid acetylation of histones and a high activity of histone deacetylase. By the 16th day the activity of this enzyme lowers almost to zero (Boffa *et al.*, 1971). The role of histone deacetylase is not clear, nor is the whole meaning of histone acetylation and deacetylation. It is obvious, however, that these processes are necessarily associated with the participation of histones in the regulation of the nuclear function in the differentiating cells.

At still later stages of muscle development in the prenatal and postembryonic periods, one observes a further rise in activity of the specific muscle enzymes: 5'-adenylate deaminase, creatine kinase, adenylate kinase, and muscle aldolase (Young, 1970). In animals born fully prepared for independent existence the activity of these enzymes increases many times over a few days before hatching (chick) or birth (guinea pig). At the same time, a newborn rat is absolutely immature, and the activity of its muscle enzymes increases gradually during the first 10 days after birth.

This relationship of the activity of muscle enzymes with their function is also manifested when we compare various organs. In the rabbit's heart muscle, which functions actively already in the embryo, an increase in activity of the muscle enzymes occurs a week before birth, whereas in the diaphragm the same process begins only after birth in accordance with its function (Young, 1970).

The diversity of proteins synthesized during myogenesis is reflected in the number of active single-copy DNA sequences (genes). Experiments on RNA-DNA hybridization show that in the primordial muscles of chick embryos the diversity is particularly large (up to 12-14% of the total single-copy DNA) at the beginning of fusion of the myoblasts into myotubes, then drops off to 7% and remains at this level in the adult muscles (Ordahl and Caplan, 1976).

In a myogenic cell culture the synthesis of muscle enzymes and myosin is stimulated by using a medium which does not promote proliferation (lower embryonic extract content, addition of insulin and somatotropin, etc.) and activates the myoblast fusion. But if the myoblast fusion and the formation of myotubes are arrested artificially, say, by removing calcium from the medium, the same accumulation of muscle enzymes as in normal differentiation begins (Delain and Wahrmann, 1975; Turner *et al.*, 1976).

Liver differentiation is characterized by many specific enzymes; among them are enzymes associated with the ornithine cycle, aldolase B, a specific isozyme composition of lactate dehydrogenase, and enzymes associated with the metabolism of aromatic amino acids — tryptophan and tyrosine. The appearance of, or a sharp increase in, the activity of these enzymes precedes the beginning of, or a change in, liver function, which occurs at birth and at hatching in mammals and birds, and during the tadpole-to-frog metamorphosis in amphibians.

Thus, the activity of tyrosine transaminase in the rat liver is detected immediately after birth, then it increases rapidly, and almost reaches a maximum within 12 h. This process is due to gene activation, since it can be repressed by actinomycin D, but also depends on the adrenal function, probably via steroid synthesis (see Herrmann and Tootle, 1964).

Three enzymes, which effect the successive processing of tyrosine — tyrosine-α-ketoglutarate aminotransferase, hydroxyphenylpyruvate hydrolase, and homogentisate oxygenase — also appear for the first time after birth. Then the activity of all the three increases strictly simultaneously during the first 24 h and, more slowly, on subsequent days (Goswami *et al.*, 1973).

The results of an investigation where one of the liver enzymes, ethanolamine kinase, was studied in the rat's embryonic and postembryonic development, are highly instructive (Weinhold and Rethy, 1972). The activity of this enzyme begins to rise sharply 5 days before birth, and by the moment of birth exceeds the initial level many times over. During the first 2 postnatal days its activity drops to the initial level and then increases gradually again for 10 days. A number of assumptions could be made as to the cause for these changes. However, when determining the enzyme activity in an extract freed from the low-molecular-weight components it was found that there was no sharp decrease in ethanolamine kinase activity after birth but it increased monotonically throughout the period under investigation. The low-molecular-weight fraction, however, contained a strong inhibitor of this enzyme, choline, which appears in the liver after birth. Thus the cause for the temporary drop in the enzyme activity was found, and it was no enzyme degradation with subsequent resynthesis. It remains unclear, however, whether the choline and the enzyme are located in the same compartments of the cell or whether their interaction is an artefact caused by homogenization. In the latter case choline does not act as regulator of metabolism, and no actual decrease in the enzyme activity in the cells takes place.

In addition to the appearance of new enzymes, liver differentiation is characterized by a change in isozyme composition. For instance, in early rat embryos the activity of the specific liver aldolase B increases to replace the activity of the muscle aldolase A (Weber and Rutter, 1964). This transition is accompanied by a temporary appearance of hybrid forms of aldolase consisting of A- and B-subunits. In chick embryos, the initial embryonic form is aldolase C characteristic of the brain tissue: from day 5 to day 8 of incubation aldolase C is replaced by aldolase B, and the hybrid C/B molecules form in the liver in this period (Lebherz, 1972). This is one more evidence that isozymes are not replaced through cell replacement but by their true differentiation, i.e. by a gradual repression of the transcription activity of one gene and switching-on of another within each cell.

The overall activity of glucose-6-phosphatase in the chick liver gradually decreases in the course of development (Wang, 1968). But intracellular replacement of the mitochondrial enzyme by its isozyme localized in the microsomal fraction occurs at the same stages. While the maximum of glucose-6-phosphatase activity in the mitochondria falls on the 11th day of incubation (the microsomal activity at the time constitutes no more than 15%), in the microsomes it reaches a maximum prior to hatching (day 20) (when in the mitochondria it is 15-20%) (Simbonis and MacBride, 1966). A number of liver enzymes [tyrosine aminotransferase, tryptophan pyrrolase, UDP-glucuronyltransferase (UGT), etc.] are not present, or their activity is very low, before the birth of the fetus in mammals or before the hatching of a chick. The mechanism of this inhibition of expression has been investigated for UGT (Skea and Nemeth, 1969). It was found that the repression factor is the embryonic medium: on explantation of embryonic liver cells the UGT activity in them immediately begins to rise rapidly. However, when an embryonic liver is transplanted to chorioallantois, i.e. to the same embryonic medium, no enzyme appears in the liver cells.

The complexity of calculating the enzyme activity and its correlation with the enzyme synthesis are evident from the changes in arginase activity in the liver and kidneys of a mouse embryo (Bradley, 1973) (Fig. 77). The liver grows from the time of the anlage formation (day 3) and up to birth. But the total arginase activity per whole liver, which is very low at these stages, does not rise until day 11 or 12.

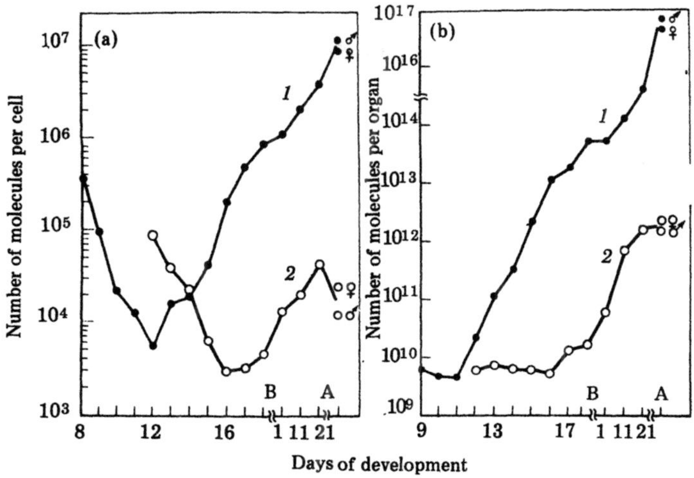

Fig. 77. Activity of arginase in mouse development. The arginase activity was determined in the liver (*1*) and kidneys (*2*) during the embryonic and postembryonic mouse development and calculated as the number of the enzyme molecules per cell (*a*) and per organ (*b*). *B* represents birth; *A* is the adult. At the early stages the organ grows rapidly and the per-cell activity decreases; at the subsequent stages the enzyme synthesis rate greatly exceeds the proliferation rate (Bradley, 1973).

As a result, within 3-4 days the arginase activity decreases 100-fold per 1 μg DNA and 50-fold per cell (the cell size slightly diminishes). In the next period, an intensive arginase synthesis begins, and its amount increases 1000-fold both per cell and per DNA unit. But since an intensive growth of the liver (and of the number of its cells) continues simultaneously, the arginase activity per whole liver increases almost million-fold from the prenatal day 11. The arginase activity in the kidneys changes similarly, but at later dates: up to day 16 of development it remains constant as calculated per whole kidney, but decreases by tens of times as calculated per one DNA unit or one cell. In the next period, it increases 10-fold per one DNA unit or one cell, and 500-fold per whole kidney.

These data show to what extent conclusions depend on the right statement of the problem. If one is interested in the arginase activity, it should be expressed in specific activity units, i.e. per 1 mg of protein or per cell. If, however, the object of investigation is the synthesis of enzyme protein, it would be more correct to calculate it per whole organ. Finally, if an attempt is made to evaluate the intensity of the gene function by the enzyme activity, the latter may be expressed as calculated per one DNA unit.

Rutter with coworkers (1968, 1973) and Wessels and Cohen (1967) have carried out most comprehensive studies of the enzymatic aspect of development of the pancreas in combination with the embryological approach. This object turned out to be extremely suitable because of the synthesis of a broad range of strictly tissue-specific and readily determinable exocrine enzymes, and also of insulin and glucagon, important protein hormones which enter the blood. Some morphogenetic processes in the pancreas have also proved accessible for analysis.

The development of this organ in mammalian embryos begins at the 20- to 25-somite stage (day 9 in the mouse and day 10.5 in the rat) in the form of an outgrowth of the primordial gut, which gradually forms a system of lobules and acinous follicles characteristic of exocrine glands. They are formed by adenose epithelium and are connected by ducts which open into the gut. The epithelium is surrounded by mesenchymal cells which subsequently form the connective stroma of the gland. Type-B cell islets, which synthesize the insulin, form rather early among the exocrine acini, and after birth one can detect there glucagon-synthesizing type-A cells. The determination of the pancreas occurs apparently at the 10-somite stage (day 8 in the mouse), when the gut and mesenchyme explantation already leads to the normal development of the gland. The differentiation of the adenose epithelium, however, requires a continuous control of mesenchymal cells or the presence of some factors produced in these cells. The nature of these factors has not yet been established, but they are associated with the pellet of the homogenate of the mesenchymal cells: they are thermolabile and sensitive to the trypsin action (Rutter *et al.*, 1968). The mesenchymal factor determines not only the morphological differentiation and synthesis of enzymes, but also the proliferative activity (synthesis of DNA, mitoses).

By the 14th day of development the morphological differentiation of the pancreas as an organ is completed, and the mitotic activity in the cells greatly diminishes. At this time cytodifferentiation begins, that is the formation of endoplasmic reticulum, zymogen granules, etc. The specific enzymes of the pancreas are revealed at very early stages of its development, already in 12-day rat embryos (Fig. 78). Their activity is not high, it varies from 10^3 (lipase) to 10^5 (amylase) expressed in terms of the number of enzyme molecules per cell. Reliably detectable accumulation of insulin in the primordial gut is revealed already on the 11th day of rat development (one molecule per cell), and on the 12th day the gland anlage already contains over 10^4 insulin molecules per cell.

This primary level of activity, which Rutter *et al.* named protodifferentiation, persists for several days, after which (on days 14-16 of development) there is a sharp increase in the concentration of the enzymes and hormones. The synthesis of insulin, amylase, and carboxylase A is the first to accelerate (day 14), followed by lipase A, chymotrypsin, and ribonuclease (day 15), and then, on day

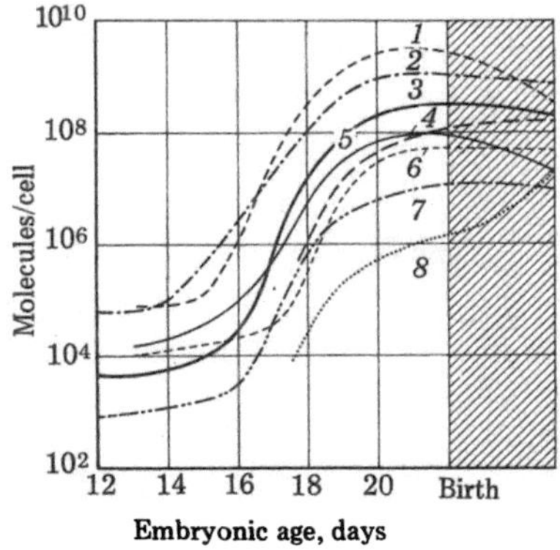

Fig. 78. Accumulation of enzymes by the developing pancreas in the mouse. *1*, Amylase; *2*, chymotrypsin; *3*, ribonuclease; *4*, trypsin; *5*, *6*. carboxypeptidases B and A; *7*, *8*, lipases A and B. Small amounts of enzymes are present in the cells prior to their visible differentiation, then their number increases 10^4- to 10^5-fold (Rutter *et al.*, 1968).

16-17, by trypsin, carboxypeptidase B, and lipase B. Within 3 to 4 days the concentration of the enzyme molecules in the cells increases by about 10,000 times and reaches a maximum level of 10^6 (lipase B) to 10^9 (chymotrypsin) molecules per cell. The insulin concentration increases from 10^4 on days 12-14 to 10^7 by day 18. Thus the protodifferentiated state changes to the differentiated one. In the period of postembryonic development and in the adult animal the enzyme concentration in the cells of a differentiated pancreas may vary considerably (10-fold), depending on the conditions, mainly on the diet.

The pancreas anlage consists of two lobes, the gut outgrowths. The dorsal lobe appears at the 23- to 25-somite stage, and the ventral, 12 h later (29- to 30-somite stage). The increase in activity of the specific enzymes — amylase and carboxypeptidases A and B — occurs, however, at a time specific to each enzyme but strictly synchronously in the two lobes (Spooner *et al.*, 1970). This indicates that in the pancreas cells a strict time-sequential synthesis of enzymes occurs, which, however, is not associated directly and unambiguously with the morphological processes of organ development.

Using the primordial pancreas cultivated with labeled amino acids, it was possible to investigate the synthesis of some enzymes directly, i.e. to separate the proteins electrophoretically, identify the fractions as different enzymes, and determine their radioactivity. The synthesis of different enzymes proceeds at a different rate. In a tissue culture the protein synthesis in a 13-day pancreas continues all the time (7 days) at a constant rate. At this level the incorporation of labeled amino acids into two enzymes — amylase and chymotrypsin — begins to increase on the 2nd day of explantation (day 15 of development) and attains a maximum by the 6th day (day 19), reaching 15% of the overall protein synthesis.

As far as the enzymatic activity is concerned Rutter *et al.* distinguish three states: the predifferentiated state, when there is no enzyme in the tissue or when there is less than one enzyme molecule per cell, the protodifferentiated state, when the enzyme activity is very low (10^3-10^5 molecules per cell), and the differentiated state with a high enzyme activity (10^6-10^9 molecules per cell) (Fig. 78). Accordingly, three transitional states are distinguished: the first, from the pre- to the protodifferentiated state (it evidently coincides with the beginning of morphogenesis), the second, from the protodifferentiated to the differentiated state (end of proliferation and beginning of cytodifferentiation), and the third which determines the fluctuations in the activity of the functioning gland.

Similar situations have been described above, although not so explicitly, with respect to the differentiation of other tissues: early appearance of myosin in myoblasts, followed by its rapid accumulation in muscular fibers, appearance of arginase in the embryonic liver and its rapid accumulation in the next period, etc. The two stages of organ development, which are reflected in the different levels of enzyme activity, are apparently common to many processes of cytodifferentiation. These processes may not be directly related to the morphogenesis of an organ as a whole.

In conclusion, we shall consider two other rather peculiar cases of differentiation of exocrine glands, which are characterized by a specific enzyme set. One of them is the venom gland of the poisonous snake, which begins to differentiate, morphologically, very early, on the 8th day of embryonic development, and is completely functionally competent by the moment of hatching on the 40th day of development (at 30°C). In distinction to the pancreas, a considerable time elapses between the morphological formation of the venom gland (days 8-14) and the beginning of the venom enzymes synthesis (day 25). The activity of phosphodiesterase in the venom increases relatively slowly, by a factor of 10 in 15 to 20 days of the second half of development. The activities of benzoylarginine ethylesterase and oxydase of amino acids is detected on the 26th-30th day and increases by almost 100 times within the next 10 days (Fein *et al.*, 1971).

Kafatos (1972) investigated much more comprehensively the formation of the proteolytic hatching enzyme cocoonase, which is synthesized at the end of pupal development and serves to dissolve the fibroin threads of the silk moth cocoon. At the early stages of metamorphosis the epithelial cells synthesize the cuticular layer of the pupa, and this is followed by a protracted (18- to 21-day) differentiation of the cocoonase-secreting zymogen cells. This process, like many others in insect development, is controlled by the antagonistic juvenile hormone and ecdysone. The short-term action of ecdysone includes differentiation, which begins with DNA synthesis and is followed (development days 3-4) by a series of mitoses. As a result of the last two cell divisions, a secreting tissue is formed, each unit of this tissue consisting of one channel cell and two zymogen cells connected to it. During the following days (days 5-12) the cytoplasm of the zymogen cells grows, and their nuclei undergo multiple DNA replication, which results in 100- to 200-fold polyploidization. The number of ribosomes and the density of endoplasmic reticulum in the cytoplasm increase, and polysomes, consisting of 9 to 12 ribosomes each, are formed. This corresponds to a synthesis of proteins with a molecular weight of $2.7\text{-}3.6 \times 10^4$ daltons, which coincides with that of zymogen, the inactive cocoonase precursor (3.3×10^4 daltons). A cavity is formed at the apical end of the zymogen cells, which is filled with zymogen protein within 9-16 days of development. In the following days the enzyme is secreted outside, after which the zymogen cells degenerate.

Autoradiographic study of the migration of labeled proteins shows that the fraction of secreted proteins increases gradually: on the 10th day of development the zymogen synthesis constitutes 20% of the overall protein synthesis, while on the 14th day it makes up already 70%. The zymogen does not yet show any proteolytic activity: it becomes an active enzyme, cocoonase, during its emergence from the channel into the cocoon cavity. This activation is probably effected by partial zymogen proteolysis, since it can easily be induced *in vitro* by treatment with diluted solutions of trypsin or pupation protease. but not by the cocoonase itself.

The zymogen synthesis continues for only 3 or 4 days, but within this time the synthesis rate increases from 3 μg on the 11th to 15 μg on the 17th day (in all, 45 μg zymogen is formed in one pupa) Measurements of protein synthesis at different moments following treatment with actinomycin D enabled Kafatos to determine the half-life of mRNA's for these proteins. For all the nonzymogen proteins this time was 2.5 h, on the average, while for the zymogen mRNA, it was almost 100 h.

If proteins synthesized in normal zymogen cells are labeled with ^3H-leucine, and those in actinomycin D-treated cells, with ^{14}C-leucine, with subsequent electrophoretic separation of their mixture, the $^3H : {}^{14}C$ ratio will indicate the sensitivity of the different proteins to actinomycin D, i.e. the life span of their mRNA's. It turned out that in the zymogen cells, in 48 of the 50 fractions investigated the half-life varied from 1 to 4.5 h, in one it was 10 h, and in the zymogen fraction, 100 h. The life span of the proteins themselves also varied from 5 to 90 h, and there was no perceptible correlation between the life span of the mRNA's and that of the proteins encoded by them.

Calculations show that only 10% of the total mRNA transcribed is the cocoonase (zymogen) mRNA. But owing to the high stability of the zymogen mRNA it accumulates rapidly and increases by 27 times from the 9th to the 14th day, while the synthesis of zymogen itself increases from 9 to 70% within this time (Kafatos, 1972). Considering that the average mRNA synthesis rate is from 3-4 (chymotrypsin, cocoonase) to 13-20 (fibroin, hemoglobin, ovalbumin) molecules per minute per gene, it can be calculated that one gene of cocoonase per one chromosome set is sufficient to ensure the necessary synthesis of the zymogen RNA, provided the nuclei are polyploid. This means that even for such specialized syntheses as is that of cocoonase, no gene amplification is required.

Kafatos' series of investigations probably surpasses all studies of this kind in the perfection of cytological and molecular analysis of differentiation. At the same time, his work shows how little is known about the mechanisms of regulation of transcription, translation, and those of cytodifferentiation as a whole.

In summarizing the section on enzymes one can see that only in a few cases are there sufficient grounds to identify the activity of an enzyme with its amount and the increase in its activity with the level of its synthesis. Nevertheless, the data obtained in studying enzymes in developing systems have provided abundant new information, which would have been hard to amass in other ways.

CONCLUSION

The data presented in this book indicate that the molecular events occurring in developmental processes consist of ordinary phenomena: replication, transcription, and translation, which form the basis of general molecular biology. But in developing systems these processes take an unusual form at times, and this brings about certain phenomena, which are specific only to oocytes or to differentiating cells. An extremal case of this kind is, presumably, oogenesis with its highly peculiar biological features. Indeed, a single cell must, within a relatively short time, produce an egg, which often exceeds it in size by thousands or millions of times and which carries huge reserves of nutrients and a protein-synthesizing system capable of sustaining the processes of early development without direct involvement of their own nuclei. To meet these requirements, the oocyte undergoes a number of unusual processes. At the level of DNA structure and replication it is the formation of lampbrushes and DNA amplification. At the transcription level, the oocyte experiences an uncoordinated synthesis of 5S rRNA, which considerably precedes that of 28S and 18S rRNA. Of special importance is the deposition of 4S and 5S RNA's and mRNA's in the form of RNP, which are stored for many months before being utilized. Other cells participate in providing the oocyte with yolk proteins, and, among other processes, vitellogenin is transferred from the site of its synthesis into the oocyte. Such a transfer of a protein into the cell is probably an exceptional phenomenon specific only to oogenesis.

The molecular events in cells of the embryo are less peculiar than those occurring in oogenesis, but they, too, differ from those that take place in differentiated cells. Thus, during fast cleavages replication takes up almost the whole interphase period, and many animals undergo almost no transcription. But then an activation period follows when the transcription level in the nuclei of embryonic cells reaches its highest values and only decreases in subsequent development.

In embryonic cells, during the differentiation of highly specialized cells, of great importance are the changes in the life span of mRNA's which ensure the preferential synthesis of particular proteins despite the fact that the transcription rates of different active genes differ only slightly. Processing may play an important, and possibly the decisive, role in determining the rate of transfer of RNA's into the cytoplasm and even in the choice of the mRNA species to be transferred.

The rate of translation is the next level of regulation of the protein synthesis on pretranscribed templates. It is not only at the stages where the nuclei practically do not function, but also in other periods of development, that the overall protein synthesis rate and the selective synthesis of protein species are regulated by specific translation of different templates rather than by the differences in transcription.

It is difficult to establish as yet any general laws in molecular biology of development. A comparison of data obtained on 3-5 most widely used objects (some species of invertebrates, amphibians, fishes, and mammals) showed that the differences at the molecular level may be very significant. In some cases oogenesis may last long and the transcription occurs in the oocyte's own nucleus, while in others it proceeds faster and the transcription takes place in the nuclei of the nurse cells. In some animals, the eggs are small and the RNA synthesis begins immediately upon fertilization, in others the eggs are large and the activation of transcription takes place only in a multicellular embryo. In echinoderms, amphibians, and fishes, cleavages follow one another very rapidly, and almost no time remains for transcription during the short interphase. At the same time, in mammals and in the ascarid each cleavage takes hours, and the nuclei are capable of transcribing sufficient quantity of RNA even during early cleavage to supply the small volume of cytoplasm around the nuclei. The role of the nuclei in the course of early development of these animals must differ substantially from their role in species with a faster development or with larger eggs.

There are also differences in such a feature of early development as the similarity between the oocyte and the embryo mRNA in various animals. In sea urchins and fishes, at the blastula stage many mRNA's are synthesized which were transcribed in oogenesis, while in amphibians there is no such coincidence, and genes transcribed at late blastula are quite different from those functioning at the lampbrush stage in oogenesis. These examples only illustrate the diversity of molecular mechanisms of development. To evaluate their actual diversity and find out which characters are general and which are individual, the range of thoroughly studied biological objects must be greatly extended. The number of possible objects for molecular biology and of available techniques is ever increasing, but much effort must still be expended in accumulating data on many species.

It is very important to know to what extent the molecular processes depend on the biology of reproduction in the species under investigation. It is obvious that the egg size (the yolk reserve), the oogenesis rate, the duration of embryogenesis, the degree of development of the larva or of the newborn organism are all direct objects of natural selection. At the same time molecular processes, such as DNA amplification or polyploidization of the nurse cell nuclei, the time of beginning of transcription or the sequence of switching-on different genes are not selected directly, but are reflected in some way or other in the number or viability of the offspring. Hence, the above-mentioned diversity of molecular mechanisms of development is due to the differences in reproductive biology. It can, in turn, be expected that the evolution itself, or its rate, in any case, will depend on whether the molecular mechanisms of development are flexible enough, i.e. capable of changing through mutations, creating varieties better adapted to the changing environment.

Molecular biology of development, which is about 30 years old, has not yet completed its descriptive period; we do not yet know the basic parameters for many processes. Progress in this field, in distinction, for instance, to molecular biology of viruses, is impeded not only by the high complexity of the system, but also by the complexity and insufficient diversity of methods.

BIBLIOGRAPHY

Abe, H., and K. Yamana, "The synthesis of 5S RNA and its regulation during early embryogenesis of *Xenopus laevis*," *Biochim. Biophys. Acta,* **240**, 392-406 (1971).

Abramova, N. B., O. S. Klyachko, and A. A. Neyfakh, "Expression of paternal genes controlling the activity of cytochrome oxidase in the development of fish hybrids," *Ontogenez,* **9**, 291-296 (1978).

Abramova, N. B., and A. A. Neyfakh, "A study of mechanisms for the respiratory rate elevation during embryonic development of fishes" *Zh. Evol. Biokhim. Fiziol.*, **1**, 227-233 (1965).

Abramova, N. B., and A. A. Neyfakh, "A study of mechanisms for aspartate-aminotransferase activity changes in early development of the sea urchin," *Ontogenez*, **2**, 71-78 (1971).

Abramova, N. B., and A. A. Neyfakh, "Migration of newly synthesized RNA during mitosis. III. Nuclear RNA in the cytoplasm of metaphase cells," *Exp. Cell Res.*, **77**, 136-142 (1973).

Abramova, N. B., and M. N. Vasilyeva, "Some properties of embryonic mitochondria of the loach," *Ontogenez,* **4**, 288-293 (1973).

Adamson, E. D., and S. E. Ayers, "The localization and synthesis of some collagen types on developing mouse embryos," *Cell,* **16**, 953-965 (1979).

Adamson, E. D., and H. R. Woodland, "Histone synthesis in early amphibian development. Histone and DNA synthesis are not co-ordinated," *J. Mol. Biol.*, **88**, 263-285 (1974).

Adamson, E. D., and H. R. Woodland, "Changes of the rate of histone synthesis during oocyte maturation and very early development of *Xenopus laevis*," *Develop. Biol.,* **57**, 136-149 (1977).

Agrell, I., and H. Persson, "Changes in the amount of nucleic acids and free nucleotides during embryonic development of sea urchins," *Nature*, **178**, 1398-1399 (1956).

Aitkhozhin, M. A., A. U. Akhanov, and Kh. I. Doshanov, "Informosomes of germinating wheat embryos," *FEBS Letters*, **31**, 104-107 (1973).

Aitkhozhin, M. A., N. V. Belitsina, and A. S. Spirin, "Nucleic acids in early embryonic development of fishes (an example of the loach *Misgurnus fossilis*)," *Biokhimiya,* **29**, 169-175 (1964).

Aizenshtadt, T. B., "Some peculiarities of the oocyte ultrastructure as related to yolk synthesis," *Zh. Obshch. Biol.*, **26**, 230-236 (1965).

Aizenshtadt, T. B., and T. A. Dettlaff, "The oocyte ultrastructure of *Acipenser stellatus* during maturation. I. Dictyosomes and the Golgi apparatus," *Ontogenez*, **3**, 280-288 (1972).

Alfageme, C. R., and A. A. Infante, "Nuclear RNA in sea urchin embryos. I. Some characteristics of ribonucleoprotein complexes," *Exp. Cell Res.*, **96**, 255-262 (1975).

Alfert, M., "A cytochemical study of oogenesis and cleavage in the mouse," *J. Cell Comp. Physiol.*, **36**, 381-409 (1950).

Allen, R. E., M. H. Stromer, D. E. Goll, and R. M. Robson, "Accumulation of myosin, actin, tropomyosin and α-actin in cultural muscle cells," *Develop. Biol.*, **69**, 655-660 (1979).

Allen, W. R., and F. H. Wilt, "The amount of nuclear protein synthesis in sea urchin blastulae," *Exp. Cell Res.*, **97**, 151-163 (1975).

Altman, S., "Transfer RNA biosynthesis," in: *International Review of Biochemistry,* Vol. 17 (B. F. C. Clark, ed.),

Baltimore University Park Press, Baltimore (1978) pp. 19-44.

Anderson, R. V., and J. A. Lengyel, "Rates of synthesis of major classes of RNA in *Drosophila* embryos," *Develop. Biol.*, **70**, 217-231 (1979).

Anderson, W. A., "Nuclear and cytoplas mic DNA synthesis during early embryogenesis of *Paracentrotus lividus*," *J. Ultrastructur. Res.*, **26**, 95-107 (1969).

Araki, M., M. Yanagida, and T. S. Okada, "Crystallin synthesis in lens differentiation in cultures of neural retinal cells of chick embryos," *Develop. Biol.*, **69**, 170-181 (1979).

Araya, A., M. Krauskopf, and M. A. Q. Siddiqui, "Selective changes in methionine tRNA species during development of the posterior silkglands of *Bombix mori*," *Biochem. Biophys. Res. Commun.*, **67**, 924-934 (1975).

Arceci, R. J., D. R. Senger, and P. R. Gross, "The programmed switch in lysine-rich histone synthesis at gastrulation," *Cell*, **9**, 171-178 (1976).

Arking, R., "Tissue-, age-, and stage-specific patterns of protein synthesis during the development of *Drosophila melanogaster*," *Develop. Biol.*, **63**, 118-127 (1978).

Aronson, A. I., and F. H. Wilt, "Properties of nuclear RNA in sea urchin embryos," *Proc. Nat. Acad. Sci. USA*, **62**, 186-193 (1969).

Arthur, C. G., M. C. Weide, W. S. Vincent III, and E. S. Goldstein, "mRNA sequence diversity during early embryogenesis in *Drosophila melanogaster*," *Exp. Cell Res.*, **121**, 87-94 (1979).

Asao, T., "Behaviour of histones and cytoplasmic basic proteins during embryogenesis of the Japanese newt *Triturus pyrrhogaster*," *Exp. Cell Res.*, **58**, 243-252 (1969).

Asao, T., "Regional histone changes in embryos of the newt *Triturus pyrrhogaster* during early development," *Exp, Cell Res.*, **61**, 255-265 (1970).

Attardi, G., Y. Aloni, B. Attardi, D. Ojala, L. Pica-Mattoccia, D. L. Robberson, and B. Storrie, "Transcription of mitochondrial DNA in HeLa cells," *Cold Spring Harbor Symp. Quant. Biol.*, **35**, 599-619 (1970).

Attardi, B., and G. Attardi, "Expression of the mitochondrial genome in HeLa cells. I. Properties of the discrete RNA components from the mitochondrial fraction," *J. Mol. Biol.*, **55**, 231-245 (1971).

Attardi, G., P. C. Huang, and S. Kabat, "Recognition of ribosomal RNA sites in DNA. II. The HeLa cell system," *Proc. Nat. Acad. Sci. USA*, **54**, 185-192 (1965).

Attardi, G., C. Tu, and B. Keeley, "Photosensitivity and heat resistance conferred by BrdU incorporation upon a thymidine kinase-deficient mouse cell line with persistent mitochondrial enzyme activity," *J. Cell Sci.*, **19**, 55-63 (1975).

Auriault, C., "Approche moléculaire de la spermatogenèse (synthèse de l'ARN et modification des nucléoprotéines basiques)," *Simbioses*, **7**, 127-138 (1975).

Avadhani, N. G., "Messenger ribonucleic acid metabolism in mammalian mitochondria: relationship between the decay of mitochondrial mRNA and their poly(A)," *Biochemistry*, **18**, 2673-2678 (1979).

Bachvarova, R., E. H. Davidson, V. G. Allfrey, and A. E. Mirsky, "Activation of RNA synthesis associated with gastrulation," *Proc. Nat. Acad. Sci. USA*, **55**, 358-365 (1966).

Bachvarova, R., and V. De Leon, "Polyadenylated RNA of mouse ova and loss of maternal RNA in early development," *Develop. Biol.*, **74**, 1-8 (1980).

Bäckström, S., "Activity of glucose-6-phosphate dehydrogenase in sea urchin embryos of different developmental trends," *Exp. Cell Res.*, **18**, 347-356 (1959).

Bäckström, S., "6-Phosphogluconate dehydrogenase in sea urchin embryos," *Exp. Cell Res.*, **32**, 566-569 (1963).

Badman, W. S., and J. W. Brookbank, "Serological studies of two hybrid sea urchins," *Develop. Biol.*, **21**, 243-256 (1970).

Bakken, A. H., "A cytological and genetical study of oogenesis in *Drosophila melanogaster*," *Develop. Biol.*, **33**, 100-122 (1973).

Balmain, A., D. Loehren, A. Alonso, and K. Goerttler, "Protein synthesis during fetal development of mouse epidermis. II. Biosynthesis of histidin-rich and cystein-rich proteins *in vitro* and *in vivo*," *Develop. Biol.*, **73**, 338-344 (1979).

Baltus, E., and J. Brachet, "Le dosage de l'acide désoxyribonucléique dans les oeufs de batraciens," *Biochim. Biophys. Acta*, **61**, 157-163 (1962).

Baltus, E., J. Hanocq-Quertier, and J. Brachet, "Isolation of DNA from the yolk platelets of *Xenopus laevis* oocyte," *Proc. Nat. Acad. Sci. USA*, **61**, 469-476 (1968).

Baltus, E., J. Quertier, A. Ficq, and J. Brachet, "Biochemical studies of nucleate and anucleate fragments isolated from sea urchin eggs. A comparison between fertilization and parthenogenetic activation," *Biochim. Biophys. Acta*, **95**, 408-417 (1965).

Bamburg, J. R., E. M. Shooter, and L. Wilson, "Developmental changes in microtubule protein of chick brain," *Biochemistry*, **12**, 1476-1482 (1973).

Barbehenn, E. K., R. G. Wales, and O. H. Lowry, "Measurement of metabolites in preimplantation embryos. A new means to study metabolic control in early embryos," *J. Embryol. Exp. Morphol.*, **43**, 29-46 (1978).

Barrett, D., "Similar accumulation and distribution patterns of two purine-oxidizing enzymes in *Rana catesbiana* development," *Develop. Biol.*, **10**, 289-304 (1964).

Barrett, D., and G. M. Angelo, "Maternal characteristics of hatching enzymes in hybrid sea urchin embryos," *Exp. Cell Res.*, **57**, 159-166 (1969).

Battey, J., and D. A. Clayton, "The transcription map of mouse mitochondrial DNA," *Cell*. **14**, 143-156 (1978).

Beams, H. W., and R. G. Kessel, "Intracisternal granules of the endoplasmic reticulum in the crayfish oocyte," *J. Cell Biol.*, **13**, 158-162 (1962).

Bechtol, K. B., S. C. Brown, and R. H. Kennett, "Recognition of differentiation antigens of spermatogenesis in the mouse by using antibodies from spleen cell-myeloma hybrids after syngenic immunization," *Proc. Nat. Acad. Sci. USA*, **76**, 363-367 (1979).

Beermann, S., "A quantitative study of chromatin diminution in embryonic mitosis of *Cyclops furcifer*," *Genetics*, **54**, 567-576 (1968).

Belitsina, N. V., M. A. Aitkhozhin, L. P. Gavrilova, and A. S. Spirin, "Informational ribonucleic acids of differentiating animal cells," *Biokhimiya*, **29**, 363-374 (1964).

Bell, W. J., and R. H. Barth, "Initiation of yolk deposition by juvenile hormone," *Nature New Biol.*, **230**, 220-222 (1971).

Bell, E., and R. Reeder, "The effect of fertilization on protein synthesis in the egg of the surf clam *Spisula solidissima*," *Biochim. Biophys. Acta*, **142**, 500-511 (1967).

Benbow, R. M., and C. C. Ford, "Cytoplasmic control of nuclear DNA synthesis during early development of *Xenopus laevis*: a cell free assay," *Proc. Nat. Acad. Sci. USA*, **72**, 2437-2441 (1975).

Benne, R., J. Edman, R. R. Trout, and J. W. B. Hershey, "Phosphorylation of eukaryotic protein synthesis initiation factors," *Proc. Nat. Acad. Sci. USA*, **75**, 108-112 (1978).

Benson, S. C., and E. L. Triplett, "The synthesis and activity of tyrosinase during development of the frog *Rana pipiens*," *Develop. Biol.*, **40**, 270-282 (1974).

Benttinen, L. C., and D. G. Comb, "Early and late histones during sea urchin development," *J. Mol. Biol.*, **57**, 355-358 (1971).

Berg, W. E., "Rates of protein synthesis in whole and half embryos of the sea urchin," *Exp. Cell Res.*, **40**, 469-489 (1965).

Berg, W. E., "Kinetics of uptake and incorporation of valine in the sea urchin embryo," *Exp. Cell Res.*, **49**, 379-395 (1968).

Berg, W. E., and D. H. Mertes, "Rates of synthesis and degradation of protein in the sea urchin embryos," *Exp. Cell Res.*, **60**, 218-224 (1970).

Bergami, M., T. E. Mansour, and E. Scarano, "Properties of glycogen phosphorylase before and after fertilization in the sea urchin eggs," *Exp. Cell Res.*, **49**, 650-655 (1968).

Berger, S. L., and H. L. Cooper, "Very short-lived and stable mRNAs from resting human lymphocytes," *Proc. Nat. Acad. Sci. USA*, **72**, 3873-3877 (1975).

Berns, A. J. M., M. Van Kraaikamp, H. Bloemendal, and C. D. Lane, "Calf crystallin synthesis in frog cells: the translation of lens-cell 14S RNA in oocytes," *Proc. Nat. Acad. Sci. USA*, **69**, 1606-1609 (1972).

Bester, A. J., D. S. Kennedy, and S. M. Heywood, "Two classes of translational control RNA: their role in the regulation of protein synthesis," *Proc. Nat. Acad. Sci. USA*, **72**, 1523-1527 (1975).

Bibring, T., J. Brachet, and F. S. Gaeta, "Some physical properties of cytoplasmic deoxyribonucleic acid in unfertilized eggs of *Arbacia lixula*," *Biochim. Biophys. Acta*, **108**, 644-651 (1965).

Bieber, S., J. A. Spence, and G. H. Hitchings, "Nucleic acids and their derivatives in the development of *Rana pipiens*. I. Oogenesis," *Exp. Cell Res.*, **16**, 202-213 (1959).

Bielka, H., J. Schultz, and M. Bottger, "Isolation and properties of DNA from eggs and gastrulae of *Ascaris lumbricoides*," *Biochim. Biophys. Acta*, **157**, 209-212 (1968).

Bier, K., "Synthese, interzellulärer Transport, und Abbua von Ribonukleinsäure im Ovar der Stubenfliege *Musca domestica*," *J. Cell Biol.*, **16**, 436-440 (1963).

Bird, A. P., "Light microscope autoradiography of ribosomal DNA amplification in *Xenopus laevis*," *Chromosoma*, **46**, 421-433 (1974).

Bird, A. P., and M. L. Birnstiel, "The relationship between protein synthesis and ribosomal DNA amplification in *Xenopus laevis*," *Biochim. Biophys. Acta*, **247**, 157-163 (1971).

Bird, A., E. Rogers, and M. L. Birnstiel, "Is genes amplification RNA-directed?", *Nature New Biol.*, **242**, 226-230 (1973).

Birnstiel, M. L., J. Speirs, J. Purdom, K. Jones, and U. E. Loening, "Properties and composition of the isolated ribosomal DNA satellite of *Xenopus laevis*," *Nature*, **219**, 454-463 (1968).

Birnstiel, M. L., H. Wallace, J. Sirlin, and M. Fischberg, "Localization of the ribosomal DNA complements in the nucleolar organizer region of *Xenopus laevis*," *Nat. Cancer Inst. Monograph*, **23**, 431-447 (1966).

Black, J. O., and E. Bresnick, "Ontogenetic changes of proteins of endoplasmic reticulum," *J. Cell Biol.*, **52**, 733-742 (1972).

Blackler, A. W., "The integrity of the reproductive cell line in the Amphibia," *Current Topics Develop. Biol.*, **5**, 71-93 (1970).

Blau, H. M., and F. C. Kafatos, "Secretory kinetics in the follicular cells of silkmoths during eggshell formation," *J. Cell Biol.*, **78**, 131-151 (1978).

Blau, H. M., and F. C. Kafatos, "Morphogenesis of silkmoth chorion. Pattern of distribution and insolubilization of structural proteins," *Develop. Biol.*, **72**, 211-226 (1979).

Bloom, A. M., and B. B. Mukherjee, "RNA synthesis in maturing mouse oocytes," *Exp. Cell Res.*, **74**, 577-582 (1972).

Blose, S. H., "Ten-nanometer filaments and mitosis: maintenance of structural continuity in dividing endothelial cells," *Proc. Nat. Acad. Sci. USA*, **76**, 3372-3376 (1979).

Boedtker, H., R. B. Crkvenjakov, J. A. Last, and P. Doty, "The identification of collagen messenger RNA," *Proc. Nat. Acad. Sci. USA*, **71**, 4208-4212 (1974).

Boffa, L. C., E. L. Gershey, and G. Vidali, "Changes of the histone deacetylase activity during chick embryo muscle development," *Biochim. Biophys. Acta*, **254**, 135-143 (1971).

Bogdanov, A. A., and R. K. Ledneva, "Nucleic-protein recognition," in: *Science and Technology Achievements, Mol. Biol. Series* [in Russian], Vol. 5, VINITI, Moscow (1975).

Bolotin, M., D. Coch, J. Deutsch, P. Netter, E. Petrochilo, and P. P. Slonimski, "La recombinaison des mitochondries chez *Saccharomyces cerevisiae*," *Bull. Inst. Pasteur*, **69**, 215-239 (1971).

Bonaldo, M. F., R. V. Santelli, and F. J. S. Lara, "The transcript from a DNA puff of *Rhynchosciara* and its migration to the cytoplasm," *Cell*, **17**, 827-833 (1979).

Bonhag, P. F., "Histochemical studies of the ovarian nurse tissues and oocytes of the milkweed bug *Oncopeltus fasciatus* (Dallas)," *J. Morphol.*, **96**, 381-411 (1955).

Borst, P., "Mitochondrial DNA," *Ann. Rev. Biochem.*, **41**, 333-376 (1972).

Borst, P., and A. M. Kroon, "Mitochondrial DNA, physicochemical properties, replication and genetic function," *Int. Rev. Cytol.*, **26**, 107-190 (1969).

Botvinnik, N. M., and A. A. Neyfakh, "Changes in aspartate-aminotransferase activity during early development of the sea urchin," *Biokhimiya*, **34**, 643-647 (1969).

Botvinnik, N. M., and A. A. Neyfakh, "Expression of esterase-controlling genes during embryonic development of fish hybrids," *Ontogenez*, **8**, 217-222 (1977).

Bourassa, W. L., and C. Terner, "Studies of metabolism in embryonic development. VII. Changes of RNA and protein synthesis in mitochondria of trout eggs after fertilization. Effects of ethidium bromide," *Exp. Cell Res.*, **118**, 261-267 (1979).

Brachet, J., "Effects of actinomycin, puromycin and cycloheximide upon the maturation of amphibian oocytes," *Exp. Cell Res.*, **48**, 233-236 (1967).

Brachet, J., "Cytochemical changes in frog liver during gametogenesis," *Currents Mod. Biol.*, **2**, 298-306 (1969).

Brachet, J., and A. Ficq, "Remarques à propos du rôle biologique des acides nucléiques," *Arch. Biol.*, **67**, 431-446 (1956).

Brachet, J., and A. Ficq, "Binding sites of ^{14}C-actinomycin in amphibian ovocytes and an autoradiography technique for the detection of cytoplasmic DNA," *Exp. Cell Res.*, **38**, 153-159 (1965).

Brachet, J., A. Ficq, and R. Tencer, "Amino acid incorporation into proteins of nucleate and anucleate fragments of sea urchin eggs: effect of parthenogenetic activation," *Exp. Cell Res.*, **32**, 168-170 (1963).

Bradley, M. O., "Arginase activity patterns and their regulation during embryonic development in liver and kidney," *Develop. Biol.*, **33**, 1-17 (1973).

Brahmachary, R. L., "Molecular embryology of invertebrates," *Adv. Morphogenesis*, **10**, 115-173 (1974).

Brahmachary, R. L., D. Chosal P. K. Tapaswi, and T. Basu, "Rhythmic incorporation of ^{35}S during early embryonic mitosis in *Limnaea* (mollusc)," *Exp. Cell Res.*, **65**, 325-328 (1971).

Brandhorst, B. P., "Simultaneous synthesis, translation, and storage of mRNA including histone mRNA in sea urchin eggs," *Develop. Biol.*, **79**, 139-148 (1980).

Brandhorst, B. P., and M. Bannet, "Terminal completion of poly(A) synthesis in sea urchin embryos," *Develop. Biol.*, **63**, 421-431 (1978).

Brandhorst, B. P., and T. Humphreys, "Stability of nuclear and messenger RNA molecules in sea urchin embryos," *J. Cell Biol.*, **53**, 474-482 (1972a).

Brandhorst, B. P., and T. Humphreys, "Synthesis and decay rates of major classes of DNA-like RNA in sea urchin embryos," *Biochemistry*, **10**, 877-881 (1972b).

Brandhorst, B. P., D. P. S. Verma, and D. Fromson, "Polyadenylated and nonadenylated mRNA fractions from sea urchin embryos code for the same abundant proteins," *Develop. Biol.*, **71**, 128-141 (1979).

Brasiello, A. R., "Autoradiographic study of ribonucleic acid synthesis during spermatogenesis of *Asellus aquaticus*," *Exp. Cell Res.*, **53**, 252-260 (1968).

Brasiello, A. R., "Protein synthesis during spermatogenesis of *Asellus aquaticus*," *Exp. Cell Res.*, **67**, 279-286 (1971).

Bravo, R., and J. E. Allende, "Conditions affecting protein synthesis in amphibian oocytes," *Arch. Biochem. Biophys.*, **172**, 648-653 (1976).

Bravo, R., and J. Knowland, "Classes of proteins synthesized in oocytes, eggs, embryos and differentiated tissues of *Xenopus laevis*," *Differentiation*, **13**, 101-108 (1979).

Brawerman, G., "Characteristics and significance of the polyadenylate sequence in mammalian messenger RNA," *Progr. Nucl. Acid Res. Mol. Biol.*, **17**, 117-148 (1976).

Brawerman, G., and J. Diez, "Metabolism of the polyadenylate sequence of nuclear RNA and messenger RNA in mammalian cells," *Cell*, **5**, 271-280 (1975).

Bresch, H., "Mitochondrial DNA metabolism in young *Psammechinus miliaris* embryos," *FEBS Letters*, **31**, 233-236 (1973).

Briggs, R., and G. Cassens, "Accumulation in the oocyte nucleus of a gene product essential for embryonic development beyond gastrulation," *Proc. Nat. Acad. Sci. USA*, **55**, 1103-1109 (1966).

Brinster, R. L., "Parental glucose phosphate isomerase activity in three-day mouse embryos," *Biochem. Genetics*, **9**, 187-191 (1973).

Brinster, R. L., J. L. Wiebold, and S. Brunner, "Protein metabolism in preimplanted mouse ova," *Develop. Biol.*, **51**, 215-224 (1976).

Bristow, D. A., and E. M. Deuchar, "Changes in nucleic acid concentration during development of *Xenopus laevis* embryos," *Exp. Cell Res.*, **35**, 580-589 (1964).

Brock, H. W., and R. Reeves, "An investigation of *de novo* protein synthesis in the South African clawed frog, *Xenopus laevis*," *Develop. Biol.*, **66**, 128-141 (1978).

Brotherton, T. W., D. H. K. Chui, J. Gauldie, and M. Patterson, "Hemoglobin ontogeny during normal mouse fetal development," *Proc. Nat. Acad. Sci. USA*, **76**, 2853-2857 (1979).

Brown, D. D., "The nucleolus and synthesis of ribosomal RNA during oogenesis and embryogenesis of *Xenopus laevis*," *Nat. Cancer Inst. Monograph*, **23**, 297-313 (1966).

Brown, D. D., and A. W. Blackler, "Gene amplification proceeds by a chromosome copy mechanism," *J. Mol. Biol.*, **63**, 75-83 (1972).

Brown, D. D., and I. B. Dawid, "Specific gene amplification in oocytes," *Science*, **160**, 272-280 (1968).

Brown, D. D., I. B. Dawid, and R. H. Reeder, "*Xenopus borealis* misidentified as *Xenopus mulleri*," *Develop. Biol.*, **59**, 266-267 (1977).

Brown, D. D., and J. B. Gurdon, "Absence of ribosomal RNA synthesis in the anucleate mutant of *Xenopus laevis*," *Proc. Nat. Acad. Sci. USA*, **51**, 139-146 (1964).

Brown, D. D., and E. Littna, "RNA synthesis during the development of *Xenopus laevis*, the South African clawed toad," *J. Mol. Biol.*, **8**, 669-678 (1964).

Brown, D. D., and E. Littna, "The synthesis and accumulation of low molecular weight RNA during embryogenesis of *Xenopus laevis*," *J. Mol. Biol.*, **20**, 95-112 (1966).

Brown, D. D., and C. S. Weber, "Gene linkage by RNA-DNA hybridization. I. Unique DNA sequences homologous to 4S, 5S RNA and ribosomal RNA. II. Arrangement of the redundant gene sequences for 28S and 18S ribosomal RNA," *J. Mol. Biol.*, **34**, 661-697 (1968).

Brown, D. D., P. C. Wensink, and E. Jordan, "A comparison of the ribosomal DNA's of *Xenopus laevis* and *Xenopus mulleri*: the evolution of tandem genes," *J. Mol. Biol.*, **63**, 57-73 (1972).

Brown, I. R., and R. B. Church, "RNA transcription from nonrepetitive DNA in the mouse," *Biochem. Biophys. Res. Commun.*, **42**, 850-856 (1971).

Brown, I. R., and R. B. Church, "Transcription of nonrepeated DNA during mouse and rabbit development," *Develop. Biol.*, **29**, 73-84 (1972).

Brown, R. D., and G. P. Tocchini-Valentini, "On the role of RNA in gene amplification," *Proc. Nat. Acad. Sci. USA*, **69**, 1746-1748 (1972).

Brown, R. D., and G. P. Tocchini-Valentini, "The preparation of RNA-directed DNA polymerase from ovaries of *Xenopus laevis*," in: *Methods in Enzymology*, Vol. 29E, Acad. Press, New York (1973), pp. 173-177.

Bruns, R. R., D. J. S. Hulmes, S. F. Therrien, and J. Gross, "Procollagen segment-long-spacing crystallites. Their role in collagen fibrillogenesis," *Proc. Nat. Acad. Sci. USA*, **76**, 313-317 (1979).

Buckingham, M. E., D. Caput, A. Cohen, R. C. Whalen, and F. Gross, "The synthesis and stability of cytoplasmic messenger RNA during myoblast differentiation in culture," *Proc. Nat. Acad. Sci. USA*, **71**, 1466-1470 (1974a).

Buckingham, M. E., A. Cohen, F. Gross, D. Lizzati, D. Charmot, and G. Drugeon, "Expression of the myosin gene in a hybrid cell derived from a rat myoblast and a mouse fibroblast," *Biochimie*, **56**, 1571-1573 (1974b).

Buckingham, M. E., and F. Gross, "The use of metrizamide to separate cytoplasmic ribonucleoprotein particles in muscle cell cultures: a method for the isolation of messenger RNA, independent of its polyA content," *FEBS Letters*, **53**, 355-359 (1975).

Burakova, T. A., and A. A. Kostomarova, "An autoradiographic study of protein synthesis in the loach blastoderm," *Ontogenez*, **6**, 225-233 (1975).

Burdon, R. H., "Ribonucleic acid maturation in animal cells," *Progr. Nucl. Acid Res. Mol. Biol.*, **11**, 33-79 (1971).

Burgos, M., "The Feulgen reaction in mature unfertilized sea urchin eggs," *Exp. Cell Res.*, **9**, 360-363 (1955).

Burny, A., G. Marbaix, J. Quertier, and J. Brachet, "Demonstration of functional polyribosomes in nucleate and anucleate fragments of sea urchin eggs following parthenogenetic activation," *Biochim. Biophys. Acta*, **103**, 526-528 (1965).

Burt, M. A. "Glucose metabolism and chick neurogenesis. 1. Glucose-6-phosphate dehydrogenase activity in the embryonic brachial spinal cord," *Develop. Biol.*, **12**, 213-232 (1965).

Byrd, E. W., Jr., "Phospholipid metabolism following fertilization in sea urchin eggs and embryos," *Develop. Biol.*, **46**, 309-316 (1975).

Byrd, E. W., Jr., and H. E. Kasinsky, "Histone synthesis during early embryogenesis in *Xenopus laevis*," *Biochemistry*, **12**, 246-253 (1973).

Cabada, M. O., C. Darnbrough, P. J. Ford, and P. C. Turner, "Differential accumulation of two size classes of poly(A) associated with messenger RNA during oogenesis in *Xenopus laevis*," *Develop. Biol.*, **57**, 427-439 (1977).

Cahn, D. R., "Developmental changes in embryonic enzyme patterns: the effect of oxidative substrates on lactic dehydrogenase in beating chick embryonic heart cell cultures," *Develop. Biol.*, **9**, 327-346 (1964).

Callan, H. G., "Chromosomes and nucleoli of the axolotl *Ambystoma mexicanum*," *J. Cell Sci.*, **1**, 85-98 (1966).

Candelas, G. G., and R. M. Iverson, "Evidence for translational level control of protein synthesis in the development of sea urchin eggs," *Biochem. Biophys. Res. Commun.*, **24**, 867-871 (1966).

Cape, M., and M. Decroly, "Mesure de la capacité 'template' des acides ribonucléiques des oeufs de *Xenopus laevis* au cours du development," *Biochim. Biophys. Acta*, **174**, 99-107 (1969).

Carroll, A., W. R. Eckberg, and H. Ozaki, "A comparison of protein synthesis patterns in normal and animalized sea urchin embryos," *Exp. Cell Res.*, **90**, 328-332 (1975).

Carroll, A. G., and H. Ozaki, "Changes in the histones of the sea urchin *Strongylocentrotus purpuratus* at fertilization," *Exp. Cell Res.*, **119**, 307-319 (1979).

Cassidy, D. A., and A. W. Blackler, "A developmental analysis of the histone patterns of two species of *Xenopus* and their hybrids," *Develop. Biol.*, **63**, 224-232 (1978).

Castaneda, M., "The activity of ribosomes of sea urchin eggs in response to fertilization," *Biochim. Biophys. Acta*, **179**, 381-388 (1969).

Cave, M. D., "Localization of ribosomal DNA within oocytes of the house cricket, *Acheta domesticus* (Orthoptera: Gryllidae)," *J. Cell Biol.*, **55**, 310-321 (1972).

Cave, M. D., "Synthesis and characterization of amplified DNA in oocytes of the house cricket, *Acheta domesticus* (Orthoptera: Gryllidae)," *Chromosoma*, **42**, 1-22 (1973).

Cave, M. D., "Absence of rDNA amplification in a meroistic (telotrophic) ovary," *J. Cell Biol.*, **63**, 53a (1974).

Cave, M. D., "Absence of ribosomal RNA gene amplification in nurse cells of polytrophic meroistic ovaries," *J. Cell Biol.*, **70**, 119a (1976).

Chamberlain, J. P., "RNA synthesis in anucleate egg fragments and normal embryos of the sea urchin *Arbacia punctulata*," *Biochim. Biophys. Acta*, **213**, 183-193 (1970).

Chamberlain, J., and C. B. Metz, "Mitochondrial RNA synthesis in sea urchin embryos," *J. Mol. Biol.*, **64**, 593-607 (1972).

Chapman, B. S., and A. J. Tobin, "Distribution of developmentally regulated hemoglobins in embryonic erythroid populations, *Develop. Biol.*, **69**, 375-387 (1979).

Chapman, V. M., W. K. Whitten, and F. H. Ruddle, "Expression of paternal glucose phosphate isomerase-1 (Gpi-1) in preimplantation stages of mouse embryos," *Develop. Biol.*, **26**, 153-158 (1971)

Chase, J. H., and I. B. Dawid, "Biogenesis of mitochondria during *X. laevis* development," *Develop. Biol.*, **27**, 504-518 (1972).

Chen, P. S., "Changes in DNA and RNA during embryonic Urodela development," *Exp. Cell Res.*, **21**, 523-534 (1960).

Chentsov, Yu. S., "The structure and chemistry of nucleolus as a site for ribosome synthesis in the cell," *Usp. Sovr. Biol.*, **62**, 324-344 (1966).

Chi, J. C., S. A. Fellini, and H. Holtzer, "Differences among myosins synthesized in non-myogenic cells, presumptive myoblasts, and myoblasts," *Proc. Nat. Acad. Sci. USA*, **72**, 4999-5003 (1975).

Chia, W. K., and P. E. Morrison, "Autoradiographic and ultrastructural studies on origin of yolk protein in house fly, *Musca domestica* L.," *Canad. J. Zool.*, **50**, 1569-1576 (1972).

Childs, G., R. Maxson, and L. H. Kedes, "Histone gene expression during sea urchin embryogenesis: isolation and characterization of early and late messenger RNA's of *Strongylocentrotus purpuratus* by gene specific hybridization and template activity," *Develop. Biol.*, **73**, 153-173 (1979).

Chmilevsky, D. A., "RNA synthesis and development of the nucleolar apparatus in the ruff oogenesis," *Tsitologiya*, **13**, 1233-1256 (1971).

Choi, W. C., and W. Nagl, "Patterns of DNA and RNA synthesis during the development of ovarian nurse cells in *Gerris najas* (Heteroptera)," *Develop. Biol.*, **61**, 262-272 (1977).

Chuguyev, Yu. P., A. Ya. Strongin, and V. V. Sovs, "The isoelectric focusing of the plasma membrane proteins in the embryonic cells of the sea urchin *Strongylocentrotus intermedius*," *Biokhimiya*, **41**, 1713-1715 (1976).

Church, R. B., and I. R. Brown, "Tissue specificity of genetic transcription," in: *Results and Problems in Cell Differentiation*, Vol. 3 (H. Ursprung, ed), Springer-Verlag, Berlin, New York (1972), pp. 1-12.

Church, R. B., and G. A. Schultz, "Differential gene activity in the pre- and postimplantation mammalian embryo," *Current Topics Develop. Biol.*, **8**, 179-202 (1974).

Clandinin, M. T., and G. A. Schultz, "Levels and modification of methionyl-transfer RNA in preimplantation rabbit embryos," *J. Mol. Biol.*, **93**, 517-523 (1975).

Clavert, J., "La biochimie de l'ovogénèse," *Arch. Neerl. Zool.*, **10**, Suppl., 1-18 (1953).

Clawson, G. A., M. Koplitz, B. Castler-Schechter, and E. A. Smuckler, "Energy utilization and RNA transport: their interdependence," *Biochemistry*, **17**, 3747-3752 (1978).

Claycomb, W. C., and C. A. Villee, "Lactate dehydrogenase isozymes of *X. laevis*. Factors affecting their appearance during early development," *Develop. Biol.*, **24**, 413-427 (1971).

Clayton, D. A., R. W. Davis, and J. Vinograd, "Homology and structural relationships between dimeric and monomeric circular forms of mitochondrial DNA from human leukemic leucocytes," *J. Mol. Biol.*, **47**, 137-153 (1970).

Clegg, J. S., and A. L. Golub, "Protein synthesis in *Artemia salina* embryos. II. Resumption of RNA and protein synthesis upon cessation of dormancy in the encysted gastrula," *Develop. Biol.*, **19**, 178-200 (1969).

Clegg, K. B., and P. C. Denny, "Synthesis of rabbit globin in a cell-free protein synthesis system utilizing sea urchin egg and zygote ribosomes," *Develop. Biol.*, **37**, 263-272 (1974).

Cognetti, G., R. D. Platz, M. L. Meistrich, and T. Di Liegro, "Studies on protein synthesis during sea urchin oogenesis. 1. Synthesis of histone F2b," *Cell Different.*, **5**, 283-291 (1977).

Cohen, L. H., K. M. Newrock, and A. Zweidler, "Stage-specffic switches in histone synthesis during embryogenesis of the sea urchin," *Science*, **190**, 994-997 (1975).

Cohen, P. P., "Biochemical differentiation during metamorphosis," *Science*, **168**, 533-543 (1970).

Coleman, J. R., and A. W. Coleman, "Muscle differentiation and macromolecular synthesis," *J. Cell. Physiol.*, **72**, Suppl., 19-34 (1968).

Collier, J. R., "The effect of removing the polar lobe on the protein synthesis of the embryo of *Ilyanassa obsoleta*," *Acta Embryol. Morphol. Exp.*, **4**, 70-76 (1961).

Collier, J. R., "The transcription of genetic information in the spiralian embryo," *Current Topics Develop. Biol.*, **1**, 39-63 (1966).

Collier, J. R., "Polyadenylation of nascent RNA during the embryogenesis of *Ilyanassa obsoleta*," *Exp. Cell Res.*, **95**, 263-268 (1975).

Collier, J. R., and R. Schwartz, "Protein synthesis during *Ilyanassa* embryogenesis," *Exp. Cell Res.*, **54**, 403-406 (1969).

Colman, A., "Synthesis of RNA in oocytes of *Xenopus laevis* during culture *in vitro*," *J. Embryol. Exp. Morphol.*, **32**, 515-532 (1974).

Comb, D. G., S. Katz, R. Branda, and C. J. Pinzino, "Characterization of RNA species synthesized during early development of sea urchins," *J. Mol. Biol.*, **14**, 195-213 (1965).

Craig, N., and R. P. Perry, "Persistent cytoplasmic synthesis of ribosomal proteins during selective inhibition of ribosomal RNA synthesis," *Nature New Biol.*, **223**, 75-80 (1971).

Craig, S. P., "Synthesis of RNA in non-nucleate fragments of sea urchin eggs," *J. Mol. Biol.*, **47**, 615-618 (1970).

Craig, S. P., and J. Piatigorsky, "Protein synthesis and development in the absence of cytoplasmic RNA synthesis in non-nucleate egg fragments and embryos of sea urchins: effect of ethidium bromide," *Develop. Biol.*, **24**, 214-231 (1971).

Craig, S. P., and J. Piatigorsky, "Cell elongation and δ-crystallin synthesis without RNA synthesis in cultured early embryonic chick lens epithelia," *Biochim. Biophys. Acta*, **299**, 642-653 (1973).

Crippa, M., E. Davidson, and A. Mirsky, "Persistence in early embryos of information RNA's from the lampbrush chromosome stage of oogenesis," *Proc. Nat. Acad. Sci. USA*, **57**, 885-892 (1967).

Crippa, M., and P. R. Gross, "Maternal and embryonic contributions to the functional messenger RNA of early development," *Proc. Nat. Acad. Sci. USA*, **62**, 120-127 (1969).

Crippa, M., and G. P. Tocchini-Valentini, "Synthesis of amphibian DNA that codes for ribosomal RNA," *Proc. Nat. Acad. Sci. USA*, **68**, 2769-2773 (1971).

Czizhak, G., "Evidences for inductive properties of the micromere-RNA in sea-urchin embryos," *Naturwissenschaften*, **52**, 141-142 (1965).

Daentl, D. L., and C. J. Epstein, "Developmental interrelationships of uridine uptake, nucleotide formation and incorporation into RNA by early mammalian embryos," *Develop. Biol.*, **24**, 428-442 (1971).

Daneholt, B., "Transcription in polytene chromosomes (review)," *Cell*, **4**, 1-9 (1975).

Darnbrough, C., and P. J. Ford, "Turnover and processing of poly(A) in full-grown oocytes and during progesterone-induced oocyte maturation in *Xenopus laevis*," *Develop. Biol.*, **71**, 323-340 (1979).

Das, C. C., B. P. Kaufmann, and H. Gay, "Autoradiographic evidence of synthesis of an arginin-rich histone during spermiogenesis in *Drosophila melanogaster*," *Nature*, **204**, 128-129 (1964).

Das, N. K., P. Luykx, and M. Alfert, "The nucleolus and RNA metabolism in *Urechis* eggs," *Develop. Biol.*, **12**, 72-78 (1965).

Das, N. K., J. Micou-Eastwood, G. Ramamurthy, and M. Alfert, "Sites of synthesis and processing of ribosomal RNA precursors within the nucleolus of *Urechis caupo* eggs," *Proc. Nat. Acad. Sci. USA*, **67**, 968-975 (1970).

Davidson, E. H., *Gene Activity in Early Development*, Acad. Press, New York, London (1969).

Davidson, E. H., V. G. Allfrey, and A. E. Mirsky, "On the RNA synthesized during the lampbrush phase of amphibian oogenesis," *Proc. Nat. Acad. Sci. USA*, **52**, 501-508 (1964).

Davidson, E. H., and R. J. Britten, "Regulation of gene expression: possible role of repetitive sequences," *Science*, **204**, 1052-1059 (1979).

Davidson, E. H., M. Crippa, F. R. Kramer, and A. E. Mirsky, "Genomic function during the lampbrush chromosomes stage of amphibian oogenesis," *Proc. Nat. Acad. Sci. USA*, **56**, 856-863 (1966).

Davidson, E. H., M. Crippa, and A. E. Mirsky, "Evidence for the appearance of novel gene products during amphibian blastulation," *Proc. Nat. Acad. Sci. USA*, **60**, 152-159 (1968).

Davidson, E. H., G. W. Haslett, R. J. Finney, V. G. Allfrey, and A. E. Mirsky, "Evidence for prelocalization of cytoplasmic factors affecting gene activation in early embryogenesis," *Proc. Nat. Acad. Sci. USA*, **54**, 696-704 (1965).

Davidson, E., and B. R. Hough, "Synchronous oogenesis in *Engystomops pustulosus*, a neotropic anuran, suitable for laboratory studies. Localization in the embryo of RNA synthesized at the lampbrush stage," *J. Exp. Zool.*, **172**, 25-48 (1969).

Davidson, E. H., and B. R. Hough, "Genetic information in oocyte RNA," *J. Mol. Biol.*, **56**, 491-506 (1971).

Davidson, E. H., and A. E. Mirsky, "Gene activity in oogenesis," in: *Genetic Control of Differentiation*, Brook Haven Symp. Biol., Vol. 18, pp. 77-98 (1965).

Davis, F. C., and R. W. Davis, "Polyadenylation of RNA in immature oocytes, mature oocytes and early embryos of *Urechis caupo*," *J. Cell Biol.*, **70**, 278a (1976).

Davis, F. C., and R. W. Davis, "Polyadenylation of RNA in immature oocytes and early cleavage of *Urechis caupo*," *Develop. Biol.*, **66**, 86-96 (1978).

Davis, F. C., and F. H. Wilt, "RNA synthesis during oogenesis in the echiuroid worm *Urechis caupo*," *Develop. Biol.*, **27**, 1-12 (1972).

Dawid, I. B., "Deoxyribonucleic acid in amphibian eggs," *J. Mol. Biol.*, **12**, 581-599 (1965).

Dawid, I., "Evidence for the mitochondrial origin of frog egg cytoplasmic DNA," *Proc. Nat. Acad. Sci. USA*, **56**, 269-276 (1966).

Dawid, I., "The nature of mitochondrial RNA in oocytes of *Xenopus laevis* and its relation to mitochondrial DNA," in: *Control of Organelle Development*, Symp. Soc. Exp. Biol., **24**, 227-246, Cambridge University Press (1970).

Dawid, I. B., "Mitochondrial RNA in *Xenopus laevis*. I. The expression of the mitochondrial genome," *J. Mol. Biol.*, **63**, 201-216 (1972a).

Dawid, I. B., "Evolution of mitochondrial DNA sequences in *Xenopus*," *Develop. Biol.*, **29**, 139-151 (1972b).

Dawid, I., and A. W. Blackler, "Maternal and cytoplasmic inheritance of mitochondrial DNA in *Xenopus*," *Develop. Biol.*, **29**, 152-161 (1972).

Dawid, I., and D. Brown, "The mitochondrial and ribosomal DNA components of oocytes of *Urechis caupo*., *Develop. Biol.*, **22**, 1-14 (1970).

Dawid, I. B., D. D. Brown, and R. H. Reeder, "Composition and structure of chromosomal and amplified ribosomal DNA's of *Xenopus laevis*," *J. Mol. Biol*, **51**, 341-360 (1970).

Dawid, I., and D. R. Wolstenholme, "Ultracentrifuge and electron microscope studies on the structure of mitochondrial DNA," *J. Mol. Biol.*, **28**, 233-245 (1967).

Dawid, I., and D. R. Wolstenholme, "Renaturation and hybridization studies of mitochondrial DNA," *Biophys. J.*, **8**, 65-81 (1968).

Decroly, M., M. Cape, and J. Brachet, "Studies on the synthesis of RNA in embryonic stages of *Xenopus laevis*," *Biochim. Biophys. Acta*, **87**, 34-39 (1964).

Dehm, P., and N. A. Kefalides, "The collagenous component of lens basement membrane," *J. Biol. Chem.*, **253**, 6680-6686 (1978).

Dehn, P. F., and R. A. Wallace, "Sequestered and injected vitellogenin. Alternative routes of protein processing in *Xenopus laevis* oocytes," *J. Cell Biol.*, **58**, 721-724 (1973).

Delain, D., and J. P. Wahrmann, "Is fusion a trigger for myoblast differentiation?", *Exp. Cell Res.*, **93**, 495-498 (1975).

Delaunay, J., M. Wegnez, and H. Denis, "Biochemical research on oogenesis. Comparison between the proteins of the ribosomes and the proteins of the 42S particles from small oocytes of *Xenopus laevis*," *Develop. Biol.*, **42**, 379-387 (1975).

Denis, H., "Gene expression in amphibian development. I. Validity of the method used: interspecific and intraspecific hybridization between nucleic acids, properties of messenger RNA synthesized by developing embryos," *J. Mol. Biol.*. **22**, 269-284 (1966a).

Denis, H., "Gene expression in amphibian development. II. Release of the genetic information in growing embryos," *J. Mol. Biol.*, **22**, 285-304 (1966b).

Denis, H., "Role of messenger RNA in embryonal development," *Adv. Morphogenesis*, **7**, 115-150 (1968).

Denis, H., *Précis d'embryologie moléculaire*, Presses universitaires de France, Vendôme (1974).

Denis, H., and M. Mairy, "Recherches biochimiques sur l'oogénèse. II. Distribution intracellulaire du RNA dans les petits oocytes de *Xenopus laevis*," *Europ. J. Biochem.*, **25**, 524-534 (1972).

Denis, H., A. Mazabraud, and M. Wegnez, "Biochemical research on oogenesis. Comparison between transfer RNAs from somatic cells and from oocytes in *Xenopus laevis*," *Europ. J. Biochem.*, **58**, 43-50 (1975).

Denis, H., B. Picard, M. Mairy, and J. C. Clerot, "Biochemical research on oogenesis. The storage particles of the teleost fish *Tinca tinca*," *Develop. Biol.*, **77**, 218-223 (1980).

Denis, H., and M. Wegnez, "Biochemical research on oogenesis. Oocytes and liver cells of the teleost fish *Tinca tinca* contain different kinds of 5S RNA," *Develop. Biol.*, **59**, 228-236 (1977).

Denis, H., M. Wegnez, and R. Willem, "Recherches biochimiques sur l'oogénèse. V. Comparaison entre le RNA 5S somatique et le RNA 5S des oocytes de *Xenopus laevis*," *Biochimie*, **54**, 1189-1196 (1972).

Denny, P. C., and A. Tyler, "Activation of protein biosynthesis in non-nucleate fragments of sea urchin eggs," *Biochem. Biophys. Res. Commun.*, **14**, 245-249 (1964).

De Petrocellis, B., and E. Parisi, "Changes in alkaline deoxyribonuclease activity in sea urchin during embryonic development," *Exp. Cell Res.*, **73**, 496-500 (1972).

De Petrocellis, B., and E. Parisi, "Deoxyribonuclease in sea urchin embryos (comparison of the activity present in developing embryos, in nuclei and mitochondria)," *Exp. Cell Res.*, **79**, 53-62 (1973a).

De Petrocellis, B., and E. Parisi, "Effect of actinomycin and puromycin on the deoxyribonuclease activity in *Paracentrotus lividus* at various stages of development," *Exp. Cell Res.*, **82**, 351-356 (1973b).

Dettlaff, T. A., "Action of actinomycin and puromycin upon frog oocyte maturation," *J. Embryol. Exp. Morphol.*, **16**, 183-195 (1966).

Dettlaff, T. A., L. A. Nikitina, and O. G. Stroyeva, "The role of the germinal vesicle in oöcyte maturation in anurans as revealed by the removal and transplantation of nuclei," *J. Embryol. Exp. Morphol.*, **12**, 851-873 (1964).

Dettlaff, T. A., L. A. Nikitina, and O. G. Stroyeva, "An analysis of the role and specificity of the germinal vesicle in oocyte maturation in anurans as revealed by its removal and replacement with somatic cell nuclei," *Dokl. Akad. Nauk SSSR*, **160**, 1441-1443 (1965).

Dettlaff, T. A., and M. N. Skoblina, "The role ot germinal vesicle in the process of oocyte maturation in Anura and Acipenseridae," *Ann. d'Embryol. Morphol.*, 133-151 (1969).

Deuchar, E. M., "Amino acids in developing tissues of *Xenopus laevis*," *J. Embryol. Exp. Morphol.*, **4**, 327-346 (1956).

Deuchar, E. M., "Amino acids and differentiation in animal embryos," *Cell Different., Symp. Soc. Exp. Biol.*, **17**, 58-73 (1963a).

Deuchar, E. M., "Tracing amino acids from yolk protein into tissue protein. I. Incorporation of tritiated leucine into oocytes and its distribution in the early embryo of *Xenopus laevis*," *Acta Embryol. Morphol. Exp.*, **6**, 311-323 (1963b).

Deuchar, E. M., "Tracing amino acids from yolk protein into tissue protein. II. l-proline," *Acta Embryol. Morphol. Exp.*, **7**, 49-60 (1964).

Diacumakos, J. G., J. Garnjobski, and E. L. Tatum, "A cytoplasmic character in *Neurospora crassa*. The role of nuclei and mitochondria," *J. Cell Biol.*, **26**, 427-443 (1965).

Dobner, P., and R. A. Flickinger, "Repetitive RNA transcripts in frog embryo cytoplasm," *Biochim. Biophys. Acta*, **432**, 401-403 (1976).

Dolecki, G. J., and L. D. Smith, "Poly(A) RNA metabolism during oogenesis in *Xenopus laevis*," *Develop. Biol.*, **69**, 217-236 (1979).

Donohoo, P., and F. C. Kafatos, "Differences in the proteins synthesized by the progeny of the first two blastomeres of *Ilyanassa*, a "mosaic' embryo," *Develop. Biol.*, **32**, 224-229 (1973).

Dontsova, G. V., T. M. Ignatyeva, N. N. Rott, and I. I. Tolstorukov, "Nucleic acids in the early embryogenesis of the trout," *Ontogenez*, **1**, 471-481 (1970).

Douvas, A. S., C. A. Harrington, and J. Bonner, "Major nonhistone proteins of rat liver chromatin: preliminary identification of myosin, actin, tubulin and tropomyosin," *Proc. Nat. Acad. Sci. USA*, **72**, 3902-3906 (1975).

Dubroff, L. M., and M. Nemer, "Developmental shifts in the synthesis of heterogeneous nuclear RNA classes in the sea urchin embryo," *Nature*, **260**, 120-124 (1976).

Dumont, J. N., "Oogenesis in the annelid *Enchytraeus abbidus* with special reference to the origin and cytochemistry of yolk," *J. Morphol.*, **129**, 317-328 (1969).

Durante, M. C., "L'acide désoxyribonucléique des gamétes de *Gryllus domesticus*," *C. R. Acad. Sci. Paris*, **241**, 1340-1343 (1955).

Durante, M., "The alanineglycine dipeptidase activity during the development of *Ascidia malaca*," *Acta Embryol. Morphol. Exp.*, **4**, 352-355 (1961).

Dworkin, M. B., and I. B. Dawid, "Use of a cloned library for the study of abundant poly $(A)^+$ RNA during *Xenopus laevis* development," *Develop. Biol.*, **76**, 449-464 (1980).

Dworkin, M. B., and A. A. Infante, "Relationship between the mRNA of polysomes and free ribonucleoprotein particles in the early sea urchin embryo," *Develop. Biol.*, **53**, 73-90 (1976).

Dworkin, M. B., and A. A. Infante, "RNA synthesis in unfertilized sea urchin eggs," *Develop. Biol.*, **62**, 247-257 (1978).

Dym, H. P., D. S. Kennedy, and S. M. Heywood, "Sub-cellular distribution of the cytoplasmic myosin heavy chain mRNA during myogenesis," *Differentiation*, **12**, 145-155 (1979).

Easton, D. P., and A. H. Whiteley, "The relative contributions of newly synthesized and stored messages to H1 histone synthesis in interspecies hybrid echinoid embryos," *Differentiation*, **12**, 127-133 (1979).

Ecker, R. E., and L. D. Smith, "The kinetics of protein synthesis in early amphibian development," *Biochim. Biophys. Acta*, **129**, 186-192 (1966).

Ecker, R. E., and L. D. Smith, "Protein synthesis in amphibian oocytes and early embryos," *Develop. Biol.*, **18**, 232-249 (1968).

Ecker, R. E., and L. D. Smith, "The nature and fate of *Rana pipiens* proteins synthesized during maturation and early cleavage," *Develop. Biol.*, **24**, 559-576 (1971).

Ecker, R. E., L. D. Smith, and S. Subtelny, "Kinetics of protein synthesis in enucleate frog oocytes," *Science*, **160**, 1115-1117 (1968).

Edds, I., "Development of collagen in the frog embryo," *Proc. Nat. Acad. Sci. USA*, **44**, 296-305 (1958).

Edström, J. E., "Microelectrophoretic determination of deoxyribonucleic acid content and base composition in microscopic tissue samples," *Biochim. Biophys. Acta*, **80**, 399-410 (1964).

Edström, J. e., and J. G. Gall, "The base composition of ribonucleic acid in lampbrush chromosomes, nucleoli, nuclear sap and cytoplasm of *Triturus* oocytes," *J. Cell Biol.*, **19**, 279-284 (1963).

Egrie, J. C., and F. H, Wilt, "Changes in poly(adenylic acid) polymerase activity during sea urchin embryogenesis," *Biochemistry*, **18**, 269-274 (1979).

Ekizashvili, V. K., L. L. Gause, and K. A. Kafiani, "Some characteristics of the DNA-dependent RNA polymerase from a developing embryo of the loach," in: *Transactions of the Tbilisi State Univers.*, A8 (153), 211-215 (1974).

Ellem, K. A. O., and R. B. L. Gwatkin, "Pattern of nucleic acid synthesis in the early mouse embryo," *Develop. Biol.*, **18**, 311-330 (1968).

Ellis, C. H., "The genetic control of sea urchin development: a chromatographic study of protein synthesis in the *Arbacia punctulata* embryo," *J. Exp. Zool.*, **163**, 1-22 (1966).

Elsdale, T. R., M. Fischberg, and S. Smith, "A mutation that reduces nucleolar number in *Xenopus laevis*," *Exp. Cell Res.*, **14**, 642-644 (1958).

Elson, D., and E. Chargaff, "On the desoxyribonucleic acid content of sea urchin gametes," *Experientia*, **8**, 143-145 (1952).

Emerson, C. P., Jr., and S. K. Beckner, "Activation of myosin synthesis in fusing and mononucleated myoblasts," *J. Mol. Biol.*, **93**, 431-447 (1975).

Emerson, C. P., and T. Humphreys, "Regulation of DNA-like RNA and the apparent activation of ribosomal synthesis in sea urchin embryos: quantitative measurements of newly synthesized RNA," *Develop. Biol.*, **23**, 86-112 (1970).

Emerson, C. P., Jr., and T. Humphreys, "Ribosomal RNA synthesis and the multiple atypical nucleoli in cleaving embryos," *Science*, **171**, 898-901 (1971).

Emmons, S. W., M. R. Klass, and D. Hirsh, "Analysis of the constancy of DNA sequences during development and evolution of the nematode *Caenorhabditis elegans*," *Proc. Nat. Acad. Sci. USA*, **76**, 1333-1337 (1979).

Enger, M. D., and R. A. Walters, "Isolation of low molecular weight methylated ribonucleic acids from 10S to 30S particles of Chinese hamster cell fractions," *Biochemistry*, **9**, 3551-3562 (1970).

England, M. C., and D. T. Mayer, "The comparative nucleic acid content of liver nuclei, ova nuclei and spermatozoa of the frog," *Exp. Cell Res.*, **12**, 249-253 (1957).

Epel, D., "A primary metabolic change of fertilization: interconversion of pyridine nucleotides," *Biochem. Biophys. Res. Commun.*, **17**, 62-68 (1964a).

Epel, D., "Simultaneous measurement of TPNH formation and respiration following fertilization of sea urchin egg," *Biochem. Biophys. Res. Commun.*, **17**, 69-73 (1964b).

Epel, D., "Protein synthesis in sea urchin eggs: a 'late' response to fertilization," *Proc. Nat. Acad. Sci. USA*, **57**, 899-906 (1967).

Epel, D., "Activation of an N^+-dependent amino acid transport system upon fertilization of sea urchin eggs," *Exp. Cell Res.*, **72**, 74-89 (1972).

Epel, D., B. C. Pressman, S. Elsaesser, and A. M. Weaver, "The program of structural and metabolic changes following fertilization of sea urchin eggs," in: *The Cell Cycle: Gene-Enzyme Interactions* (G. M. Padilla *et al.*, eds.), Acad. Press, New York (1969b), pp. 279-298.

Epel, D., A. M. Weaver, A. V. Muchmore, and R. T. Schimke, "β-1,3-Glucanase of sea urchin eggs: release from particles at fertilization," *Science*, **163**, 294-296 (1969a).

Eppenberger, H. M., M. Eppenberger, R. Richterich, and H. Aebi, "The ontogeny of creatine kinase isozymes," *Develop. Biol.*, **10**, 1-16 (1964).

Eppig, J. J., and J. N. Dumont, "Oogenesis in *Xenopus laevis*. II. The induction and subcellular localization of tyrosinase activity in developing oocytes," *Develop. Biol.*, **36**, 330-342 (1974).

Epstein, C. I., "Expression of the mammalian X chromosome before and after fertilization," *Science*, **175**, 1467-1468 (1972).

Epstein, C. I., and S. A. Smith, "Amino acid uptake and protein synthesis in preimplantation mouse embryos," *Develop. Biol.*, **33**, 171-184 (1973).

Epstein, C. I., and S. A. Smith, "Electrophoretic analysis of proteins synthesized by preimplantation mouse embryos," *Develop. Biol.*, **40**, 233-244 (1974).

Ernst, S. G., B. R. Hough-Evans, R. J. Britten, and E. H. Davidson, "Limited complexity of the RNA in micromeres of sixteen-cell sea urchin embryos," *Develop. Biol.*, **79**, 119-127 (1980).

Evans, D., and M. Birnstiel, "Localization of amplified ribosomal DNA in the oocyte of *Xenopus laevis*," *Biochim. Biophys. Acta*, **166**, 274-276 (1968).

Evans, L. E., and H. Ozaki, "Nuclear histones in unfertilized sea urchin eggs," *Exp. Cell Res.*, **79**, 228-231 (1973).

Failly-Crepin, C., and G. R. Martin, "Protein synthesis and differentiation in a clonal line of teratocarcinoma and in preimplantation mouse embryo," *Cell Different.*, **8**, 61-73 (1979).

Fan, H., and S. Penman, "Regulation of synthesis and processing of nucleolar components in metaphase-arrested cells," *J. Mol. Biol.*, **59**, 27-38 (1971).

Fansler, B., and L. A. Loeb, "Sea urchin nuclear DNA polymerase. II. Changing localization during early development," *Exp. Cell Res.*, **57**, 305-310 (1969).

Fantoni, A., S. Bordin, and M. Lunadei, "Control of ribosomal RNA maturation in differentiating yolk sac erythroid cells," *Cell Different.*, **1**, 219-228 (1972).

Faulhabe, I., and L. Lyra, "Studies of activity and isoenzyme patterns of LDH during development of *Xenopus laevis, Triturus alpestris* and *vulgaris*, and *Ambystoma mexicanum*," *Wilh. Roux.' Arch. Entwickl. Org.*, **176**, 191-205 (1975).

Fedecka-Bruner, B., M. Anderson, and D. Epel, "Control of enzyme synthesis in early sea urchin development: aryl-sulfatase activity in normal and hybrid embryos," *Develop. Biol.*, **25**, 655-671 (1971).

Fedoroff, N. V., "On spacers," *Cell*, **16**, 697-710 (1979).

Fein, A., A. Bdolah, and E. Kochva, "Developmental pattern of enzyme secretion in the embryonic venom glands of *Vipera palaestina* (Ophidia, Reptilia)," *Develop. Biol.*, **24**, 520-532 (1971).

Feldherr, C. M., "The uptake of endogeneous proteins by oocyte nuclei," *Exp. Cell Res.*, **93**, 411-419 (1975).

Feldherr, C. M., and P. A. Richmond, "The localization of the nuclear protein pool in *Xenopus* oocytes," *Exp. Cell Res.*, **100**, 417-419 (1976).

Fessler, J. H., and L. I. Fessler, "Biosynthesis of procollagen," *Ann. Rev. Biochem.*, **47**, 129-162 (1978).

Ficq, A., "Synthesis and detection of DNA in early oogenesis," *Exp. Cell Res.*, **53**, 691-693 (1968).

Ficq, A., F. Aiello, and E. Scarano, "Métabolism des acides nucléiques dans l'oeuf d'oursin en développement," *Exp. Cell Res.*, **29**, 128-136 (1963).

Ficq, A., and J. Brachet, "RNA-dependent DNA polymerase: possible role in the amplification of ribosomal DNA in *Xenopus* oocytes," *Proc. Nat. Acad. Sci. USA*, **68**, 2774-2776 (1971).

Filipowicz, W., J. M. Sierra, C. Nombela, S. Ochoa, W. C. Merrick, and W. F. Anderson, "Polypeptide chain initiation in eukaryotes: initiation factor requirements for translation of natural messengers," *Proc. Nat. Acad. Sci. USA*, **73**, 44-48 (1976).

Filipowicz, W., J. M. Sierra, and S. Ochoa, "Polypeptide chain initiation in eukaryotes: initiation factor MP in *Artemia salina* embryos," *Proc. Nat. Acad. Sci. USA*, **72**, 3947-3951 (1975).

Finamore, F. J., and J. S. Clegg, "Biochemical aspects of morphogenesis in the brine shrimp, *Artemia salina*," in: *Cell Cycle: Gene-Enzyme Interaction* (G. M. Padilla *et al.*, eds.), Acad. Press, New York (1969), pp. 249-281.

Finamore, F. J., D. J. Thomas, G. T. Grouse, and B. Llouyd, "Biochemistry of amphibian oocytes. I. Method of isolation and nucleic acid content of nuclei," *Arch. Biochem.*, **88**, 10-15 (1960).

Finamore, J., and E. Volkin, "Nucleotide and nucleic acid metabolism in developing amphibian embryos. II. Composition and metabolic activity of avarian egg nucleic acids," *Exp. Cell Res.*, **15**, 405-411 (1958).

Firtel, R. A., and A. Monroy, "Polysomes and RNA synthesis during early development of the surf clam, *Spisula solidissima*," *Develop. Biol.*, **21**, 87-104 (1970).

Fischmann, D. A., "The synthesis and assembly of myofibrils in embryonic muscle," *Current Topics Develop. Biol.*, **5**, 235-280 (1970).

Fitzmaurice, L. C., and R. F. Bauer, "Transcription ability of nuclear DNAs from different development stages of the sea urchin," *Biochim. Biophys. Acta*, **272**, 510-517 (1972).

Flickinger, R. A., "Evidence from sea urchin-sand dollar hybrid embryos for a nuclear control of alkaline phosphatase activity," *Biol. Bull.*, **112**, 21-28 (1957).

Flickinger, R. A., "Comparison of DNA and RNA synthesis in dorsal and ventral frog embryos ectoderm during gastrulation," *Wilh. Roux' Arch. Entwickl. Org.*, **171**, 256-258 (1972).

Flickinger, R., R. Greene, D. Kohl, and M. Miyagi, "Patterns of synthesis of DNA-like RNA in parts of developing frog embryos," *Proc. Nat. Acad. Sci. USA*, **56**, 1712-1718 (1966).

Flynn, J. M., and H. R. Woodland, "The synthesis of histone H1 during early amphibian development," *Develop. Biol.* **75**, 222-230 (1980).

Follett, B. K., T. J. Nicholls, and M. R. Redshaw, "The vitellogenic response in the South African clawed toad (*Xenopus laevis* Daudin)," *J. Cell Physiol.*, **72**, 91-102 (1968).

Ford, P. J., "Non-coordinated accumulation and synthesis of 5S RNA ribonucleic acid by ovaries of *Xenopus laevis*," *Nature*, **233**, 561-564 (1971).

Ford, P. J., T. Mathieson, and M. Rosbash "Very long-lived messenger RNA in ova ries of *Xenopus laevis*," *Develop. Biol.* **57**, 417-426 (1977).

Ford, P. J., and E. M. Southern, "Different sequences for 5S RNA in kidney cells and ovaries of *X. laevis*," *Nature New Biol.*, **241**, 7-12 (1973).

Frederiksen, S., and P. Hellung-Larsen, "Small molecular weight RNA components in sea urchin eggs," *Exp. Cell Res.*, **71**, 289-291 (1972).

Frederiksen, S., and P. Hellung-Larsen, "Synthesis of small molecular weight RNA components during early stages of sea urchin embryo development," *Exp. Cell Res.*, **89**, 217-227 (1974).

Frederiksen, S., P. Hellung-Larsen, and J. Engberg, "Small molecular weight RNA components in sea urchin embryos," *Exp. Cell Res.*, **78**, 287-294 (1973).

Freeman, S. B., "A comparison of certain isozyme patterns in lobeless and normal embryos of the snail *Ilyanassa obsoleta*," *J. Embryol. Exp. Morphol.*, **26**, 339-349 (1971).

Friedman, P. A., E. G. Platzer, and E. Y. Carroll, Jr., "Tubulin characterization during embryogenesis of *Ascaris suum*," *Develop. Biol.*, **76**, 47-57 (1980).

Fromson, D., and D. P. S. Verma, "Translation of nonadenylated messenger RNA of sea urchin embryos," *Proc. Nat. Acad. Sci. USA*, **73**, 148-151 (1976).

Fry, B. J., and P. R. Gross, "Patterns and rates of protein synthesis in sea urchin embryos. I. Uptake and incorporation of amino acids during the first cleavage cycle," *Develop. Biol.*, **21**, 105-124 (1970a).

Fry, B. J., and P. R. Gross, "Patterns and rates of protein synthesis in sea urchin embryos. II. The calculation of absolute rates," *Develop. Biol.*, **21**, 125-146 (1970b).

Furuichi, Y., M. Morgah, A. J. Shatkin, W. Jelinek, M. Salditt-Georgieff, and J. E. Darnell, "Methylation, blocked 5′ termini in HeLa cell mRNA," *Proc. Nat. Acad. Sci. USA*, **72**, 1904-1908 (1975).

Gabrielli, F., and C. Baglioni, "Maternal messenger RNA and histone synthesis in embryos of surf clam *Spisula solidissima*," *Develop. Biol.*, **43**, 254-263 (1975).

Gahmberg, C. G., K. Nilsson, and L. C. Andersson, "Specific changes in the surface glycoprotein pattern of human propyelocytic leukemic cell line HL-60 during morphologic and functional differentiation," *Proc. Nat. Acad. Sci. USA*, **76**, 4087-4091 (1979).

Galau, G. A., R. J. Britten, and E. H. Davidson, "A measurement of the sequence complexity of polysomal messenger RNA in sea urchin embryos," *Cell*, **2**, 9-22 (1974).

Galau, G. A., W. H. Klein, R. J. Britten, and E. H. Davidson, "Functional significance of mRNA sequences in the least abundant concentration class," *Arch. Biochem. Biophys.*, **179**, 584-599 (1977).

Galau, G. A., W. H. Klein, M. M. Davis, B. I. Wold, R. J. Britten, and E. H. Davidson, "Structural gene sets active in

embryos and adult tissues of the sea urchin," *Cell*, **7**, 485-505 (1976).
Gall, J. G., and H. G. Callan, "^{3}H-uridine incorporation in lampbrush chromosomes," *Proc. Nat. Acad. Sci. USA*, **48**, 562-570 (1962).
Gall, J. G., "Kinetics of deoxyribonuclease action on chromosomes," *Nature*, **198**, 36-38 (1963).
Gall, J. G., "Differential synthesis of the genes for r-RNA during amphibian oogenesis," *Proc. Nat. Acad. Sci. USA*, **60**, 553-560 (1968).
Gall, J. G., and M. Pardue, "Formation and detection of RNA-DNA hybrid molecules in cytological preparations," *Proc. Nat. Acad. Sci. USA*, **63**, 378-382 (1969).
Gambino, R., S. Metafora, L. Felicetti, and J. Raisman, "Properties of the ribosomal salt wash from unfertilized sea urchin eggs and its effect on natural mRNA translation," *Biochim. Biophys. Acta*, **312**, 377-391 (1973).
Ganion, L. R., and R. G. Kessel, "Intracellular synthesis, transport and packaging of proteinaceous yolk in oocytes of *Orconectes immunis*," *J. Cell Biol.*, **52**, 420-437 (1972).
Garrels, J., "Changes in protein synthesis during myogenesis in a clonal cell line," *Develop. Biol.*, **73**, 134-152 (1979).
Gasaryan, K., and N. Andreeva, "Changes in the template activity of chromatin from pigeon erythrocytes and erythroblasts induced by partial salt deproteinization," *FEBS Letters*, **27**, 263-266 (1972).
Gause, G. G., *The Mitochondrial DNA* [in Russian], Nauka, Moscow (1977).
Gause, G. G., and V. S. Mikhailov, "State of the DNA synthesizing system in isolated mitochondria from mature eggs of the loach (*Misgurnus fossilis*)," *Biochim. Biophys. Acta*, **324**, 189-198 (1973).
Gause, G. G., and V. S. Mikhailov, "The content of D-loop molecules in mitochondrial DNA of unfertilized oocytes and developing embryos of *Misgurnus fossilis*," *Dokl. Akad. Nauk SSSR*, **217**, 1197-1200 (1974).
Geyer-Duszynska, J., "Genetic factors in oogenesis and spermatogenesis in Cecidyidae," *Chromosomes Today*, **1**, 174-192 (1966).
Gilbert, W., and D. Dressler, "DNA replication: the rolling circle model," *Cold Spring Harbor Symp. Quant. Biol.*, **33**, 473-484 (1968).
Gineitis, A. A., H. H. Nivinskas, and V. I. Vorobyov (Vorobyev), "Non-histone chromatine proteins from the sea urchin *Strongylocentrotus droebachiensis* sperm and embryo," *Exp. Cell Res.*, **98**, 248-252 (1976).
Girard, M., and D. Baltimore, "The effect of HeLa cell cytoplasm on the rate of sedimentation of RNA," *Proc. Nat. Acad. Sci. USA*, **56**, 999-1002 (1966).
Giudice, G., *Developmental Biology of Sea Urchin Embryos*, Acad. Press, New York (1973).
Giudice, G., and V. Mutolo, "Synthesis of ribosomal RNA during sea urchin development," *Biochim. Biophys. Acta*, **138**, 276-285 (1967).
Giudice, G., and V. Mutolo, "Synthesis of ribosomal RNA during sea urchin development. II. Electrophoretic analysis of nuclear and cytoplasmic RNA," *Biochim. Biophys. Acta*, **179**, 341-347 (1969).
Giudice, G., G. Sconzo, I. Albanese, G. Ortolani, and M. Camurat, "Cytoplasmic giant RNA in sea urchin embryos. I. Proof that it is not derived from artifactual nuclear leakage," *Cell Different.*, **3**, 287-294 (1974).
Giudice, G., G. Sconzo, A. Bono, and I. Albanese, "Studies on sea urchin oocytes. I. Purification and cell fractionation," *Exp. Cell Res.*, **72**, 90-94 (1972a).
Giudice, G., G. Sconzo, F. Ramirez, and I. Albanese, "Giant RNA is also found in the cytoplasm in sea urchin embryos," *Biochim. Biophys. Acta*, 262, 401-403 (1972b).
Glišin, V R., and M. V. Glišin, "RNA metabolism following fertilization in sea urchin eggs," *Proc, Nat. Acad. Sci. USA*, **52**, 1548-1553 (1964).
Glišin, V. R., M. V. Glišin, and P. Doty,

"The nature of mRNA in the early stages of sea urchin development," *Proc. Nat. Acad. Sci. USA*, **56**, 285-294 (1966).

Glišin, V., A. Savic, S. Ruzdijić, N. Bajković, and R. Mileusnić, "Old and new informational macromolecules during early development of sea urchin embryos," *Studia Biophys.*, **31-32**, 345-357 (1972).

Goldsmith, M. R., and G. Basehoar, "Organization of the chorion genes of *Bombyx mori*, a multigene family. I. Evidence for linkage to chromosome 2," *Genetics*, **90**, 291-310 (1978).

Goldsmith, M. R., and E. Clermont-Rattner, "Organization of the chorion genes of *Bombyx mori*, a multigene family. II. Partial localization of three gene clusters," *Genetics*, **92**, 4 (1979).

Goldstein, E. S., "Translated and sequestered untranslated message sequences in *Drosophila* oocytes and embryos," *Develop. Biol.*, **63**, 59-66 (1978).

Goldstein, L., M. V. N. Rao, and D. M. Prescott, "The migration of RNA from cytoplasm to nucleus in *Amoeba proteus*," *Ann. Embryol. Morphol.*, Suppl. 1, 189-197 (1969).

Goldstein, P., and N. A. Straus, "Molecular characterization of *Ascaris suum* DNA and chromatin diminution," *Exp. Cell Res.*, **116**, 462-466 (1978).

Goswami, M. N. O., A. J. Rosenberg, and F. Menry, "A comparative analysis of the ontogenetic development of rat liver sequential enzymes — tyrosine α-ketoglutarate aminotransferase hydroxyphenylpyruvate hydroxylase and homogentisate oxygenase," *Develop. Biol.*, **30**, 129-136 (1973).

Gould, M., "RNA and protein synthesis in the unfertilized eggs of *Urechis caupo*," *Develop. Biol.*, **19**, 460-481 (1969a).

Gould, M., "A comparison of RNA and protein synthesis in fertilized and unfertilized eggs of *Urechis caupo*," *Develop. Biol.*, **19**, 482-497 (1969b).

Govaert, J., "Etude quantitative de la teneur en acide desoxyribonucléique des noyaux des cellules somatiques et germinatives chez *Fasciola hepatica*," *Arch. Biol.*., **68**, 165 (1957).

Graham, C. F., "The regulation of DNA synthesis and mitosis in multinucleate frog eggs," *J. Cell Sci.*, **1**, 363 (1966).

Graham, C. F., *Regulation of Mammalian Reproduction* (S. Segal, ed.), Thomas Publ. Co., Springfield (1973), pp. 286-301.

Graham, C. F., K. Arms, and J. B. Gurdon, "The induction of DNA synthesis by frog egg cytoplasm," *Develop. Biol.*, **14**, 349-381 (1966).

Graham, C. F., and R. W. Morgan, "Changes in the cell cycle during early amphibian development," *Develop. Biol.*, **14**, 439-460 (1966).

Grainger, R. M., and F. H. Wilt, "Incorporation of C^{13}, N^{15}-labeled nucleosides and measurement of RNA synthesis and turnover in sea urchin embryos," *J. Mol. Biol.*, **104**, 589-601 (1976).

Grainiger, J. L., M. M. Winkler, S.S. Shen, and R. A. Steinhardt, "Intracellular pH control of protein synthesis rate in the sea urchin egg and early embryo," *Develop. Biol.*, **68**, 396-406 (1979).

Grant, P., "The synthesis of deoxyribonucleic acid during early development of *Rana pipiens*.," *J. Cell Comp. Physiol.*, **52**, 227-248 (1958).

Grant, P., "Informational molecules and embryonic development," in: *Biochemistry of Animal Development*, Vol. 1 (R. Weber, ed.) Acad. Press, New York (1965), pp. 483-594.

Graziosi, G., and D. R. Robertis, "Molecular anisotropy of the early *Drosophila* embryo," *Nature*, **258**, 1577-159 (1975).

Green, H., B. Goldberg, M. Schwartz, and D. D. Brown, "The synthesis of collagen during the development of *Xenopus laevis*., *Develop. Biol.*, **18**, 391-400 (1968).

Greenberg, J. R., "Synthesis and properties of ribosomal RNA in *Drosophila*," *J. Mol. Biol.*, **46**, 85-98 (1969).

Greene, R. F., and R. A. Flickinger, "Qualitative changes in DNA-like RNA during development in *Rana pipiens*," *Biochim. Biophys. Acta*, **217**, 447-460 (1970).

Gregg, J. R., and S. Løvtrup, "Synthesis of desoxyribonucleic acid in lethal amphibian hybrids," *Biol. Bull.*, **108**, 29-34 (1955).

Grosfeld, H., H. Sorq, and U. Z. Littauer, "Membrane associated cytoplasmic mRNA in *Artemia salina;* functional and physical changes during development," *Nucleic Acids Res.*, **4**, 2109-2121 (1977).

Gross, K. W., M. Jacobs-Lorena, C. Baglioni, and P. R. Gross, "Cell-free translation of maternal messenger RNA from sea urchin eggs," *Proc. Nat. Acad. Sci. USA*, **70**, 2614-2618 (1973a).

Gross, K. W., J. Ruderman, M. Jacobslo, C. Baglioni, and P. R. Gross, "Cell-free synthesis of histone directed by messenger RNA from sea urchin embryos," *Nature New Biol.*, **241**, 272-274 (1973b).

Gross, P. R., "The immediacy of genomic control during early development," *J. Exp. Zool.*, **157**, 21-38 (1964).

Gross, P., "The control of protein synthesis in embryonic development and differentiation," *Current Topics Develop. Biol.*, **2**, 1-45 (1967).

Gross, P. R., and G. H. Cousinou, "Macromolecule synthesis and the influence of actinomycin on early development," *Exp.* Cell. Res., 33, 368—395 (1964).

Gross, P. R., and B. J. Fry, "Continuity of protein synthesis through cleavage metaphase," *Science,* **153**, 749-751 (1966).

Gross, P. R., L. J. Malkin, and M. Hubbard, "Synthesis of RNA during the oogenesis in the sea urchin," *J. Mol. Biol.*, **13**, 463-469 (1965).

Gross, P. R., L. J. Malkin, and W. A. Moyer, "Templates for the first proteins of embryonic development," *Proc. Nat. Acad. Sci. USA*, **51**, 407-414 (1964).

Grossmann, A., S. Inoue, and L. Fischman, "Release of particulate structure and TAME hydrolase activity from sea urchin eggs after fertilization," *Nature New Biol.*, **243**, 279-280 (1973b).

Grossman, A., M. Levy, W. Troll, and G. Weissmann, "Redistribution of tosylarginin methylester hydrolase activity after fertilization of sea urchin *Arbacia punctulata* eggs," *Nature New Biol.*, **243**, 277-278 (1973a).

Gurdon, J. B., "On the origin and persistence of a cytoplasmic state inducing nuclear DNA synthesis in frog's eggs," *Proc. Nat. Acad. Sci. USA*, **58**, 545-552 (1967).

Gurdon, J. B., "Nucleic acids synthesis in embryos," *Essays Biochem.*, **4**, 25-68 (1969).

Gurdon, J. B., "Nuclear transplantation and the control of gene activity in animal development," *Proc. Roy. Soc. B.*, **176**, 303-314 (1970).

Gurdon, J. B., *The Control of Gene Expression in Animal Development*, Oxford Univ. Press, London, New York (1974).

Gurdon, J. B., "Gene expression in early animal development: the study of its control by the microinjection of amphibian eggs," *Harvey Lecture,* 1973-1974, Ser. 69, New York (1975), pp. 49-69.

Gustafson, T., "Enzymatic aspects of embryonic differentiation," *Int. Rev. Cytol.*, **3**, 277-327 (1954).

Gustafson, T., "Morphogenetic significance of biochemical patterns in sea urchin embryos," in: *The Biochemistry of Animal Development*, Vol. 1 (R. Weber, ed.), Acad. Press, New York (1965).

Gustafson, T., and P. Lenicque, "Studies on mitochondria in the developing sea urchin egg," *Exp. Cell Res.*, **3**, 251-274 (1952).

Gustafson, T., and P. Lenicque, "Studies on mitochondria in early cleavage stages of the sea urchin egg," *Exp. Cell Res.*, **8**, 114-117 (1955).

Haba, G., G. W. Cooper, and V. Elting, "Hormonal requirements for myogenesis of striated muscle *in vitro:* insulin and somatotropin," *Proc. Nat. Acad. Sci. USA*, **56**, 1719-1723 (1966).

Hadje-Azimi, I., "Studies on *Xenopus laevis* immunoglobulins," *Immunology*, **21**, 463-473 (1971).

Hadje-Azimi, I., and M. Michea-Hamzehpour, "*Xenopus laevis* 19S immunoglobulin: ultrastructure and J-chain isolation," *Immunology*, **30**, 587-591 (1976).

Hadjiolov, A. A., "Patterns of ribosome biogenesis in eukaryotes," *Trends Biochem. Soc.*, **2**, 84-86 (1977).

Hagenmaier, N. E., "Der Nucleinsäure- bzw. Ribonucleoproteid-Status während der Früentwicklung von Fischkeimen (*Salmo irideus* und *Salmo trutto fario*)," *Wilh. Roux' Arch. Entwickl. Org.*, **162**, 19-40 (1969).

Haggis, A., "Deoxyribonucleic acid in germinal vesicles of oocytes of *Rana pipiens*," *Science*, **154**, 670 (1966).

Hall, T. J., and M. R. Cummings, "*In vitro* synthesis and processing of ribosomal RNA in the housefly ovary," *Develop. Biol.*, **46**, 233-242 (1975).

Hallberg, R. L., "Mitochondrial DNA in *Xenopus laevis* oocytes. I. Displacement loop occurrence," *Develop. Biol.*, **38**, 346-351 (1974).

Hallberg, R. L., and D. D. Brown, "Coordinated synthesis of some ribosomal proteins and ribosomal RNA in embryos of *Xenopus laevis*," *J. Mol. Biol.*, **46**, 393-411 (1969).

Hallberg, R. L., and D. C. Smith, "Ribosomal protein synthesis in *Xenopus laevis*," *Develop. Biol.*, **42**, 40-52 (1975).

Hand, G. S., "Stimulation of protein synthesis in unfertilized sea urchin and sand dollar eggs treated with trypsin," *Exp. Cell Res.*, **64**, 204-208 (1971).

Hanocq, F., M. Kirsch-Volders, J. Hanocq-Quertier, E. Baltus, and G. Steinert, "Characterization of yolk DNA from *Xenopus laevis* oocytes ovulated *in vitro*," *Proc. Nat. Acad. Sci. USA*, **69**, 1322-1326 (1972).

Hanocq-Quertier, J., E. Baltus, A. Ficq, and J. Brachet, "Studies on the DNA of *Xenopus laevis* oocytes," *J. Embryol. Exp. Morphol.*, **19**, 273-282 (1968).

Hansen-Delkeskamp, E., "Satelliten-Deoxyribonucleinsäuren in Gonaden und somatischen Gewebe von *Acheta domestica*," *Z. Naturforschung*, **24b**, 1331-1335 (1969).

Harding, C. V., D. Harding, and P. Perlmann, "Antigens in sea urchin hybrid embryos," *Exp. Cell Res.*, **6**, 202-210 (1954).

Harris, S. E., and H. S. Forrest, "RNA and DNA synthesis in developing eggs of the milkweed bug, *Oncopeltus fasciatus*," *Science*, **156**, 1613-1615 (1967).

Hartmann, J. F., and D. G. Comb, "Transcription of nuclear cytoplasmic genes during early development of sea urchin embryos," *J. Mol. Biol.*, **41**, 155-158 (1969).

Hartmann, J. F., M. H. Ziegler, and D. G. Comb, "Sea urchin embryogenesis. I. RNA synthesis by cytoplasmic and nuclear genes during development," *Develop. Biol.*, **25**, 209-231 (1971).

Hassel, J., and N. W. Klein, "Quantitative analysis of ovalbumin utilization by the cultured chick embryo and its relationship to growth regulation during development," *Develop. Biol.*, **26**, 380-392 (1971).

Hatlen, L., and G. Attardi, "Proportion of the HeLa cell genome complementary to tRNA and 5S RNA," *J. Mol. Biol.*, **56**, 535-553 (1971).

Hausman, S. J., S. M. Anderson, and W. H. Telfer, "The dependence of *Cecropia* yolk formation *in vitro* on specific blood proteins," *J. Cell Biol.*, **48**, 303-313 (1971).

Heideman, S. R., J. Townsend, and R. Tompkins, "Synthesis of soluble protein in oocytes of *Xenopus laevis*," *J. Exp. Zool.*, **191**, 253-261 (1975).

Hennig, W., "RNA synthesis of the Y-chromosomes of *Drosophila hydei*," *J. Mol. Biol.*, **38**, 227-239 (1968).

Hentschel, C. C., R. M. Kay, and J. G. Williams, "Analysis of *Xenopus laevis* globins during development and erythroid cell maturation and the construction of recombinant plasmids containing sequences derived from adult globin mRNA," *Develop. Biol.*, **72**, 350-363 (1979).

Herrmann, H., S. M. Heywood, and A. C. Marchok, "Reconstruction of muscle development as a sequence of macromolecular synthesis," *Current Topics Develop. Biol.*, **5**, 181-234 (1970).

Herrmann, H., and M. L. Tootle, "Specific and general aspects of the development of enzymes and metabolic pathways," *Physiol. Rev.*, **44**, 289-371 (1964).

Hess, O., "The function of the lampbrush loops formed by the Y-chromosome of *Drosophila hydei* spermatocyte nuclei," *Mol. Genet.*, **103**, 58-71 (1968).

Heywood, S. M., and A. Rich, "*In vitro* synthesis of native myosin, actin and tropomyosin from embryonic chick polyribosomes," *Proc. Nat. Acad. Sci. USA*, **59**, 590-597 (1968).

Hickey, E. D., L. A. Weber, and C. Baglioni, "Inhibition of initiation of protein synthesis by 7-methylguanosine-5-monophosphate," *Proc. Nat. Acad. Sci. USA*, **73**, 19-23 (1976).

Hinegardner, R. T., "The DNA content of isolated sea urchin nuclei," *Exp. Cell Res.*, **25**, 341-347 (1961).

Hinegardner, R. T., B. Rao, and D. E. Feldman, "The DNA synthetic period during early development of the sea urchin egg," *Exp. Cell Res.*, **36**, 53-61 (1964).

Hirama, M. N., and Y. Mano, "Polysomes of the sea urchin embryo," *Exp. Cell Res.*, **86**, 15-24 (1974).

Hnilica, L. S., and A. W. Johnson, "Fractionation and analysis of nuclear proteins in sea urchin embryos," *Exp. Cell Res.*, **63**, 261-270 (1970).

Hoff-Jørgensen E., "Deoxynucleic acid in some gametes and embryos," *Proc. Sympos. Colston Res. Soc.*, **7**, 79-90 (1954).

Hoff-Jørgensen, E., and E. Zeuthen, "Evidence of cytoplasmic desoxyribosides in the frog's egg," *Nature*, **169**, 245-246 (1952).

Hogan, B., and P. R. Gross, "Nuclear RNA synthesis in sea urchin embryos," *Exp. Cell Res.*, **72**, 101-114 (1972).

Holmes, R. S., and G. S. Whitt, "Developmental genetics of the esterase isozymes of *Fundulus heteroclitus*," *Biochem. Genet.*, **4**, 471-480 (1970).

Holt, C. E., and E. G. Gurney, "Minor components of the DNA of *Physarum polycephalum*," *J. Cell Biol.*, **40**, 484-496 (1969).

Hough, B. R., and E. H. Davidson, "The RNA of the mature *Xenopus laevis*," in: *Handbook of Genetics*, Vol. 4, New York (1976), pp. 51-55.

Hough, B. R., M. J. Smith, R. J. Britten, and E. H. Davidson, "Sequence complexity of heterogeneous nuclear RNA in sea urchin embryos," *Cell*, **5**, 291-299 (1975).

Hough, B. R., P. H. Yancey, and E. H. Davidson, "Persistence of maternal RNA in *Engystomops* embryos," *J. Exp. Zool.*, **185**, 357-368 (1973).

Hough-Evans, B. R., S. G. Ernst, R. J. Britten, and E. H. Davidson, "RNA complexity in developing sea urchin oocytes," *Develop. Biol.*, **69**, 258-269 (1979).

Hough-Evans, B. R., M. Jacobs-Lorena, M. R. Cummings, R. J. Britten, and E. H. Davidson, "Complexity of RNA in eggs of *Drosophila melanogaster* and *Musca domestica*," *Genetics*, **95**, 81-94 (1980).

Hough-Evans, B. R., B. J. Wold, S. G. Ernst, R. J. Britten, and E. H. Davidson, "Appearance and persistence of maternal RNA sequences in sea urchin development," *Develop. Biol.*, **60**, 258-277 (1977).

Hourcade, D., D. Dressler, and J. Wolfson, "The amplification of ribosomal RNA genes involves a rolling circle intermediate," *Proc. Nat. Acad. Sci. USA*, **70**, 2926-2930 (1973a).

Hourcade, D., D. Dressler, and J. Wolfson, "The nucleolus and the rolling circle," *Cold Spring Harbor Symp. Quant. Biol.*, **38**, 537-549 (1973b).

Hsieh, B., M. M.-Y. Chi, J. Knor, and O. H. Lowry, "Enzymes of glycogen metabolism and related metabolites in preimplantation mouse embryos," *Develop. Biol.*, **72**, 342-349 (1979).

Hultin, T., "The protein metabolism of sea urchin eggs during early development studied by means of ^{15}N-labeled ammonia," *Exp. Cell Res.*, **1**, 599-602 (1950).

Hultin, T., "Activation of ribosomes in sea urchin eggs in response to fertilization," *Exp. Cell Res.*, **25**, 405-417 (1961).

Hultin, T., and J. Morris, "The ribosomes of encysted embryos of *Artemia salina* during cryptobiosis and resumption of development," *Develop. Biol.*, **17**, 143-164 (1968).

Hultin, T., and G. Wessel, "Incorporation of ^{14}C-labeled carbon dioxide into the protein of developing sea urchin eggs," *Exp. Cell Res.*, **3**, 613-616 (1952).

Humphrey, R. R., "A recessive factor (O, for ova deficient) determining a complex of abnormalities in the Mexican axolotl (*Ambystoma mexicanum*)," *Develop. Biol.*, **13**, 57-76 (1966).

Humphreys, T., "Efficiency of translation of mRNA before and after fertilization in sea urchin," *Develop. Biol.*, **20**, 435-458 (1969).

Humphreys, T., "Measurements of messenger RNA entering polysomes upon fertilization of sea urchin eggs," *Develop. Biol.*, **26**, 201-208 (1971).

Hynes, R.O., G.A. Greenhouse, R. Minkoff, and P.R. Gross, "Properties of the three cell types in sixteen-cell sea urchin embryos: RNA synthesis," *Develop. Biol.*, **27**, 457-478 (1972b).

Hynes, R. O., and P. R. Gross, "Informational RNA sequences in early sea urchin embryos," *Biochim. Biophys. Acta*, **259**, 104-111 (1972).

Hynes, R.O., A. Raff, and P.R. Gross, "Properties of the three cell types in sixteen-cell sea urchin embryos: aggregation and microtubule protein synthesis," *Develop. Biol.*, **27**, 150-164 (1972a).

Iatrou, K., A.W. Spira, and G.H. Dixon, "Protamine messenger RNA: evidence for early synthesis and accumulation during spermatogenesis in rainbow trout," *Develop. Biol.*, **64**, 82-98 (1978).

Ilan, Ju., and J. Ilan, "Translation of maternal messenger ribonucleoprotein particles from sea urchin in a cell-free system from unfertilized eggs and product analysis," *Develop. Biol.*, **66**, 375-385 (1978).

Illmensee, K., and A.P. Mahowald, "The autonomous function of germ plasm in a somatic region of the *Drosophila* egg," *Exp. Cell Res.*, **97**, 127-140 (1976).

Imoh, H., "Changes in H1 histone during development of newt embryos," *Exp. Cell Res.*, **108**, 57-62 (1977).

Imoh, H., "Re-examination of histone changes during development of newt embryos," *Exp. Cell Res.*, **113**, 23-29 (1978).

Imoh, H., and J. Kawakami, "Histone synthesis during development of *Triturus* embryos," *J. Embryol. Exp. Morphol.*, **30**, 73-82 (1973).

Infante, A.A., and P.N. Graves, "Stability of free ribosomes and polysomes of the sea urchin," *Biochim. Biophys. Acta*, **246**, 100-110 (1971).

Infante, A.A., and M. Nemer, "Accumulation of newly synthesized RNA templates in a unique class of polyribosomes during embryogenesis," *Proc. Nat. Acad. Sci. USA*, **58**, 681-688 (1967).

Ingwall, J.S., C.D. Weiner, M.F. Morales, E. Davis, and F.E. Stockdale, "Specificity of creatine in the control of muscle protein synthesis," *J. Cell Biol.*, **63**, 145-151 (1974).

Inman, R., "Denaturation maps of the left and right sides of the lambda DNA molecule determined by electron microscopy," *J. Mol. Biol.*, **28**, 103 (1967).

Inman, R., and M. Schnos, "Partial denaturation of thymine- and 5-bromouracil-containing λ DNA in alkali," *J. Mol. Biol.*, **49**, 93-98 (1970).

Innis, M. A., and S. P. Craig, "Mitochondrial regulation in sea urchins. II. Formation of polyribosomes within the mitochondria of 4-8 cell stage embryos of the sea urchin," *Exp. Cell Res.*, **3**, 223-230 (1978).

Ivanenkov, V.V., "The expression of the paternal genes of lactatdehydrogenase, glutamatdehydrogenase and acetylcholine esterase in the development of interfamilial hybrids between the Cobitidae and Cyprinidae (pisces)," *Ontogenez*, **7**, 579-589 (1976).

Ivanenkov, V.V., and G.N. Korobtsov, "Thermostability of acetylcholine esterase and non-specific esterases during the development of sea urchin hybrids *Strongylocentrotus droebachiensis* × *Strongylocentrotus intermedius*," *Ontogenez*, **7**, 341-349 (1976).

Izawa, M., V.G. Allfrey, and A.E. Mirsky, "The relationship between RNA synthesis and loop structure in lampbrush chromosomes," *Proc. Nat. Acad. Sci. USA*, **49**, 544-551 (1963a).

Izawa, M., V.G. Allfrey, and A.E Mirsky, "Composition of the nucleus and chromosomes in the lampbrush stage of the newt oocyte," *Proc. Nat. Acad. Sci. USA*, **50**, 811-817 (1963b).

Jacobs-Lorena, M., B.R. Hough-Evans, R.J. Britten, and E.H. Davidson, "Complexity of RNA in developing oocytes of *Drosophila melanogaster*," *Develop. Biol.*, **76**, 509-513 (1980).

Jackson, J.F., R.M. Clayton, R. Williamson, I. Thomson, D.E.S. Truman, and D.I. de Pomerai, "Sequence complexity and tissue distribution of chick lens crystallin mRNAs," *Develop. Biol.*, **65**, 383-395 (1978).

Jain, K.S., and S. Sarkar, "Poly(riboadenylate)-containing messenger ribonucleoprotein particles of chick embryonic muscles," *Biochemistry*, **18**, 745-752 (1979).

Jamamoto, K., "Studies on the formation of fish eggs. II. Changes in the nucleus of the oocyte of *Liopsetta obscura* with special reference to the activity of the nucleolus," *J. Fac. Sci. Hokkaido Univ.*, Ser. 6, **12**, 375-384 (1956).

Jared, D.W., and R.A. Wallace, "Comparative chromatography of the yolk proteins of teleosts," *Compar. Biophys. Physiol.*, **24**, 437-443 (1968).

Jaworska, H., and A. Lima-de-Faria, "Amplification of ribosomal DNA in *Acheta*. VI. Ultrastructure of two types of nuclear component associated with ribosomal DNA," *Hereditas*, **74**, 169-186 (1973a).

Jaworska, H., and A. Lima-de-Faria, "Amplification of ribosomal DNA in *Acheta*. VII. Transfer of DNA-RNA assemblies from the nucleus to the cytoplasm," *Hereditas*, **74**, 186-199 (1973b).

Jenkins, K.D., and P.C. Denny, "Effects of fertilization on the ribosomal subunit pool of sea urchin egg," *Biochim. Biophys. Acta*, **217**, 206-208 (1970).

Jenkins, N.A., J.F. Kaumeyer, E.M. Young, and R.A. Raff, "A test for masked message: the template activity of messenger ribonucleoprotein particles isolated from sea urchin eggs," *Develop. Biol.*, **63**, 279-298 (1978).

John, H., M.L. Birnstiel, and K.W. Jones, "RNA-DNA hybrids at the cytological level," *Nature*, **223**, 582-587 (1969).

Johnson, A.W., and L.S. Hnilica, "Cytoplasmic and nuclear basic protein synthesis during early sea urchin development," *Biochim. Biophys. Acta*, **246**, 141-154 (1971).

Johnson, A.W., J.A. Wilhelm, D.N. Ward, and L.S. Hnilica, "The composition of sea urchin sperm and embryo histones," *Biochim. Biophys. Acta*, **259**, 140-149 (1973a).

Johnson, A.W., J.A. Wilhelm, and L.S. Hnilica, "Nuclear basic protein acetylation during early sea urchin development," *Biochim. Biophys. Acta*, **295**, 150-158 (1973b).

Johnson, J.D., and D. Epel, "Relationship between release of surface proteins and metabolic activation of sea urchin eggs at fertilization," *Proc. Nat. Acad. Sci. USA*, **72**, 4474-4478 (1975).

Kafatos, F.C. "The cocoonase zymogen cells of silk moths: a model of terminal cell differentiation for specific protein synthesis," *Current Topics Develop. Biol.*, **7**, 125-192 (1972).

Kafatos, F.C., A. Efstratiadis, M.R. Goldsmith, C.W. Jones, T. Maniatis, J.C. Regier, G. Rodakis, N. Rosenthal, G.K. Sim, G. Thireos, and L. Villa-Komaroff, "The developmentally regulated multigene families encoding chorion protein in silk moths," in: *Differentiation and Development, Miami Winter Symp.*, Vol. 15, (F. Ahmad, J. Schultz, T. R. Russel, R. Werner, eds.), Acad. Press, New York (1978), pp. 299-315.

Kafiani, K. A., R. S. Akhalkatsi, and K. G. Gasaryan, "Nuclear RNA polymerase activity and template efficiency of developing loach (*Misgurnus fossilis*) embryos," *Biochim. Biophys. Acta*, **324**, 133-142 (1973).

Kafiani, K.A., and M.Ya. Timofeeva, "The nuclear RNA synthesis in early embryonic development," *Dokl. Akad. Nauk SSSR*, **154**, 721-724 (1964).

Kafiani, K.A., and M.Ya. Timofeeva, "The dynamics of nuclear RNA synthesis in early embryogenesis," in: *Cell Differentiation and Induction Mechanisms* [in Russian], Nauka, Moscow (1965).

Kafiani, K.A., M.Ya. Timofeeva, A.A. Neyfakh, N.L. Melnikova, and J.A. Rachkus, "RNA synthesis in the early embryogenesis of a fish (*Misgurnus fossilis*)," *J. Embryol. Exp. Morphol.*, **21**, 295-308 (1969).

Karp, G., C. Manes, and W.E. Hahn, "Ribosome production and protein synthesis in the preimplantation rabbit embryo," *Develop. Biol.*, **31**, 404-408 (1973).

Karp, G.C., and M.D. Weems, "H^3-amino acid uptake and incorporation in sea urchin gastrulae and exogastrulae: an autoradiographic study," *J. Exp. Zool.*, **194**, 535-546 (1975).

Kasamatsu, H., D. H. Robberson, and J. Vinograd, "A novel closed-circular mitochondrial DNA with properties of a replicating intermediate," *Proc. Nat. Acad. Sci. USA*, **68**, 2252-2257 (1971).

Kasinsky, N. E., N. K. Bols, and E. V. Berd, "On the histone synthesis during spermatogenesis in amphibian embryos," *Ontogenez*, **4**, 558-567 (1973).

Kaulenas, M. S., and D. Fairbairn, "Ribonuclease-stable polysomes in the egg of *Ascaris lumbricoides*," *Develop. Biol.*, **14**, 481-494 (1966).

Kaulenas, M. S., and D. Fairbairn, "RNA metabolism of fertilized *Ascaris lumbricoides* eggs during uterine development," *Exp. Cell Res.*, **52**, 233-251 (1968).

Kaulenas, M. S., W. E. Foor, and D. Fairbairn, "Ribosomal RNA synthesis during cleavage of *Ascaris lumbricoides* eggs," *Science*, **136**, 1201-1203 (1969).

Kaulenas, M. S., and B. R. Unsworth, "Synthesis of ribosomal proteins during sea urchin early embryonic development," *Mol. Gen. Genet.*, **135**, 231-243 (1974).

Kaumeyer, J. F., N. A. Jenkins, and R. A. Raff, "Messenger ribonucleoprotein particles in unfertilized sea urchin eggs," *Develop. Biol.*, **63**, 266-278 (1978).

Kedes, L. H. "Histone messengers and histone genes (review)," *Cell*, 8, 321-331 (1976).

Kedes, L. H., and M. L. Birnstiel, "Reiteration and clustering of DNA sequences complementary to histone messenger RNA," *Nature*, **230**, 165-169 (1971).

Kedes, L. H., and P. R. Gross, "Synthesis and function of messenger RNA during early embryonic development," *J. Mol. Biol.*, **42**, 559-575 (1969).

Kedes, L. H., P. R. Gross, G. Cognetti, and A. L. Hunter, "Synthesis of nuclear and chromosomal proteins on light polyribosomes during cleavage in sea urchin embryo," *J. Mol. Biol.*, **45**, 337-351 (1969a).

Kedes, L. H., B. Hogan, G. Cognetti, S. Selvig, P. Yanover, and P. R. Gross, "Regulation of translation and transcription of messenger RNA during early embryonic development," *Cold Spring Harbor Symp. Quant. Biol.*, **34**, 717-723 (1969b).

Kedes, L., and L. Stavy, "Structural and functional identity of ribosomes from eggs and embryos of sea urchin," *J. Mol. Biol.*, **43**, 337-340 (1969).

Kessel, R. G., "Mechanisms of protein yolk synthesis and deposition in crustacean oocytes," *Z. Zellforsch. Mikroskop. Anat.*, **89**, 17-38 (1968).

Khesin, R. B., "The instability of the genome," *Mol. Biol.*, **14**, 1205-1233 (1980).

Kidder, G. M., "Gene transcription in mosaic embryos. 1. The pattern of RNA synthesis in early development of the coot clam *Mulinia lateralis*," *J. Exp. Zool.*, **180**, 55-74 (1972).

Kidder, G. M., "The ribosomal RNA cistrons in clam gametes," *Develop. Biol.*, **49**, 132-142 (1976).

Kierszenbaum, A. L., and L. L. Tres, "Structural and transcriptional features of the mouse spermatid genome," *J. Cell Biol.*, **65**, 258-270 (1975).

20. Зак. 1447

Kijima, S., and F. H. Wilt, "Rate of nuclear ribonucleic acid turnover in sea urchin embryos," *J. Mol. Biol.*, **40**, 235-246 (1969).

Kiknadze, I. I., *Functional Organization of the Chromosomes* [in Russian], Nauka, Moscow, Leningrad (1972).

Kimmel, C. B., S. K. Sessions, and M. C. MacLeod, "Evidence for an association of most nuclear RNA with chromatin," *J. Mol. Biol.*, **102**, 177-192 (1976).

Kindas-Mügge, I., C. D. Lans, and G. Kreil, "Insect protein synthesis in frog cells. The translation of honey bee promilittin messenger RNA in *Xenopus* oocytes," *J. Mol. Biol.*, **87**, 451-462 (1974).

Kirby, K. S., and E. A. Cook, "Isolation of deoxyribonucleic acid from mammalian tissues," *Biochem. J.*, **104**, 254-257 (1967).

Kirnos, M. D., "The structure of mitochondrial DNA in the animals (the nucleotide composition, pyrimidine sequences, characteristics of methylation)," *Author's Abstract of Candidate's Dissertation*, Moscow (1975).

Kirschner, R. H., D. R. Wolstenholme, and N. J. Gross, "Replicating molecules of circular mitochondrial DNA," *Proc. Nat. Acad. Sci. USA*, **60**, 1466-1472 (1968).

Kistler, A., and R. Weber, "Enzyme patterns in mitochondria of eggs, liver, and skeletal muscle during larval development of *Xenopus*," *Develop. Biol.*, **37**, 236-247 (1974).

Klose, J., and H. von Wallenberg-Pachaly, "Changes of soluble protein populations during organogenesis of mouse embryos as revealed by protein mapping," *Develop. Biol.*, **51**, 324-331 (1976).

Klyachko, O. S., and A. A. Neyfakh, "Distribution of endogeneous and injected lactate dehydrogenase between blastoderm and yolk in the loach embryo," *Ontogenez*, **11**, 476-483 (1980a).

Klyachko, O. S., and A. A. Neyfakh, "Isolation and characteristics of LDH-containing complexes from the loach embryos," *Ontogenez*, **11**, 485-491 (1980b).

Knowland, J. S., "Polyacrylamide gel electrophoresis of nucleic acids synthesized during the early development of *Xenopus laevis*," *Biochim. Biophys. Acta*, **204**, 416-429 (1970).

Knowland, J., and C. Graham, "RNA synthesis at the two-cell stage of mouse development," *J. Embryol. Exp. Morphol.*, **27**, 167-176 (1972).

Koch, E. A., P. A. Smith, and R. C. King, "The division and differentiation of *Drosophila* cystocytes," *J. Morphol.*, **121**, 55-70 (1967).

Konovalov, Yu. D., and A. Ya. Mestechkina, "Peptide hydrolase activity in embryonic development of the carp," *Ontogenez*, **6**, 201-204 (1975).

Korochkin, L. I., *Genes Interactions during Development* [in Russian], Nauka, Moscow (1977).

Koster, J. G., H. E. Kasinsky, and O. H. I. Destree, "Newly synthesized histones in chromatin of early embryos of *Xenopus laevis*," *Cell Different.*, **8**, 93-104 (1979).

Kostomarova, A. A., and N. S. Korobtsova, "An autoradiographic study of nuclear RNA synthesis in intact embryos and isolated cells of the loach *Misgurnus fossilis*," *Dokl. Akad. Nauk SSSR*, **191**, 493-495 (1970).

Kostomarova, A. A., and N. V. Nechayeva, "The distribution of ribosomal cytoplasmic RNA in the blastoderm of the loach *Misgurnus fossilis* L.," *Ontogenez*, **1**, 391-397 (1970).

Kostomarova, A. A., and A. A. Neyfakh, "The method of separation of blastoderm in the loach embryos and its possible application," *Zh. Obshch. Biol.*, **25**, 386-395 (1964).

Kostomarova, A. A., and N. N. Rott, "An autoradiographic study of nuclear RNA synthesis in diploid and haploid embryos of the loach," in: *The Cell Nucleus and Its Ultrastructures* [in Russian], Nauka, Moscow (1970), pp. 258-261.

Krigsgaber, M. R., A. A. Kostomarova, T. A. Terekhova, and T. A. Burakova, "Synthesis of nuclear and cytoplasmic proteins in the early development of fishes and echinoderms," *J. Embryol. Exp. Morphol.*, **26**, 611-622 (1971).

Krigsgaber, M. R., and A. A. Neyfakh, "Protein synthesis in the blastoderm of *Misgurnus fossilis* embryo," *Dokl. Akad. Nauk SSSR*, **180**, 1259-1261 (1968).

Krigsgaber, M. R., and A. A. Neyfakh, "Investigation of the mode of nuclear control over protein synthesis in early development of loach and sea urchin," *J. Embryol. Exp. Morphol.*, **28**, 491-509 (1972).

Krigsgaber, M. R., A. A. Neyfakh, and N. P. Madatova, "Synthesis of ribosomal proteins in early development of the loach (*Misgurnus fossilis* L.)," *Ontogenez*, **7**, 47-57 (1976).

Kung, C. S., "On the size relationship between nuclear and cytoplasmic RNA in sea urchin embryos," *Develop. Biol.*, **36**, 343-356 (1974).

Kunkel, N. S., K. Hemminki, and E. S. Weinberg, "Size of histone gene transcripts in different embryonic stages of the sea urchin, *Strongylocentrotus purpuratus*," *Biochemistry*, **17**, 2591-2598 (1978).

Kunkel, N. S., and E. S. Weinberg, "Histone gene transcripts in the cleavage and mesenchyme blastula embryo of the sea urchin *S. purpuratus*," *Cell*, **14**, 313-326 (1978).

Kunz, W., "Funktions, structuren im oocytenkern von *Locusta migratoria*," *Chromosoma*, **20**, 332-370 (1967).

Kuriki, Y., and R. Okazaki, "Deoxyribonucleic acid and acid-soluble deoxyribosidic compounds in amphibian embryos," *Embryologia* (Nagoya), **4**, 337-348 (1959).

Kuzin, B. A., A. A. Aronshtam, and L. I. Korochkin, "A study of the determination of Est-6 gene expression in the genitalia of *Drosophila melanogaster* males," *Ontogenez*, **6**, 323-330 (1975).

La Marca, M. J., L. D. Smith, and K. Keem, "Hormonal effects on RNA synthesis by stage 6 oocytes of *Xenopus laevis*," *Develop. Biol.*, **47**, 384-393 (1975).

La Marca, M. J., M. C. Fidler, L. D. Smith, and M. C. Strobel, "Quantitative and qualitative analysis of RNA synthesis in stage 6 and stage 4 oocytes of *Xenopus laevis*," *Develop. Biol.*, **34**, 106-118 (1973).

La Marca, M. J., and P. M. Wasserman, "Program of early development in the mammal: changes in absolute rates of synthesis of ribosomal protein during oogenesis and early embryogenesis in the mouse," *Develop. Biol.*, **73**, 103-119 (1979).

Lamb, M. M., and C. D. Laird, "Increase in nuclear poly(A)-containing RNA at syncytial blastoderm in *Drosophila melanogaster* embryos," *Develop. Biol.*, **52**, 31-42 (1976).

Lambert, C. C., "Genetic transcription during the development and metamorphosis of the tunicate *Ascidia callosa*," *Exp. Cell Res.*, **66**, 401-409 (1971).

Landesman, R., "Ribosomal RNA synthesis in pre- and post-gastrula embryos of *Xenopus laevis*," *Cell Different.*, **1**, 209-213 (1972).

Landesman, R., and P. R. Gross, "Pattern of macromolecule synthesis during development of *Xenopus laevis*. II. Identification of the 40S precursor to ribosomal RNA," *Develop. Biol.*, **19**, 244-260 (1969).

Larsen, C. J., F. Galibert, A. Hampe, and M. Boiron, "Etude des RNAs nucléaires de faible poinds moléculaire de la cellule KB," *Bull. Soc. Chim. Biol.*, **51**, 649-668 (1969).

Larsen, C. J., P. Lebowitz, J. M. Weissman, and B. Dubuy, "Studies of the primary structure of low molecular weight ribonucleic acid other than tRNA," *Cold Spring Harbor Symp. Quant. Biol.*, **35**, 35-46 (1970).

Lebherz, H. G., "The ontogeny of fructose diphosphate aldolase B in the chick," *Develop. Biol.*, **27**, 143-149 (1972).

Lengyel, J. A., and S. Penman, "Differential stability of cytoplasmic RNA in a *Drosophila* cell line," *Develop. Biol.*, **57**, 243-253 (1977).

Leonard, D. A., and M. J. La Marca, "*In vivo* synthesis and turnover of cytoplasmic ribosomal RNA by stage 6 oocytes of *Xenopus laevis*," *Develop. Biol.*, **45**, 199-202 (1975).

Lerner, A. M., E. Bell, and J. E. Darnell, "Ribosomal RNA in the developing chick embryo," *Science*, **141**, 1187-1188 (1963).

Lerner, M. R., J. A. Boyle, S. M. Mount, S. L. Wolin, and J. A. Steitz, "Are snRNAs involved in splicing?", *Nature*, **283**, 220-224 (1980).

Lev, L., T. L. Thomes, A. S. Lee, R. C. Angerer, R. J. Britten, and E. H. Davidson, "Developmental expression of two cloned sequences coding for rare sea urchin embryo messages," *Develop. Biol.*, **76**, 322-340 (1980).

Levenson, R. G., and K. B. Marcu, "Evidence for polyadenylated histone mRNA in *Xenopus laevis* oocytes," *J. Cell Biol.*, **70**, 195a (1976).

Levey, I. L., G. B. Stull, and R. L. Brinster, "Poly(A) and synthesis of polyadenylated RNA in the preimplantation mouse embryo," *Develop. Biol.*, **64**, 140-148 (1978).

Levis, R., and S. Penman, "The metabolism of poly(A)$^+$ and poly(A)$^-$ hnRNA in cultured *Drosophila* cells studied with a rapid uridine pulse-chase," *Cell*, **11**, 105-113 (1977).

Levner, M. H., "RNA transcription in mature sea urchin eggs," *Exp. Cell Res.*, **85**, 296-302 (1974).

Levy, S., P. Wood, M. Grunstein, and L. Kedes, "Individual histone messenger RNAs: identification by template activity," *Cell*, **4**, 239-248 (1975).

Librera, E., "Effect of actinomycin D on the embryonic development of *Ascidia malaca*," *Acta Embryol. Morphol. Exp.*, **71**, 242-248 (1964).

Lievens, A., and J. Brachet, "RNA polymerase activity during development of normal and hybrid amphibian embryos," *Biochim. Biophys. Acta*, **204**, 263-266 (1970).

Lifton, R. P., and L. H. Kedes, "Size and sequence homology of masked maternal and embryonic histone messenger RNAs," *Develop. Biol.*, **48**, 47-55 (1976).

Lima-de-Faria, A., "The molecular organization of the chromosomes of *Acheta*, involved in ribosomal DNA amplification," *Cold Spring Harbor Symp. Quant. Biol.*, **38**, 559-571 (1973).

Lima-de-Faria, A., M. Birnstiel, and H. Jaworska, "Amplification of ribosomal cistrons in the heterochromatin of *Acheta*," *Genetics*, **61**, 145-161 (1969).

Lima-de-Faria, A., T. Gustafsson, and H. Jaworska, "Amplification of ribosomal DNA in *Acheta*. II. The number of nucleotide pairs of the chromosomes and chromomeres involved in amplification," *Hereditas*, **73**, 119-162 (1973a).

Lima-de-Faria, A., H. Jaworska, T. Gustafsson, and S. Daskaloff, "Amplification of ribosomal DNA in *Acheta*. III. The release of DNA copies from chromomeres," *Hereditas*, **73**, 163-191 (1973b).

Lindahl, P. E., "Zur Kenntnis des Ovarialeies bei dem Seeigel," *Wilh. Roux' Arch. Entwickl. Org.*, **127**, 300-338 (1932).

Ling, V., and G. H. Dixon, "The biosynthesis of protamine in trout testis. II. Polysome patterns and protein synthetic activities during testis maturation," *J. Biol. Chem.*, **245**, 3035-3042 (1970).

Lison, L., and J. Pasteels, "Etudes histophotométriques sur la teneur en acide desoxyribonucléique des noyaux au cours du développement embryonaire chez l'oursin *Paracentrotus lividus*," *Arch. Biol.*, **62**, 2-64 (1951).

Lockshin, R. A., "Insect embryogenesis. Macromolecular synthesis during early development," *Science*, **154**, 775-776 (1966).

Loeb, L. A., and B. Fansler, "Intracellular migration of DNA polymerase in early developing sea urchin embryos," *Biochim. Biophys. Acta*, **217**, 50-55 (1970).

Loeb, L. A., B. Fansler, R. Williams, and D. Mazia, "Sea urchin nuclear DNA polymerase. I. Localization in nuclei during rapid DNA synthesis," *Exp. Cell Res.*, **57**, 298-304 (1969).

Loening, U. E., "The mechanism of synthesis of ribosomal RNA," *Symp. Soc. Gen. Microbiol.*, **20**, 77-106 (1970).

Løvtrup-Rein, H., "Nuclear RNA and protein synthesis during early sea urchin development," *Exp. Cell Res.*, **72**, 188-194 (1972).

Lozovskaya, E. R., S. I. Slezinger, and A. A. Prokofyeva-Belgovskaya, "Differences in the relative content of AT- and GC-pairs of DNA bases in the centromeric heterochromatin of chromosome 1 in two types of human cells. 1. Autoradiographic analysis of ^{3}H-thymidine and ^{3}H-deoxycytidine incorporation," *Mol. Biol.*, **10**, 937-943 (1976).

Mabuchi, I., "Myosin from starfish egg: properties and interaction with actin," *J. Mol. Biol.*, **100**, 569-582 (1976).

MacGregor, H. C., "The role of lampbrush chromosomes in the formation of nucleoli in amphibian oocytes," *Quart. J. Microscop. Sci.*, **106**, 215-228 (1965).

MacGregor, H. C., "Nucleolar DNA in oocytes of *Xenopus laevis*," *J. Cell Sci.*, **3**, 437-444 (1968).

MacGregor, H. C., "The nucleolus and its genes in amphibian oogenesis," *Biol. Rev.*, **47**, 177-210 (1972).

MacKintosh, F. R., and E. Bell, "Regulation of protein synthesis in sea urchin eggs," *J. Mol. Biol.*, **41**, 365-380 (1969a).

MacKintosh, F. R., and E. Bell, "Proteins synthesized before and after fertilization in sea urchin eggs," *Science*, **164**, 961-963 (1969b).

MacLaughlin, I., and C. Terner, "Ribonucleic acid synthesis by spermatozoa from the rat and hamster," *Biochem. J.*, **133**, 635-639 (1973).

Maclean, N., and R. D. Jurd, "The control of haemoglobin synthesis," *Biol. Rev.*, **47**, 393-437 (1972).

Maden, B. E. H., and J. S. Robertson, "Demonstration of the '5.8S' ribosomal sequences in HeLa cell ribosomal precursor RNA," *J. Mol. Biol.*, **87**, 227-235 (1974).

Maggio, R., M. L. Vittorelli, A. M. Rinaldi, and A. Monroy, "*In vitro* incorporation of amino acids into proteins stimulated by RNA from unfertilized sea urchin eggs," *Biochem. Biophys. Res. Commun.*, **15**, 436-441 (1964).

Maggio, R., M. L. Vittorelli, J. Caffarelli-Mormino, and A. Monroy, "Dissociation of ribosomes of unfertilized eggs embryos of sea urchin," *J. Mol. Biol.*, **31**, 621-626 (1968).

Mahdavi, V., and M. Crippa, "An RNA-DNA complex intermediate in ribosomal gene amplification," *Proc. Nat. Acad. Sci. USA*, **69**, 1749-1752 (1972).

Mahowald, A. P., and M. Tiefert, "Fine structural changes in the *Drosophila* oocyte nucleus during a short period of RNA synthesis," *Wilh. Roux' Arch. Entwickl. Org.*, **165**, 8-25 (1970).

Mairy, M., and H. Denis, "Synthèse et accumulation du RNA pendant l'oogénèse de *Xenopus laevis*," *Arch. Int. Physiol. Biochim.*, **78**, 599-601 (1970).

Mairy, M., and H. Denis, "Recherches biochimiques sur l'oogénèse. 1. Synthèse et accumulation du RNA pendant l'oogenèse du crapaud sud-africain *Xenopus laevis*," *Develop. Biol.*, **24**, 143-165 (1971).

Mairy, M., and H. Denis, "Recherches biochimiques sur l'oogénèse. 2. Assemblage de ribosomes pendant le grand accroissement des oocytes de *Xenopus laevis*," *Europ. J. Biochem.*, **25**, 535-543 (1972).

Mairy-von-Frenckell, "Synthèse du RNA au cours de l'oogénèse," *Arch. Int. Physiol. Biochim.*, **78**, 393-397 (1970).

Malacinski, G. M., "Genetic control of qualitative changes in protein synthesis during early amphibian (Mexican axolotl) embryogenesis," *Develop. Biol.*. **26**, 442-451 (1971).

Malacinski, G. M., "Histone synthesis in amphibian oocytes and early embryos," *Experientia*, **30**, 610-612 (1974).

Malkin, L. I., P. R. Gross, and P. Romanoff, "Polyribosomal protein synthesis in fertilized sea urchin eggs: the effect of actinomycin treatment," *Develop. Biol.*, **10**, 378-394 (1964).

Man, N. T., and R. J. Cole, "Quantitative changes in chromosomal activity during chicken myogenesis *in vitro*, a DNA-

RNA hybridization study," *Exp. Cell Res.*, **83**, 328-334 (1974).

Manen, C., and D. H. Russell, "Early cyclic changes in polyamine synthesis during sea urchin development," *J. Embryol. Exp. Morphol.*, **30**, 243-256 (1973).

Manes, C., "Genetic and biochemical activities in preimplantation embryos," in: *The Developmental Biology of Reproduction* (C. L. Markert and J. Papaconstantinou, eds.), Acad. Press, New York (1975).

Manes, C., and J. R. Daniel, "Quantitative and qualitative aspects of protein synthesis in the preimplantation rabbit embryo," *Exp. Cell Res.*, **55**, 261-268 (1969).

Mano, Y., "Regulation of protein synthesis in early embryogenesis in the sea urchin," *Biochem. Biophys. Res. Commun.*, **33**, 877-882 (1968).

Mano, Y., "Factors involved in cyclic protein synthesis in sea urchin cells during early embryogenesis," *Biochemistry* (Tokyo), **65**, 483-487 (1969).

Mano, Y., "Cytoplasmic regulation and cyclic variation in protein synthesis in the early cleavage stage of the sea urchin embryo," *Develop. Biol.*, **22**, 433-460 (1970).

Mano, Y., "Mechanism of increase in the basal rate of protein synthesis in the early cleavage stage of the sea urchin," *J. Biochem.*, **69**, 1-10 (1971a).

Mano, Y., "Cell-free cyclic variation of protein synthesis associated with the cell cycle in sea urchin embryos," *J. Biochem.*, **69**, 11-25 (1971b).

Mano, Y., and H. Nagano, "Mechanism of release of maternal messenger RNA induced by fertilization in sea urchin eggs," *Biochemistry* (Tokyo), **67**, 611-618 (1970).

Maroun, L. E., "Cytoplasmic localization of DNA-dependent RNA polymerase in unfertilized sea urchin eggs," *Exp. Cell Res.*, **79**, 459-461 (1973).

Marshak, A., and C. Marshak, "Deoxyribonucleic acid in the nuclei of *Arbacia* eggs," *Biol. Bull. Mar. Biol. Lab. Woods Hole*, **103**, 295-299 (1952).

Marshak, A., and C. Marshak, "Deoxyribonucleic acid in *Arbacia* eggs," *Exp. Cell Res.*, **5**, 288-300 (1953).

Marshak, A., and C. Marshak, "Biological role of deoxyribonucleic acid," *Nature* (London), **174**, 919-920 (1954).

Marshak, A., and C. Marshak, "Thymine in the acid soluble fraction of *Arbacia* eggs," *J. Biophys. Biochem. Cytol.*, **1**, 167-171 (1955).

Marshak, A., and C. Marshak, "On the question of DNA content of sea urchin eggs," *Exp. Cell Res.*, **10**, 246-247 (1956).

Marshak, T. A., and A. A. Neyfakh, "Changes in the dry weight of nuclei in early embryos of the loach," *Ontogenez*, **4**, 28-33 (1973).

Martin, G. R., S. Smith, and C. Epstein, "Protein synthetic patterns in teratocarcinoma stem cells and mouse embryos at early stages of development," *Develop. Biol.*, **66**, 8-16 (1978).

Marushige, Y., and K. Marushige, "Transformation of sperm histone during formation and maturation of rat spermatozoa," *J. Biol. Chem.*, **250**, 39-45 (1975).

Marzluff, W. F., Jr., E. L. White, R. Benjamin, and R. C. C. Huang, "Low molecular weight RNA species from chromatin," *Biochemistry*, **14**, 3715-3724 (1975).

Mashkova, T. D., T. I. Serenkova, A. M. Mazo, T. A. Avdonina, M. Ya. Timofeeva, and L. L. Kiselev, "The primary structure of oocyte and somatic 5S rRNAs from the loach *Misgurnus fossilis*," *Nucl. Acids Res.*, **9**, 2141-2152 (1981).

Masui, Y., "Relative roles of the pituitary follicle cells and progesterone in the induction of oocyte maturation in *Rana pipiens*," *J. Exp. Zool.*, **166**, 365-376 (1967).

Masui, Y., "Hormonal and cytoplasmic control of meiotic maturation of frog oocytes," *Ontogenez*, **3**, 574-584 (1972).

Masui, Y., and C. L. Markert, "Cytoplasmic control of nuclear behavior during meiotic maturation of frog oocytes," *J. Exp. Zool.*, **177**, 129-146 (1971).

Matioli, G., S. Brody, and B. Thorell, "Microspectrophotometric determination of differentially extracted haemoglobin in single erythrocyte," *Acta Haematol.*, **28**, 73-85 (1962).

Matsumoto, L., H. Kasamatsu, and L. Pikó, "Mitochondrial DNA synthesis in sea urchin oocytes," *J. Cell Biol.*, **59**, 221a (1973).

Matsumoto, L., H. Kasamatsu, L. Pikó, and J. Vinograd, "Mitochondrial DNA replication in sea urchin oocytes," *J. Cell Biol.*, **63**, 146-159 (1974).

Maxson, R. E., Jr., and J. C. Egrie, "Expression of maternal and paternal histone genes during early cleavage stages of the echinoderm hybrid *Strongylocentrotus purpuratus* × *Lytechinus pictus*," *Develop. Biol.*, **74**, 335-342 (1980).

Mazabraud, A., M. Wegnez, and H. Denis, "Biochemical research on oogenesis. RNA accumulation in the oocytes of teleosts," *Develop. Biol.*, **44**, 326-332 (1975).

Mazia, D., "Deoxyribonucleic acid and deoxyribonuclease in development," *Growth,* **13**, Suppl., 5-31 (1949).

Mazia, D., G. Schatten, and R. Steinhardt, "Turning on of activities in unfertilized sea urchin eggs: correlation with changes of the surface," *Proc. Nat. Acad. Sci. USA*, **72**, 4469-4473 (1975).

McClay, D. R., A. F. Chambers, and R. H. Warren, "Specificity of cell-cell interactions in sea urchin embryos. Appearance of new cell-surface determinants at gastrulation," *Develop. Biol.*, **56**, 343-355 (1977).

McClean, D. K., and A. H. Warner, "Aspects of nucleic acid metabolism during development of the brine shrimp *Artemia salina*," *Develop. Biol.*, **24**, 88-105 (1971).

McLean, K. W., and A. H. Whiteley, "RNA synthesis during the early development of the Pacific oyster, *Crassostrea gigas*," *Exp. Cell Res.*, **87**, 132-138 (1974).

McMaster-Kaye, R., and J. S. Kaye, "Basic protein changes during the final stages of sperm maturation in the house cricket," *Exp. Cell Res.*, **97**, 378-386 (1976).

McMaster-Kaye, R., and J. H. Taylor, "Evidence for two metabolically distinct types of ribonucleic acid in chromatin and nucleoli," *J. Biophys. Biochem. Cytol.*, **4**, 5-11 (1958).

Meedel, T. H., and J. R. Whittaker, "Messenger RNA synthesis during early ascidian development," *Develop. Biol.*, **66**, 410-421 (1978).

Meeker, G. L., and R. M. Iverson, "Tubulin synthesis in fertilized sea urchin eggs," *Exp. Cell Res.*, **64**, 129-132 (1971).

Meistrich, M. L., P. K. Trostle, M. Frapart, and R. P. Erickson, "Biosynthesis and localization of lactate dehydrogenase X in pachytene spermatocytes and spermatids of mouse testes," *Develop. Biol.*, **60**, 428-441 (1977).

Melnikova, N. L., M. Ya. Timofeeva, N. N. Rott, and G. M. Ignatyeva, "Synthesis of ribosomal RNAs in early embryogenesis of the loach," *Ontogenez*, **3**, 85-94 (1972).

Melton, D. A., E. M. De Robertis, and R. Cortese, "Order and intracellular location of the events involved in the maturation of a spliced tRNA," *Nature*, **284**, 143-148 (1980).

Merlie, J. P., A. Sobel, J. P. Changeux, and F. Gros, "Synthesis of acetylcholine receptor during differentiation of cultured embryonic muscle cells," *Proc. Nat. Acad. Sci. USA*, **72**, 4028-4032 (1975).

Mermod, J. J., and M. Crippa, "Variations in the amount of polysomes in mature oocytes of *Drosophila melanogaster*," *Develop. Biol.*, **66**, 586-592 (1978).

Mermod, J. J., M. Jacobs-Lorena, and M. Crippa, "Changes in rate of RNA synthesis and ribosomal gene number during oogenesis of *Drosophila melanogaster*," *Develop. Biol.*, **57**, 393-402 (1977).

Merriam, R. W., "Protein synthesis in mature frog oocytes," *Exp. Cell Res.*, **41**, 614-621 (1966).

Merriam, R. W., "Movement of cytoplasmic proteins into nuclei induced to enlarge and initiate DNA or RNA synthesis," *J. Cell Sci.*, **5**, 333-349 (1969).

Metafora, S., L. Felicetti, and R. Gambino, "The mechanism of protein synthesis activation after fertilization of sea urchin eggs," *Proc. Nat. Acad. Sci. USA*, **68**, 600-604 (1971).

Mezger-Freed, L., "Evidence for synthesis of deoxyribonucleic acid in enucleated *Rana pipiens* eggs," *J. Cell Biol.*, **18**, 471-474 (1963).

Mikhailov, V. S., and G. G. Gause, "Replication of mitochondrial DNA in the early development of *Misgurnus fossilis*," *Develop. Biol.*, **41**, 57-71 (1974a).

Mikhailov, V. S., and G. G. Gause, "The DNA-synthesizing system of isolated mitochondria from unfertilized eggs of the loach and its *in vitro* activation by the antibiotic bruneomycin," *Mol. Biol.*, **8**, 108-118 (1974b).

Miller, L., "Relative amounts of newly synthesized poly(A)$^+$ and poly(A)$^-$ messenger RNA during development of *Xenopus laevis*," *Develop. Biol.*, **64**, 114-129 (1978).

Miller, O. L., "Extrachromosomal nucleolar DNA in amphibian oocytes," *J. Cell Biol.*, **23**, 60A (1964).

Miller, O. L., "Structure and composition of peripheral nucleoli of *Salamander* oocytes," *Nat. Cancer Inst. Monogr.*, **23**, 53-66 (1966).

Miller, O. L., and B. Beatty, "Extrachromosomal nucleolar genes in amphibian oocytes," *Genetics*, **61**, Suppl., 133-143 (1969a).

Miller, O. L., and B. Beatty, "Visualization of nucleolar genes," *Science*, **164**, 955-957 (1969b).

Miller, T. H., and D. Epel, "Studies of oogenesis in *Urechis caupo*. II. Accumulation during oogenesis of carbohydrate, RNA, microtubule protein, and soluble mitochondrial and lysosomal enzymes," *Develop. Biol.*, **32**, 331-344 (1973).

Milman, L. S., and Yu. G. Yurovitsky, *Mechanisms for Enzyme Regulation of Carbohydrate Metabolism in Early Embryogenesis* [in Russian], Nauka, Moscow (1973).

Milstone, L. M., and J. Piatigorsky, "Rates of protein synthesis in explanted embryonic chick lens epithelia: differential stimulation of δ-crystallin synthesis," *Develop. Biol.*, **43**, 91-100 (1975).

Milstone, L. M., P. Zelenka, and J. Piatigorsky, "δ-Crystallin mRNA in chick lens cells: mRNA accumulates during differential stimulation of δ-crystallin synthesis in cultured cells," *Develop. Biol.*, **48**, 197-204 (1976).

Minor, R. R., C. C. Clark, E. L. Strause, T. R. Koszalka, R. L. Brent, and N. A. Kefalides, "Basement membrane procollagen is not converted to collagen in organ cultures of parietal yolk sac endoderm," *J. Biol. Chem.*, **251**, 1789-1794 (1976).

Mintz, B., "Synthetic processes and early development in the mammalian egg," *J. Exp. Zool.*, **157**, 87-100 (1964a).

Mintz, B., "Gene expression in the morula stage of mouse embryos as observed during development of t_{12}/t_{12} lethal mutants *in vitro*," *J. Exp. Zool.*, **157**, 267-272 (1964b).

Mirkes, P. E., "Protein synthesis before and after fertilization in the egg of *Ilyanassa obsoleta*," *Exp. Cell Res.*, **60**, 115-118 (1970).

Mirkes, P. E., "Polysomes and protein synthesis during development of *Ilyanassa obsoleta*," *Exp. Cell Res.*, **74**, 503-508 (1972).

Mizuno, S., Y. R. Lee, A. H. Whiteley, and H. R. Whiteley, "Cellular distribution of RNA populations in 16-cell stage embryos of the sand dollar *Dendraster excentricus*," *Develop. Biol.*, **37**, 18-27 (1974).

Monesi, V., "Autoradiographic evidence of a nuclear histone synthesis during mouse spermiogenesis in the absence of detectable quantities of nuclear ribonucleic acid," *Exp. Cell Res.*, **36**, 683-688 (1964).

Monesi, V., "Synthetic activities during spermatogenesis in the mouse," *Exp. Cell Res.*, **39**, 197-224 (1965).

Monroy, A., "Fertilization," *Compreh. Biochem.*, **28**, 1-22 (1967).

Monroy, A., and P. R. Gross, "The control of gene activity during echinoderm embryogenesis," *Exp. Biol. Med.*, **1**, 37-51 (1967).

Monroy, A., R. Maggio, and A. M. Rinaldi, "Experimentally induced activation of the ribosomes of the unfertilized sea urchin egg," *Proc. Nat. Acad. Sci. USA*, **54**, 107-111 (1965).

Monroy, A., and M. L. Vittorelli, "Utilization of ^{14}C-glucose for amino acid and protein synthesis by the sea urchin embryo," *J. Cell Comp. Physiol.*, **60**, 285-287 (1962).

Monroy, A., M. L. Vittorelli, and R. Guarneri, "Investigations on the proteins of the cell fluid during the early development of the sea urchin *Paracentrotus lividus*," *Acta Embryol. Morphol. Exp.*, **4**, 77-95 (1961).

Moore, B. C., "Autoradiographic studies of ^{3}H-thymidine incorporation in normal and hybrid frog embryos (15 min.)," *Anat. Rec.*, **134**, 610-611 (1959).

Moritz, K. B., "DNS-Variation im keimbahnbegrenzten Chromatin und autoradiographische Befunde an seiner Funktion bei *Parascaris equorum*," *Verh. dt. zool. Ges.*, **64**, 36-42 (1970).

Moritz, K. B., and G. Roth, "Complexity of germinal and somatic DNA in *Ascaris*," *Nature*, **259**, 55-57 (1976).

Morris, G. E., A. Cooke, and R. J. Cole, "Isoenzymes of creatine phosphokinase during myogenesis *in vitro*," *Exp. Cell Res.*, **74**, 582-585 (1972).

Moscona, A. A., M. H. Moscona, and N. Saenz, "Enzyme induction in embryonic retina: the role of transcription and translation," *Proc. Nat. Acad. Sci. USA*, **61**, 160-167 (1968).

Moss, B. A., W. G. Joyce, and V. M. Ingram, "Histones in chick embryonic erythropoiesis," *J. Biol. Chem.*, **248**, 1025-1031 (1973).

Moss, P. S., and R. C. Strohman, "Myosin synthesis by fusion-arrested chick embryo myoblasts in cell culture," *Develop. Biol.*, **48**, 431-437 (1976).

Moyer, C., R. W. Armentrout, and R. D. Brown, "The onset of immunoglobulin synthesis during the development of *Xenopus laevis*," *Develop. Biol.*, **61**, 338-345 (1977).

Muckenthaler, F. A., and A. P. Mahowald, "DNA synthesis in the ooplasm of *Drosophila melanogaster*," *J. Cell Biol.*, **28**, 199-208 (1966).

Mulnard, J., "Etude morphologique et cytochimique de l'oogénèse chez *Acanthoscelides obtectus* Say (Bruchide — Coleoptera)," *Arch. Biol.*, **65**, 261 (1954).

Murison, G., "Synthesis and degradation of xanthine dehydrogenase during chick liver development," *Develop. Biol.*, **20**, 518-543 (1969).

Murray, K., "Interaction sites for proteins in DNA molecules," in: *Structure and Function of Chromatin, Ciba Faund. Sympos.*, Vol. 28, Elsevier (1975), pp. 338-351.

Nadel, M. R., M. R. Goldsmith, J. Goplerud, and F. C. Kafatos, "Specific protein synthesis in cellular differentiation. V. A secretory defect of chorion formation in the Gr col mutant of *Bombyx mori*," *Develop. Biol.*, **75**, 41-58 (1980).

Nadel, M. R., and F. C. Kafatos, "Specific protein synthesis in cellular differentiation. IV, The chorion proteins of *Bombyx mori* and their program of synthesis," *Develop. Biol..*, **75**, 26-40 (1980).

Nakano, E., and A. Monroy, "Incorporation of S^{35}-methionine in the cell fractions of sea urchin eggs and embryos," *Exp. Cell Res.*, **14**, 236-244 (1958).

Nass, M. M. K., "Mitochondrial DNA. Advances, problems, goals. Studies of size and structure of mitochondrial DNA related to biogenesis and function of this organelle," *Science*, **165**, 25-35 (1969a).

Nass, M. M. K., "Reversible generation of circular dimer and higher multiple forms of mitochondrial DNA," *Nature*, **223**, 1124-1129 (1969b).

Nass, S., M. M. K. Nass, and U. Hennix. "Deoxyribonucleic acid in isolated rat-liver mitochondria," *Biochim. Biophys. Acta*, **95**, 426-435 (1965).

Nemer, M., "Characteristics of the utilization of nucleosides by embryos of *Para-*

centrotus lividus.," *J. Biol. Chem.*, **237**, 143-149 (1962).

Nemer, M., "Old and new RNA in the embryogenesis of the purple sea urchin," *Proc. Nat. Acad. Sci. USA*, **50**, 230-235 (1963).

Nemer, M., "Transfer of genetic information during embryogenesis," *Progr. Nucl. Acid Res. Mol. Biol.*, **7**, 243-301 (1967).

Nemer, M., "Development changes in the synthesis of sea urchin embryo messenger RNA containing and lacking polyadenylic acid," *Cell*, **6**, 559-570 (1975).

Nemer, M., and S. G. Bard, "Polypeptide synthesis in sea urchin embryogenesis: an examination with synthetic polyribonucleotides," *Science*, **140**, 664-666 (1963).

Nemer, M., L. M. Dubroff, and M. Graham, "Properties of sea urchin embryo messenger RNA containing and lacking poly(A)," *Cell*, **6**, 171-178 (1975).

Nemer, M., and A. Infante, "Messenger RNA in early sea urchin embryos: size classes," *Science*, **150**, 214-217 (1965).

Nemer, M., and A. Infante, "Ribosomal ribonucleic acid of the sea urchin egg and its fate during embryogenesis," *J. Mol. Biol.*, **27**, 73-86 (1967).

Nemer, M., and D. T. Lindsay, "Evidence that the S-polysomes of early sea urchin embryos may be responsible for the synthesis of chromosomal histones," *Biochem. Biophys. Res. Commun.*, **35**, 156-159 (1969).

Newrock, K. M., C. R. Alfageme, R. V. Nardi, and L. H. Cohen, "Histone changes during chromatin remodeling in embryogenesis," *Cold. Spring Harbor Symp. Quant. Biol.*, **42**, 421-431 (1977).

Newrock, K. M., L. H. Cohen, M. W. Hendricks, R. J. Donnelly, and E. S. Weinberg, "Stage specific mRNAs coding for subtypes of H2A and H2B histones in the sea urchin embryo," *Cell*, **14**, 327-336 (1978).

Newrock, K. M., and R. A. Raff, "Polar lobe specific regulation of translation in embryos of *Ilyanassa obsoleta*," *Develop. Biol.*, **42**, 242-261 (1975).

Neyfakh, A. A., "X-ray inactivation of nuclei as a method for studying their function in the early development of fishes," *J. Embryol. Exp. Morphol.*, **7**, 173-192 (1959).

Neyfakh, A. A., "X-ray inactivation of nuclei as a method for studying their function in the development of respiration in fish embryos," *Biokhimiya*, **25**, 658-668 (1960).

Neyfakh, A. A., "A comparative radiation study of the morphogenetic function of the nuclei in animal development," *Zh. Obshch. Biol.*, **22**, 42-57 (1961).

Neyfakh, A. A., "Behavior of RNA during cell division," *Priroda*, **7**, 44-47 (1970).

Neyfakh, A. A., "Steps of realization of genetic information in early development," *Current Topics Develop. Biol.*, **6**, 45-77 (1971).

Neyfakh, A. A., "Morphogenetic nuclear function in the early development of the common pond snail (*Limnaea stagnalis*)," *Ontogenez*, **7**, 630-632 (1976).

Neyfakh, A. A., and N. B. Abramova, "Changes in the number and function of mitochondria in early ontogenesis," in: *Genetic Functions of the Cytoplasm Organelles* [in Russian], Nauka, Leningrad (1974), pp. 80-85.

Neyfakh, A. A., N. B. Abramova, and A. M. Bagirova, "Migration of newly synthesized RNA during mitosis. II. Fibroblasts of the Chinese hamster," *Tsitologiya*, **13**, 491-498 (1971).

Neyfakh, A. A., and E. R. Davydov, "Lack of nuclear control of cathepsin activity elevation in early embryonic development," *Biokhimiya*, **29**, 273-282 (1964).

Neyfakh, A. A., and G. V. Dontsova, "An X-ray study of the nuclear function in the elevation of cytochrome oxidase activity in fish embryos," *Biokhimiya*, **27**, 339-348 (1962).

Neyfakh, A. A., and A. A. Kostomarova, "Migration of newly synthesized RNA during mitosis. I. Embryonic cells of the loach," *Tsitologiya*, **11**, 231-235 (1971).

Neyfakh, A. A., A. A. Kostomarova, and

T. A. Burakova, "Transfer of RNA from nucleus to cytoplasm in early development of fish," *Exp. Cell Res.*, **72**, 223-232 (1972).

Neyfakh, A. A., and M. R. Krigsgaber, "The synthesis of protein in the embryonic development of the sea urchin following nuclei inactivation," *Dokl. Akad. Nauk SSSR*, **183**, 493-466 (1969).

Neyfakh, A. A., M. R. Krigsgaber, and M. I. Ilyin, "Protein synthesis during the early development of *Misgurnus fossilis* following radioinactivation of nuclei," *Dokl. Akad. Nauk SSSR*, **181**, 253-256 (1968).

Neyfakh, A. A., and N. N. Rott, "Synchronization of cell divisions in early loach embryos by exposure to low temperature," *Dokl. Akad. Nauk SSSR*, **125**, 432-434 (1959).

Nigon, V., and J. Dailler, "La synthèse de l'acide deoxyribonucléique au cours du développement de la drosophila," *Biochim. Biophys. Acta*, **29**, 246-255 (1958).

Nijhawan, P., and W. F. Marzluff, "Metabolism of low molecular weight ribonucleic acids in early sea urchin embryos," *Biochemistry*, **18**, 1353-1360 (1979).

Nilsson, M. O., and T. Hultin, "Evidence of a store of informational RNA in encysted embryos of *Artemia salina*," *Exp. Cell Res.*, **72**, 145-149 (1972).

Nonaka, M., and H. Terayama, "Thymidine kinase activation in unfertilized sea urchin eggs by homogenization and fertilization," *Develop. Biol.*, **43**, 322-332 (1975).

O'Brien, D. A., and A. R. Bellvé, "Protein constituents of the mouse spermatozoon. II. Temporal synthesis during spermatogenesis," *Develop. Biol.*, **75**, 405-418 (1980).

O'Connor, C. M., D. R. Balzer, and E. Lazarides, "Phosphorylation of subunit proteins of intermediate filaments from chicken muscle and nonmuscle cells," *Proc. Nat. Acad. Sci. USA*, **76**, 819-823 (1979).

Ogur, M., and G. Rosen, "The nucleic acids of plant tissues. I. The extraction and estimation of deoxypentose nucleic acid and pentose nucleic acid," *Arch. Biochem.*, **25**, 262-276 (1950).

Ohi, S., J. L. Ramirez, W. B. Upholt, and I. B. Dawid, "Mapping of mitochondrial 4S RNA genes in *Xenopus laevis* by electron microscopy," *J. Mol. Biol.*, **121**, 299-310 (1978).

Ohno, S., *Evolution by Gene Duplication*, Springer-Verlag, Berlin, Heidelberg, New York (1970).

Ohno, S., S. Karasaki, and K. Takata, "Histo- and cytochemical studies on the superficial layer of yolk platelets in the *Triturus* embryo," *Exp. Cell Res.*, **33**, 310-318 (1964).

Ojala, D., and G. Attardi, "Identification of discrete polyadenylate-containing RNA components transcribed from HeLa cell mitochondrial DNA," *Proc. Nat. Acad. Sci. USA*, **71**, 563-567 (1974).

Olmstead, J. B., G. B. Witman, K. Carlson, and J. L. Rosenbaum, "Comparison of the microtubule proteins of neuroblastoma cells, brain and chlamydomonas flagella," *Proc. Nat. Acad. Sci. USA*, **68**, 2273-2277 (1971).

O'Malley, B. W., S. L. C. Woo, S. E. Harris, J. M. Rosen, and A. R. Means, "Steroid hormone regulation of specific messenger RNA and protein synthesis in eucaryotic cells," *J. Cell. Physiol.*, **85**, 343-356 (1975).

O'Melia, A. F., and C. A. Villee, "*De novo* synthesis of transfer and 5S RNA in cleaving sea urchin embryos," *Nature New Biol.*, **239**, 51-53 (1972).

Ono, T., and M. J. Getz, "Levels of ovalbumin messenger RNA sequences in nonoviduct tissues of the chicken," *Develop. Biol.*, **75**, 481-485 (1980).

Oppenheimer, J. M., "Organization of the teleost blastoderm," *Quart. Rev. Biol.*, **22**, 105-118 (1947).

Ordahl, C. P., and A. I. Caplan, "Transcriptional diversity in myogenesis," *Develop. Biol.*, **54**, 61-72 (1976).

Osterman L. A., "A regulatory function of tRNA in the protein synthesis," *Usp. Sovr. Biol.*, **83**, 3-25 (1977).

Osterman, L. A., "Participation of tRNA in regulation of protein biosynthesis at the translational level in eukaryotes," *Biochimie*, **61**, 323-342 (1979).

Osterman, L. A., P. S. Sverdlova, I. A. Solovyova, and M. Ya. Timofeeva, "Methylation of tRNA in early embryogenesis of fishes," in: *The Structure and Functions of Nucleic Acids, Soviet-French Sympos.*, Pushchino (1974).

Ovchinnikov, L. P., A. Ts. Avanesov, and A. S. Spirin, "Informosomes of the loach embryos. II. Evidence for their pre-existence in the cell," *Mol. Biol.*, **3**, 465-471 (1969a).

Ovchinnikov, L. P., T. F. Bystrova, and A. S. Spirin, "Sedimentation and density characteristics of ribosomes and their subunits from the loach embryos," *Dokl. Akad. Nauk SSSR*, **185**, 210-213 (1969b).

Ovchinnikov, L. P., A. S. Spirin, B. Erni, and T. Staehelin, "RNA-binding proteins of rabbit reticulocytes contain the two elongation factors and some of the initiation factors of translation," *FEBS Letters*, **88**, 21-26 (1978).

Ovchinnikov, L. P., A. S. Voronina, A. S. Stepanov, N. V. Belitsina, and A. S. Spirin, "Informosomal complexes formed following the addition of RNA to animal cell homogenates," *Mol. Biol.*, **2**, 752-763 (1968).

Overton, G. C., and E. C. Weinberg, "Length and sequence heterogeneity of the histone gene repeat unit of the sea urchin *S. purpuratus*," *Cell*, **14**, 247-257 (1978).

Ozaki, H., and A. H. Whiteley, "l-Malat dehydrogenase in the development of sea urchin *Strongylocentrotus purpuratus*," *Develop. Biol.*, **21**, 196-215 (1970).

Ozernyuk, N. D., "Changes in the number of mitochondria during the loach oogenesis," *Dokl. Akad. Nauk SSSR*, **207**, 974-977 (1972).

Ozernyuk, N. D., and L. R. Palmbakh, "Growth and reproduction of mitochondria in the loach oocytes," *Ontogenez*, **6**, 442-449 (1975).

Paglia, L. M., S. J. Berry, and W. H. Kastern, "Messenger RNA synthesis, transport and storage in silkmoth ovarian follicles," *Develop. Biol.*, **51**, 173-181 (1976a).

Paglia, L. M., W. H. Kastern, and S. J. Berry, "Messenger ribonucleoprotein particles in silkmoth oogenesis," *Develop. Biol.*, **51**, 182-189 (1976b).

Painter, T. S., and A. N. Taylor, "Nucleic acid storage in toad's egg," *Proc. Nat. Acad. Sci. USA*, **28**, 311-317 (1942).

Palmer, W. G., and J. Papaconstantinou, "Increase in the complexity of α-crystallins during differentiation of lens cells," *Nature*, **219**, 170-172 (1968).

Palmiter, R., "Quantitation of parameters that determine the rate of ovalbumin synthesis," *Cell*, **4**, 189-197 (1975).

Pan, M. L., W. J. Bell, and W. H. Telfer, "Vitellogenic blood protein synthesis by insect fat body," *Science*, **165**, 393-394 (1969).

Papaconstantinou, J., and E. M. Yulku, "The regulation of ribosomal RNA synthesis and ribosomal assembly in the vertebrate lens," *J. Cell. Physiol.*, **72**, 161-180 (1968).

Pardue, M., and J. G. Gall, "Chromosomal localization of mouse satellite DNA," *Science*, **168**, 1356-1360 (1970).

Parsons, J. A., and R. C. Rustad, "The distribution of DNA among dividing mitochondria of *Tetrachymena periformis*," *J. Cell Biol.*, **37**, 683-691 (1968).

Pasternak, C. A., "Phospholipid synthesis in cleaving sea urchin eggs; model for specific membrane assembly," *Develop. Biol.*, **30**, 403-410 (1973).

Pasternak, J., and M. Haight, "DNA accumulation during oogenesis in the free-living nematode *Panagrellus silusiae*," *Chromosoma*, **49**, 279-298 (1975).

Paul, M., M. R. Goldsmith, J. R. Hunsley, and F. C. Kafatos, "Specific protein synthesis in cellular differentiation. Production of eggshell proteins by silkmoth follicular cells," *J. Cell Biol.*, **55**, 653-679 (1972).

Paul, M., and F. C. Kafatos, "Specific protein synthesis in cellular differentiation.

II. The program of protein synthetic changes during chorion formation by silkmoth follicles and its implementation organ culture," *Develop. Biol.*, **42**, 141-159 (1975).

Perkowska, E., H. C. MacGregor, and M. Birnstiel, "Gene amplification in the oocyte nucleus of mutant and wild type *Xenopus laevis*," *Nature*, **217**, 649-650 (1968).

Perlman, S., and M. Rosbash, "Analysis of *Xenopus laevis* ovary and somatic cell polyadenylated RNA by molecular hybridization", *Develop. Biol.*, **63**, 197-212 (1978).

Perlmann, P., "Soluble antigens in sea urchin gametes and developmental stages," *Exp. Cell Res.*, **5**, 394-399 (1953).

Pero, R., A. Lima-de-Faria, U. Stahle, H. Granström, and R. Ghatnekar, "Amplification of ribosomal DNA in *Acheta*. IV. The number of cistrons for 28S and 18S ribosomal RNA," *Hereditas*, **73**, 195-211 (1973).

Perry, R. P., F.-Y. Cheng, J. J. Freed, J. R. Greenberg, D. E. Kelley, and K. D. Tartof, "Evolution of the transcription unit of ribosomal RNA," *Proc. Nat. Acad. Sci. USA*, **65**, 609-616 (1970).

Peters, C., and W. R. Jeffery, "Postfertilization poly(A) protein complex formation on sea urchin maternal messenger RNA," *Differentiation*, **12**, 91-97 (1978).

Petri, W. H., A. R. Wyman, and F. C. Kafatos, "Specific protein synthesis in cellular differentiation. III. The eggshell proteins of *Drosophila melanogaster* and their program of synthesis," *Develop. Biol.*, **49**, 185-199 (1976).

Pfohl, R. J., "Changes in alkaline phosphatase during the early development of the sea urchin *Arbacia punctulata*," *Exp. Cell Res.*, **39**, 496-503 (1965).

Pfohl, R. J., "Alkaline phosphatase of sea urchin embryos: chromatographic and electrophoretic characterization," *Develop. Biol.*, **44**, 333-345 (1975).

Piatigorsky, J. H. Ozaki, and A. Tyler, "RNA and protein-synthesizing capacity of isolated oocytes of sea urchin *Lytechinus pictus*," *Develop. Biol.*, **15**, 1-22 (1967).

Piatigorsky, J., and A. Tyler, "Radioactive labelling of RNAs of sea urchin eggs during oogenesis," *Biol. Bull.*, **133**, 229-244 (1967).

Piatigorsky, J., and A. Tyler, "Changes upon fertilization in the distribution of RNA-containing particles in sea urchin eggs," *Develop. Biol.*, **21**, 13-28 (1970).

Piatigorsky, J., H. F. Webster, and S. P. Craig, "Protein synthesis and ultrastructure during the formation of embryonic chick fibers *in vivo* and *in vitro*," *Develop. Biol.*, **27**, 176-189 (1972).

Picard, B., M. Le Maire, M. Wegnez, and H. Denis, "Biochemical research on oogenesis. Composition of the 42S storage particles of *Xenopus laevis* oocytes," *Europ. J. Biochem.*, **109**, 359-368 (1980).

Picard, B., and M. Wegnez, "Isolation of a 7S particle from *Xenopus laevis* oocytes: a 5S RNA protein complex," *Proc. Nat. Acad. Sci. USA*, **76**, 241-245 (1979).

Pikó, L., "Synthesis of macromolecules in early mouse embryos cultured *in vitro*: RNA, DNA and a polysaccharide component," *Develop. Biol.*, **21**, 257-279 (1970).

Pikó, L., "Expression of mitochondrial and nuclear genes during early development," in: *Early Develop. Mammals, 2nd Symp. Brit. Soc. Develop. Biol.*, Cambridge (1975), pp. 167-187.

Pikó, L., and D. G. Chase, "Role of the mitochondrial genome during early development in mice. Effects of ethidium bromide and chloramphenicol," *J. Cell Biol.*, **58**, 357-378 (1973).

Pikó, L., D. G. Blair, A. Tyler, and J. J. Vinograd, "Cytoplasmic DNA in the unfertilized sea urchin egg: physical properties of circular mitochondrial DNA and the occurrence of catenated forms," *Proc. Nat. Acad. Sci. USA*, **59**, 838-845 (1968).

Pikó, L., and L. Matsumoto, "Number of mitochondria and some properties of mitochondrial DNA in the mouse egg," *Develop. Biol.*, **49**, 1-10 (1976).

Pikó, L., A. Tyler, and J. Vinograd, "Amount, location, priming capacity, circularity and other properties of cytoplasmic DNA in sea urchin eggs," *Biol. Bull.*, **132**, 68-90 (1967).

Pinon, H., M. Barat, C. Dufresne, and J. C. Mounolou, "Template specificity of the cytoplasmic DNA polymerase in *Xenopus laevis* oocytes," *Biochem. Biophys. Res. Commun.*, **65**, 692-698 (1975).

Pirrone, A. M., G. Spinelli, P. Acierno, M. Errera, and G. Giudice, "The RNA of unfertilized sea urchin eggs is 'capped'," *Cell Different.*, **5**, 335-342 (1977).

Poccia, D. L., and R. T. Hinegardner, "Developmental changes in chromatin proteins of the sea urchin from blastula to mature larva," *Develop. Biol.*, **45**, 81-94 (1975).

Portugal, F. H., "Specificity of lysine and tyrosine transfer RNA fractions for RNA codons," *Mech. Ageing and Develop.*, **1**, 23-32 (1972).

Prives, J. M., and B. M. Paterson, "Differentiation of cell membranes in cultures of embryonic chick breast muscle," *Proc. Nat. Acad. Sci. USA*, **71**, 3208-3211 (1974).

Puccia, E., C. Mansueto-Bonaccorso, N. Farinella-Feruzzia, and R. Morello, "Ribonucleic acid synthesis during development of an ascidian *Ciona intestinalis*," *Acta Embryol. Exp.*, **2**, 167-177 (1976).

Puckett, L. D., and L. A. Snyder, "Quantitative aspects of ribosomal RNA synthesis during ovarian development in two mutants of *Drosophila melanogaster*," *Biochem. Genet.*, **11**, 249-260 (1974).

Rachkus, Yu. A., K. A. Kafiani, and M. Ya. Timofeeva, "Some features of ribonucleic acids synthesized during the loach embryogenesis," *Ontogenez*, **2**, 276-285 (1971).

Rachkus, Yu. A., N. S. Kupriyanova, M. Ya. Timofeeva, and K. A. Kafiani, "Transcription of repetitive sequences in embryogenesis," *Mol Biol.*, **3**, 428-440, 617—626 (1969*a*, *b*).

Raff, E., "Microtubule proteins in axolotl eggs and developing embryos," *Develop. Biol.*, **58**, 56-75 (1977).

Raff, R. A., H. V. Colot, S. E. Selvig, and P. R. Gross, "Oogenetic origin of messenger RNA for embryonic synthesis of microtubule proteins," *Nature,* **235**, 211-214 (1972).

Raff, R. A., G. A. Greenhouse, K. W. Gross and P. R. Gross, "Synthesis and storage of microtubule proteins by sea urchin embryos," *J. Cell Biol.*, **50**, 516 (1971).

Raff, R. A., and J. F. Kaumeyer, "Soluble microtubule proteins of the sea urchin embryo: partial characterization of the proteins and behavior of the pool in early development," *Develop. Biol.*, **32**, 309-320 (1973).

Raff, R. A., K. M. Newrock, and R. D. Secrist, "Regulation of microtubule protein synthesis in embryos of the marine snail *Ilyanassa obsoleta*," *Develop. Biol.*, **44**, 369-374 (1975).

Ranzi, S., "Les changements protéiques au cours du développement embryonnaire et larvaire," *Année Biol.*, **30**, 523-550 (1957).

Rao, M. V. N., and D. M. Prescott, "Inclusion of predivision labeled nuclear RNA in postdivision nuclei in *Amoeba proteus*," *Exp. Cell Res.*, **62**, 286-292 (1970).

Raste, E., and I. B. Dawid, "Expression of the mitochondrial genome in *Xenopus laevis:* a map of transcripts," *Cell*, **18**, 501-510 (1979).

Raveh, D., M. Friedländer, and H. Eyal-Giladi, "Nucleolar ontogenesis in the uterine chick germ correlated with morphogenetic events," *Exp. Cell Res.*; **100**, 195-203 (1976).

Raven, C. P., *Oogenesis. The Storage of Developmental Information*, Pergamon Press, New York (1961).

Regier, J. C., and F. C. Kafatos, "Absolute rates of protein synthesis in sea urchins with specific activity measurements of radioactive leucine and leucyl-tRNA," *Develop. Biol.*, **57**, 270-283 (1977).

Regier, J. C., G. D. Mazur, and F. C. Kafatos, "The silkmoth chorion: morphological and biochemical characterization of four surface regions," *Develop. Biol.*, **76**, 286-304 (1980).

Reich, E., and D. J. L. Luck, "Replication and inheritance of mitochondrial DNA," *Proc. Nat. Acad. Sci. USA*, **55**, 1600-1607 (1966).

Rinaldi, A. M., and A. Monroy, "Polyribosome formation and RNA synthesis in the early post-fertilized stages of the sea urchin egg," *Develop. Biol.*, **19**, 73-86 (1969).

Rinaldi, A. M., and A. Parente, "Rate of protein synthesis in oocytes of *Paracentrotus lividus,*" *Develop. Biol.*, **49**, 260-267 (1976).

Rinaldi, A. M., G. Sconzo, I. Albanese, F. Ramirez, B. D. Bavister, and G. Giudice, "Cytoplasmic giant RNA in sea urchin embryos. III. Polysomal localisation," *Cell Different.*, **3**, 305-312 (1974).

Ritossa, F. M., and S. Spiegelman, "Localization of DNA complementary to ribosomal RNA in the nucleolus organizer region of *Drosophila melanogaster*," *Proc. Nat. Acad. Sci. USA*, **53**, 737-745 (1965).

Robberson, D. L., and D. A. Clayton, "Replication of mitochondrial DNA in mouse L cells and their thymidine kinase derivatives: displacement replication on a covalently-closed circular template," *Proc. Nat. Acad. Sci. USA*, **69**, 3810-3814 (1972).

Robberson, D. L., H. Kasamatsu, and J. Vinograd, "Replication of mitochondrial DNA. Circular replicative intermediates in mouse L cells," *Proc. Nat. Acad. Sci. USA,* **69**, 737-741 (1972).

Robbins, J., and S. M. Heywood, "Quantitation of myosin heavy chain mRNA during myogenesis," *Europ. J. Biochem.*, **82**, 601 (1978).

Rochaix, J.-D., A. Bird, and A. Bakken, "Ribosomal RNA gene amplification by rolling circles," *J. Mol. Biol.*, **87**, 473-487 (1974).

Rodgers, W. H., and P. R. Gross, "Inhomogeneous distribution of egg RNA sequences in the early embryo," *Cell,* **14**, 279-288 (1978).

Roeder, R. G., and W. J. Rutter, "Multiple RNA polymerases and ribonucleic acid synthesis during sea urchin development," *Biochemistry*, **9**, 2543-2553 (1970).

Romanoff, A. L., and A. J. Romanoff, *The Avian Egg*, J. Wiley, New York (1949).

Rosbash, M., and P. J. Ford, "Polyadenylic acid-containing RNA in *Xenopus laevis* oocytes," *J. Mol. Biol.*, **85**, 87-101 (1974).

Rosbash, M., P. J. Ford, and J. O. Bishop, "Analysis of the c-value paradox by molecular hybridization," *Proc. Nat. Acad. Sci. USA*, **71**, 3746-3750 (1974).

Roskam, W. G., M. Gruber, and G. Ab, "Estradiol-induced synthesis of vitellogenin. II. Immunochemical characterization of vitellogenin polysomes," *Biochim. Biophys. Acta,* **435**, 82-90 (1976).

Rossi, M., G. Augusti-Tocco, and A. Monroy, "Differential gene activity and segregation of cell lines: an attempt at a molecular interpretation of the primary events of embryonic development," *Quart. Rev. Biophys.*, **8**, 43-119 (1975).

Roth, T. F., and K. R. Porter, "Yolk protein uptake in the oocyte of the mosquito *Aedes aegypti*," *J. Cell Biol.*, **20**, 313-332 (1964).

Rott, N. N., and V. V. Popov, "Changes in the duration of the mitotic cycle and its stages in early embryogenesis of the loach," *9th Int. Embryol. Congress, Abstracts*, Moscow (1969), pp. 68-69.

Rott, N. N., and G. A. Shevelyova, "Changes in cell division characteristics at early developmental stages in diploid and haploid embryos of the loach," *Tsitologiya*, **9**, 1265-1275 (1967).

Rubin, G. M., "The nucleotide sequence of *Saccharomyces cerevisiae* 5.8S ribosomal ribonucleic acid," *J. Biol. Chem.*, **248**, 3860-3875 (1973).

Rubinstein, N., J. Chi, and H. Holtzer, "Coordinated synthesis and degradation of actin and myosin in a variety of myogenic and non-myogenic cells," *Exp. Cell Res.*, **97**, 387-393 (1976).

Rudack, D., and R. A. Wallace, "On the site of phosvitin synthesis in *Xenopus laevis*," *Biochim. Biophys. Acta*, **155**, 299 (1968).

Ruderman, J. V., and P. R. Gross, "Histones and histone mRNA in sea urchins," *J. Cell Biol.*, **59**, 296a (1973).

Ruderman, J. V., and P. R. Gross, "Histones and histone synthesis in sea urchin development," *Develop. Biol.*, **36**, 286-298 (1974).

Ruderman, J. V., H. R. Woodland, and E. A. Sturgess, "Histone messenger RNA during the early development of *Xenopus laevis*," *Develop. Biol.*, **71**, 71-82 (1979).

Russell, D. H., "Putrescine and spermidine biosynthesis in the development of normal and anucleolate mutants of *Xenopus laevis*," *Proc. Nat. Acad. Sci. USA*, **68**, 523-527 (1971).

Russell, D. H., and S. H. Snyder, "Amine synthesis in regenerating rat liver: extremely rapid turnover of ornithine decarboxylase," *Mol. Pharmacol.*, **5**, 253-262 (1969).

Rutter, W. J., J. D. Kemp, W. S. Bradshaw, W. R. Clark, R. A. Ronzio, and T. G. Sanders, "Regulation of specific protein synthesis in cytodifferentiation," *J. Cell. Physiol.*, **72**, Suppl. 1, 1-18 (1968).

Rutter, W. J., R. L. Pietet, and P. W. Morris, "Toward molecular mechanisms of developmental processes," *Ann. Rev. Biochem.*, **42**, 601-646 (1973).

Ruzdijič, S., and V. Glišin, "Towards a total analysis of polyribosome-associated RNP particles of sea urchin embryos," *Biochim. Biophys. Acta*, **269**, 441-449 (1972).

Sadgopal, A., and D. Kabat, "Synthesis of chromosomal proteins during the maturation of chicken erythrocytes," *Biochim. Biophys. Acta*, **190**, 486-495 (1969).

Sagata, N., K. Shiokawa, and K. Yamana, "A study on the steady-state population of poly(A)$^+$ RNA during early development of *Xenopus laevis*," *Develop. Biol.*, **77**, 431-448 (1980).

Sager, R., and R. Kitchin, "Selective silencing of eukaryotic DNA," *Science*, **189**, 426-433 (1975).

Saito, Z., W. J. Martin, and W. H. Cook, "Changes in the major macromolecular fractions of egg yolk during embryogenesis," *Canad. J. Biochem.*, **43**, 1755-1770 (1965).

Sakun, O. F., "The state of the oocyte nucleolar apparatus in teleost fishes just before meiotic divisions," *Dokl. Akad. Nauk SSSR*, **137**, 749-751 (1961).

Sala-Trepat, J. M., T. D. Sargent, S. Sell, and J. Bonner, "α-Fetoprotein and albumin genes of rats: no evidence for amplification-deletion or rearrangement in rat liver carcinogenesis," *Proc. Nat. Acad. Sci. USA*, **76**, 695-699 (1979).

Salzman, N. P., D. E. Moore, and J. Mendelsohn, "Isolation and characterization of human metaphase chromosomes," *Proc. Nat. Acad. Sci. USA*, **56**, 1449-1456 (1966).

Samarina, O. P., "Information transport from the nucleus to cytoplasm," in: *Molecular Biology of Nucleoplasmic Relationships* (S. Puiseux-Dao, ed.), Elsevier, Amsterdam (1975), pp. 219-229.

Samarina, O. P., A. S. Asriyan, and G. P. Georgiev, "Isolation of nuclear nucleoproteins containing messenger ribonucleic acid," *Dokl. Akad. Nauk SSSR*, **163**, 1510-1516 (1965).

Santerre, R. F., and A. Rich, "Actin accumulation in developing chick brain and other tissues," *Develop. Biol.*, **54**, 1-12 (1976).

Sarkar, P. K., and A. A. Moscona, "The ribonucleoprotein nature of the rapidly labeled RNA in the embryonic retina," *Develop. Biol.*, **26**, 187-200 (1971a).

Sarkar, P. K., and A. A. Moscona, "Changes in macromolecular synthesis associated with the induction of glutamine synthetase in embryonic retina (hydrocortisone/ differentiation/

polysomes/cytosine arabinoside), *Proc. Nat. Acad. Sci. USA*, **68**, 2308-2311 (1971b).

Sayles, C. D., L. W. Browder, and J. H. Williamson, "Expression of xanthine dehydrogenase activity during embryonic development of *Drosophila melanogaster*," *Develop. Biol.*, **33**, 213-217 (1973).

Scarano, E., B. De Petrocellis, and G. Augusti-Tocco, "Studies on the control of enzyme synthesis during the early embryonic development of the sea urchin," *Biochim. Biophys. Acta*, **83**, 174-176 (1964).

Scarano, E., and R. Maggio, "The enzymatic deamination of 5′-deoxycytidylic acid and of 5′-methyl-5′-deoxycytidylic acid in the developing sea urchin embryo," *Exp. Cell Res.*, **18**, 333-346 (1959).

Scarano, E., M. Rossi, and G. Geraci, "The regulation of the activity of deoxycytidylate aminohydrolase," in: *FEBS Symp. on Regulation of Enzyme Activity and Allosteric Interactions* (E. Kvamme and A. Pihl, eds.), Acad. Press, New York (1968), pp. 145-165.

Scheer, U., "Nuclear pore flow rate of ribosomal RNA and chain growth rate of its precursor during oogenesis of *Xenopus laevis*," *Develop. Biol.*, **30**, 13-28 (1973).

Scheer, U., M. F. Trendelenburg, and W. W. Franke, "Regulation of transcription of genes of ribosomal RNA during amphibian oogenesis. A biochemical and morphological study," *J. Cell Biol.*, **69**, 465-489 (1976).

Scherrer, K., M.-T. Imaizumi-Scherrer, C. A. Reynaud, and A. Therwath, "On premessenger RNA and transcriptions. A review," *Mol. Biol. Rep.*, **5**, 5-28 (1979).

Schmidt, B. A., P. T. Kelly, M. C. May, E. Davis, and G. W. Conrad, "Characterization of actin from fertilized eggs of *Ilyanassa obsoleta* during polar lobe formation and cytokinesis," *Develop. Biol.*, **76**, 126-140 (1980).

Schubert, D., H. Tarikas, S. Humphreys, S. Heinemann, and J. Patrick, "Protein synthesis and secretion in a myogenic cell line," *Develop. Biol.* **33**, 18-37 (1973).

Schultz, G. A., "Polyadenylic acid-containing RNA in unfertilized and fertilized eggs of the rabbit," *Develop. Biol.*, **44**, 270-277 (1975).

Schultz, G., and R. B. Church, "Transcriptional patterns in early mammalian development," in: *Biochemistry of Animal Development,* Vol. 3 (R. Weber, ed.), Acad. Press, New York (1973).

Schultz, J., "The relation of the heterochromatic chromosome regions to the nucleic acids of the cell," *Cold Spring Harbor Symp. Quant. Biol.*, **21**, 307-328 (1956).

Schultz, R. M., G. E. Letourneau, and P. M. Wasserman, "Program of early development in the mammal: changes in patterns and absolute rates of tubulin and total protein synthesis during oogenesis and early embryogenesis in the mouse," *Develop. Biol.*, **68**, 341-359 (1979a).

Schultz, R. M., G. E. Letourneau, and P. M. Wasserman, "Program of early development in the mammal: changes in the patterns and absolute rates of tubulin and total protein synthesis during oocyte growth in the mouse," *Develop. Biol.*, **73**, 120-133 (1979b).

Schultz, R. M., and P. M. Wasserman, "Specific changes in the pattern of protein synthesis during meiotic maturation of mammalian oocytes *in vitro*," *Proc. Nat. Acad. Sci. USA*, **74**, 538-541 (1977).

Schwartz, M. S., "Nucleic acid metabolism in oocytes and embryos of *Urechis caupo*," *Develop. Biol.*, **23**, 241-260 (1970).

Sconzo, G., I. Albanese, A. M. Rinaldi, G. Lopresti, and G. Giudice, "Cytoplasmic giant RNA in sea urchin embryos. II. Physicochemical characterization," *Cell Different.*, **3**, 297-304 (1974).

Sconzo, G., A. Bono, I. Albanese, and G. Giudice, "Studies on sea urchin oocytes. II. Synthesis of RNA during oogenesis," *Exp. Cell Res.*, **72**, 95-100 (1972).

Sconzo, G., and G. Giudice, "Synthesis of ribosomal RNA in sea urchin embryos.

V. Further evidence for an activation following the hatching blastula stage," *Biochim. Biophys. Acta*, **254**, 447-451 (1971).

Sconzo, G., A. M. Pirrone, V. Mutolo, and G. Giudice, "Synthesis of ribosomal RNA during sea urchin development. III. Evidence for an activation of transcription," *Biochim. Biophys. Acta*, **199**, 435-440 (1970).

Sconzo, G., E. Vitrano, A. Bono, L. D. Giovanni, V. Mutolo, and G. Giudice, "Synthesis of ribosomal RNA in sea urchin embryos. IV. Maturation of rRNA precursor," *Biochim. Biophys. Acta*, **232**, 132-139 (1971).

Seale, R. L., and A. I. Aronson, "Chromatin-associated proteins of the developing sea urchin embryo. I. Kinetics of synthesis and characterization of non-histone proteins," *J. Mol. Biol.*, **75**, 633-645 (1973).

Selman, G. G., and G. J. Pawsey, "The utilization of yolk platelets by tissues of *Xenopus* embryos studied by a safranine staining method," *J. Embryol. Exp. Morphol.*, **14**, 191-212 (1965).

Selvig, S. E., G. A. Greenhouse, and P. R. Gross, "Cytoplasmic synthesis of RNA in the sea urchin embryo. II. Mitochondrial transcription," *Cell Different.* **1**, 5-14 (1972).

Selvig, S. E., P. R. Gross, and A. Hunter, "Cytoplasmic synthesis of RNA in the sea urchin embryo," *Develop. Biol.*, **22**, 343-365 (1970).

Senger, D. R., R. J. Arceci, and P. R. Gross, "Histones of sea urchin embryos transients in transcription, translation and the composition of chromatin," *Develop. Biol.*, **65**, 416-425 (1978).

Sevaljevič, L., "Immunochemical evidence for the cytoplasmic origin of some chromosomal nonhistone proteins of sea urchin embryos," *Biochim. Biophys. Acta*, **335**, 102-108 (1974).

Seybold, W. D., and D. T. Sullivan, "Protein synthetic patterns during differentiation of imaginal discs *in vitro*," *Develop. Biol.*, **65**, 69-80 (1978).

Shaklee, J. B., M. J. Champion, and G. S. Whitt, "Developmental genetics of teleosts: a biochemical analysis of lake chubsucker ontogeny," *Develop. Biol.*, **38**, 356-382 (1974).

Sharma, O. K., L. L. Mays, and E. Borek, "Enhancement of the synthesis of a hormone-induced protein by transfer ribonucleic acids," *J. Biol. Chem.*, **248**, 7622-7624 (1973).

Shaver, J., "Mitochondrial populations during development of the sea urchin," *Exp. Cell Res.*, **11**, 548-559 (1956).

Shepherd, G. W., and R. Flickinger, "Heterogeneous nuclear RNA (hnRNA). Size in developing frog embryos," *Exp. Cell Res.*, **120**, 73-77 (1979).

Shih, R. J., C. M. O'Connor, K. Keem, and L. D. Smith, "Kinetic analysis of amino acid pools and protein synthesis in amphibian oocytes and embryos," *Develop. Biol.*, **66**, 172-182 (1978).

Shih, R. J., D. Smith, and K. Keen, "Rates of histone synthesis during early development of *Rana pipiens*," *Develop. Biol.*, **75**, 329-342 (1980).

Shinohara, T., and J. Piatigorsky, "Age-dependent differences in the ratio of synthesis of δ-crystallin polypeptides in cultured vitreous-free lenses of chick embryos," *Develop. Biol.*, **72**, 385-389 (1979).

Shiokawa, K., A. Kawahara, Y. Misumi, Y. Yasuda, and K. Yamana, "Inhibitor of ribosomal RNA synthesis in *Xenopus laevis* embryos. VI. Occurrence in mature oocytes and in unfertilized eggs," *Develop. Biol.*, **57**, 210-214 (1977).

Shiokawa, K., Y. Misumi, Y. Yasuda, Y. Nishio, S. Kurata, M. Sameshima, and K. Yamana, "Synthesis and transport of various RNA species in developing embryos of *Xenopus laevis*," *Develop. Biol.*, **68**, 503-514 (1979).

Shmerling, Zh. G., "Isolation and properties of DNA from oocytes of the sturgen," *Biokhimiya*, **30**, 113-123 (1965).

Shmerling, Zh. G., "Genetic functions of the mitochondrial DNA," *Izv. Akad. Nauk SSSR, Ser. Biol.*, **6**, 847-862 (1970).

Shutt, R. H., and L. H. Kedes, "Synthesis of histone mRNA sequences in isolated nuclei of cleavage stage sea urchin embryos," *Cell*, **3**, 283-294 (1974).

Siekevitz, P., R. Maggio, and C. Catalano "Some properties of a rapidly labelled ribonucleic acid species in *Sphaerechinus granularis*," *Biochim. Biophys. Acta*, **129**, 145-156 (1966).

Sierra, J. M., M. Meier, and S. Ochoa, "Effect ot development on the translation of messenger RNA in *Artemia salina* embryos," *Proc. Nat. Acad. Sci. USA*, **71**, 2693-2697 (1974).

Silver, D. J., and D. G. Comb, "Free amino acid pools in the developing sea urchin *Lytechinus variegatus*," *Exp. Cell Res.*, **43**, 699-700 (1966).

Silver, D. J., and D. G. Comb, "Acetic acid-soluble proteins in the developing sea urchin," *Develop. Biol.*, **16**, 107-117 (1967).

Sim, G. K., F. C. Kafatos, C. W. Jones, M. D. Koehler, A. Efstratiadis, and T. Mamatis, "Use of a cDNA library for studies on evolution and developmental expression of the chorion multigene families," *Cell*, **18**, 1303-1316 (1979).

Simbonis, S. S., and R. A. McBride, "Intracellular distribution of glucose-6-phosphatase and glycogen content in the developing chick liver," *Develop. Biol.*, **12**, 347-357 (1966).

Simmel, E. B., and D. A. Karnofsky, "Observations on the uptake of tritiated thymidine in the pronuclei of fertilized sand dollar embryos," *J. Biophys. Biochem. Cytol.*, **10**, 59-68 (1961).

Simon, Z., "Messenger RNA abundance and gene regulation in eukaryotes," *Nature*, **255**, 171-172 (1975).

Singer, R. H., and S. Penman, "Messenger RNA in HeLa cells: kinetics of formation and decay," *J. Mol. Biol.*, **78**, 321-334 (1973).

Singh, U. N., "Rate of flow of rapidly labelled RNA from nucleus to cytoplasm during embryonic development of sea urchin," *Exp. Cell Res.*, **53**, 537-543 (1968).

Siracusa, G., "RNA polymerase during early development in mouse embryo," *Exp. Cell Res.*, **78**, 460-462 (1973).

Skea, B. R., and A. M. Nemeth, "Factors influencing premature induction of UDP-glucuronyltransferase activity in cultured chick embryo liver cells," *Proc. Nat. Acad. Sci. USA*, **64**, 795-802 (1969).

Skoblina, M. N., "Experimental study of the role of the nucleus in maturation of oocytes in Amphibia and Acipenseridae," *Author's Abstract of Candidate's Dissertation*, Moscow (1970).

Skoultchi, A., and P. R. Gross, "Maternal histone messenger RNA: detection by molecular hybridization," *Proc. Nat. Acad. Sci. USA*, **70**, 2840-2844 (1973).

Slater, D. W., I. Slater, and F. J. Bollum, "Cytoplasmic poly(A) polymerase from sea urchin eggs, merogons and embryos," *Develop. Biol.*, **63**, 94-110 (1978).

Slater, D. W., I. Slater, and D. Gillespie, "Post-fertilization synthesis of polyadenylic acid in sea urchin embryos," *Nature*, **240**, 333-337 (1972).

Slater, D. W., I. Slater, D. H. Gillespie, and S. Gillespie, "Post-fertilization polyadenylation during transcriptive and translational inhibition," *Biochem. Biophys. Res. Commun.*, **60**, 1222-1228 (1974).

Slater, D. W., and S. Spiegelman, "An estimation of genetic messages in the unfertilized egg," *Proc. Nat. Acad. Sci. USA*, **56**, 164-171 (1966).

Slater, D. W., and S. Spiegelman, "Template capabilities and size distribution of echinoid RNA during early development," *Biochim. Biophus. Acta*, **166**, 82-93 (1968).

Slater, D. W., and S. Spiegelman, "Transcriptive expression during sea urchin embryogenesis," *Biochim. Biophys. Acta*, **213**, 194-207 (1970).

Slater, I., D. Gillespie, and D. Slater, "Cytoplasmic adenylation and processing of maternal RNA," *Proc. Nat. Acad. Sci. USA*, **70**, 406-411 (1973).

Slater, I., and D. W. Slater, "Polyadenylation and transcription following fertilization," *Proc. Nat. Acad. Sci. USA*, **71**, 1103-1107 (1974).

Slater, I., and D. W. Slater, "Cell-free cytoplasmic polyadenylation of oogenic RNA," *Differentiation*, **13**, 109-115 (1979).

Smith, K. D., "Genetic control of macromolecular synthesis during development of an ascidian: *Ascidia nigra*," *J. Exp. Zool.*, **164**, 393-406 (1967).

Smith, L., "Spectrophotometric assay of cytochrome-c-oxidase," *Methods Biochem. Anal.*, **2**, 427-435 (1955).

Smith, L. D., and R. E. Ecker, "Role of the oocyte nucleus in physiological maturation in *Rana pipiens*," *Develop. Biol.*, **19**, 281-309 (1969).

Smith, L. D., and R. E. Ecker, "Regulatory processes in the maturation and early cleavage of amphibian eggs," *Current Topics Develop. Biol.*, **5**, 1-38 (1970).

Smith, L. D., R. E. Ecker, and S. Subtelny, "The initiation of protein synthesis in eggs of *Rana pipiens*," *Proc. Nat. Acad. Sci. USA*, **56**, 1724-1728 (1966).

Smith, R. L., and H. S. Forrest, "Variations in isoaccepting species of transfer RNA during embryogenesis of *Oncopeltus fasciatus* (Dallas)," *Develop. Biol.*, **33**, 123-129 (1973).

Sofer, W. H., J. F. George, and R. M. Iverson, "Rate of protein synthesis during first division cycle of sea urchin eggs," *Science*, **153**, 1644-1645 (1966).

Solomon, J., "Nucleic acid content of the egg of the domestic fowl," *Biochim. Biophys. Acta*, **23**, 211-212 (1957a).

Solomon, J., "Nucleic acid content of early chick embryos and the hen's egg," *Biochim. Biophys. Acta*, **24**, 584-591 (1957b).

Solovyova (Solovjeva), I. A., N. G. Shostak, and M. Ya. Timofeeva, "Transfer RNA in the early embryogenesis. 3. Some aspects of the regulation of tRNA synthesis at early stages of the development of a loach *Misgurnus fossilis* (L.)," *Ontogenez*, **7**, 18-26 (1976).

Solovyova, I. A., and M. Ya. Timofeeva, "Low-molecular-weight RNAs in the loach embryogenesis," *Mol. Biol.*, **7**, 908-915 (1973).

Solovyova, I. A., M. Ya. Timofeeva, and I. E. Sosinskaya, "Transfer RNAs in early embryogenesis of fishes. I. Content of tRNA in unfertilized eggs and developing embryos of the loach," *Ontogenez*, **4**, 340-347 (1973a).

Solovyova, I. A., M. Ya. Timofeeva, and I. E. Sosinskaya, "Transfer RNAs in early embryogenesis of fishes. II. Synthesis of tRNA at early stages of development of *Misgurnus fossilis*," *Ontogenez*, **4**, 549-557 (1973b).

Sommerville, J., "Ribonucleoprotein particles derived from the lampbrush chromosomes of newt oocytes," *J. Mol. Biol.*, **78**, 487-503 (1973).

Sonenshein, G. E., T. E. Geoghegan, and G. Brawerman, "A major species of mammalian messenger RNA lacking a polyadenylate segment," *Proc. Nat. Acad. Sci. USA*, **73**, 3088-3092 (1976).

Speirs, J., and M. Birnstiel, "Arrangement of the 5.8S RNA cistrons in the genome of *Xenopus laevis*," *J. Mol. Biol.*, **87**, 237-256 (1974).

Spiegel, M., H. Ozaki, and A. Tyler, "Electrophoretical examination of soluble proteins synthesized in early sea urchin development," *Biochem. Biophys. Res. Commun.*, **21**, 135-140 (1965).

Spiegel, M., E. S. Spiegel, and P. S. Meltzer, "Qualitative changes in the basic protein fraction of developing embryos," *Develop. Biol.*, **21**, 73-86 (1970).

Spirin, A. S., "On 'masked' forms of messenger RNA in early embryogenesis and in other differentiating systems," *Current Topics Develop. Biol.*, **1**, 2-38 (1966).

Spirin, A. S., "Informosomes," *Europ. J. Biochem.*, **10**, 20-35 (1969).

Spirin, A. S., "Non-ribosomal ribonucleoprotein particles (informosomes) of animal cells," in: *Mechanism of Protein Biosynthesis and Its Regulation* (L. Bosch, ed.), North Holland, Amsterdam, London (1972), pp. 515-537.

Spirin, A. S., "Eukaryotic messenger RNA and informosomes: *omnia mea mecum porto*," *FEBS Letters*, **88**, 15-17 (1978).

Spirin A. S., and L. P. Gavrilova, *The Ribosome* [in Russian], Nauka, Moscow (1971).

Spirin, A. S., and M. Nemer, "Messenger RNA in early sea urchin embryos: cytoplasmic particles," *Science,* **150,** 214-217 (1965).

Spirin, A. S., N. V. Belitsina, and M. A. Aitkhozhin, "Messenger RNAs in early embryogenesis," *Zh. Obshch. Biol.,* **25,** 331-338 (1964).

Spooner, B. S., B. T. Walther, and W. J. Rutter, "The development of the dorsal and ventral mammalian pancreas *in vivo* and *in vitro*," *J. Cell Biol.,* **47,** 235-246 (1970).

Stavy, L., and P. R. Gross, "The protein-synthetic lesion in unfertilized eggs," *Proc. Nat. Acad. Sci. USA,* **57,** 735-742 (1967).

Stavy, L., and P. R. Gross, "Protein synthesis *in vitro* with fractions of sea urchin eggs and embryos," *Biochim. Biophys. Acta,* **182,** 193-202 (1969).

Stearns, L. W., *Sea Urchin Development: Cellular and Molecular Aspects,* Dowden, Hutchinson and Ross, Inc., Stroudsburg (1974).

Steinert, M., "La synthèse de l'acide ribonucléique au cours du développement embryonnaire des batraciens," *Bull. Soc. Chim. Biol.,* **33,** 549-554 (1951).

Steinert, M., and P. Van Gansen, "Binding H^3-actinomycin to vitelline platelets of amphibian oocytes. A high resolution autoradiography investigation," *Exp. Cell Res.,* **66,** 355-367 (1971).

Stepanov, A. S., and A. S. Voronina, "Formation of stable informosome-like particles at physiological temperatures," *Dokl. Akad. Nauk SSSR,* **203,** 1418-1421 (1972).

Stepanov, A. S., A. S. Voronina, L. P. Ovchinnikov, and A. S. Spirin, "The protein RNA-binding factor of animal cell extracts. I. Detection and characterization of the protein RNA-binding factor of animal cell extracts," *Biokhimiya,* **37,** 3-9 (1972).

Stephens, R. E. "Differential protein synthesis and utilization during cilia formation in sea urchin embryos," *Develop. Biol.,* **61,** 311-329 (1977).

Storti, R. V., and A. Rich, "Chick cytoplasmic actin and muscle actin have different structural genes," *Proc. Nat. Acad. Sci. USA,* **73,** 2346-2350 (1976).

Strelkov, L. A., and G. M. Ignatyeva, "Early activation of ribosomal RNA synthesis in the axolotl embryos," *Ontogenez,* **6,** 519-523 (1975).

Strom, C. M. and A. Dorfman, "Amplification of moderately repetitive DNA sequences during chick cartilage differentiation," *Proc. Nat. Acad. Sci. USA,* **73,** 3428-3432 (1976).

Stubblefield, E., R. Klevecz, and L. Deaven, "Synchronized mammalian cell culture. I. Cell replication cycle and macromolecular synthesis following brief colcemid arrest of mitosis," *J. Cell. Physiol.,* **69,** 345-354 (1967).

Sugino, Y., N. Sugino, R. Okazaki, and T. Okazaki, "Deoxynucleosidic compounds. I. Modified microbioassay method and its application to sea urchin eggs and several other materials," *Biochim. Biophys. Acta,* **40,** 417-424 (1960).

Sullivan, B., "Amphibian haemoglobins," in: *Chemical Zoology,* Vol. 9 (M. Florkin and B. T. Sheer, eds.), Acad. Press, New York (1974), pp. 77-122.

Svetlov, P. G., V. D. Bystrov, and T. F. Korsakova, "On the morphology and physiology of early developmental stages of teleost fishes," *Arkhiv Anat. Gistol. Embriol.,* **42,** 22-37 (1962).

Swanson, R. F., and I. B. Dawid, "The mitochondrial ribosome of *Xenopus laevis*," *Proc. Nat. Acad. Sci. USA,* **66,** 117-124 (1970).

Swift, H., and R. Kleinfeld, "DNA in grasshopper spermatogenesis, oogenesis and cleavage," *Physiol. Zool.,* **26,** 301-311 (1953).

Sy, J., and L. McCarty, "Characterization of 5.8S RNA from a complex with 26S ribosomal RNA from *Arbacia punctulata*," *Biochim. Biophys. Acta,* **199,** 86-94 (1970).

Sze, L. C., "Changes in the amount of desoxyribonucleic acid in the development of *Rana pipiens*," *J. Exp. Zool.*, **122**, 577-601 (1953).

Takemoto, T., Y. Nagamatsu, and T. Oka, "Casein and α-lactalbumin messenger RNAs during the development of mouse mammary gland," *Develop. Biol.*, **78**, 247-257 (1980).

Tata, J. R., "Regulation of protein synthesis by growth and developmental hormones," in: *Biochemical Actions of Hormones*, Vol. 1 (G. Litwack, ed.), Acad. Press, New York (1970), pp. 89-133.

Tata, J. R., "Cell structures and biosynthesis during hormone-mediated growth and development," in: *Hormones in Development*, Vol. 1 (M. Hamburgh and E. J. Barrington, eds.), Appleton-Century-Grofts, New York (1971), pp. 19-41.

Tata, J. R., "The expression of the vitellogenin gene," *Cell*, **9**, 1-14 (1976).

Telfèr, W. H., "The mechanism and control of yolk formation," *Ann. Rev. Entomol.*, **10**, 161-184 (1965).

Telfer, W. H., and M. E. Melins, "The mechanism of blood protein uptake by insect oocytes," *Amer. Zool.*, **3**, 185-193 (1963).

Teng, C. S., "Nuclear acidic protein of developing *oncopeltus* embryos," *Biochim. Biophys. Acta*, **366**, 385-395 (1974).

Terman, S. A., "Relative effect of transcription-level and translation-level control of protein synthesis during early development of the sea urchin," *Proc. Nat. Acad. Sci. USA*, **65**, 985-992 (1970).

Terman, S. A., and P. R. Gross, "Translation-level control of protein synthesis during early development," *Biochem. Biophys. Res. Commun.*, **21**, 595-600 (1965).

Thaler, M. M., M. C. L. Cox, and C. A. Villee, "Histones in early embryogenesis," *J. Biol. Chem.*, **245**, 1479-1483 (1970).

Theriault, Y., and R. Landesman, "Analysis of acidic nuclear proteins during development of *Xenopus laevis*," *Cell Different.*, **3**, 249-257 (1974).

Thieband, C. H., "Quantitative determination of amplified rDNA and its distribution during oogenesis in *Xenopus laevis*," *Chromosoma*, **73**, 37-44 (1979).

Thireos, G., and F. C. Kafatos, "Cell-free translation of silkmoth chorion mRNAs: identification of protein precursors and characterization of cloned DNAs by hybrid-selected translation," *Develop. Biol.*, **78**, 36-46 (1980).

Thomas, C., "Étude cytochimique au microscope électronique des structures à ARN et du glycogène dans le cytoplasme des oocytes de *Xenopus laevis*," *J. Embryol. Exp. Morphol.*, **21**, 165-176 (1969).

Thomas, C., "Ribonucleic acids and ribonucleoproteins from small oocytes of *Xenopus laevis*," *Biochim. Biophys. Acta*, **224**, 99-113 (1970).

Thomas, C., "Correlation between ultrastructural aspect of nucleoli and inhibition of ribosomal RNA synthesis in *Xenopus laevis*," *Exp. Cell Res.*, **74**, 547-551 (1972).

Thomas, C., F. Hanocq, and V. Heilporn, "Persistence of oocyte amplified rDNA during early development of *Xenopus laevis* eggs," *Develop. Biol.*, **57**, 226-229 (1976).

Thomas, R. J., "Yolk distribution and utilization during early development of a teleost embryo (*Brachydanio rerio*)," *J. Embryol. Exp. Morphol.*, **19**, 203-215 (1968).

Thomson, I., D. I. de Pomerai, J. F. Jackson, and R. M. Clayton, "Lens-specific mRNA in cultures of embryonic chick neural retina and pigmented epithelium," *Exp. Cell Res.*, **122**, 73-81 (1979).

Thomson, I., C. E. Wilkinson, J. F. Jackson, D. I. de Pomerai, R. M. Clayton, D. E. S. Truman, and R. Williamson, "Isolation and cell-free translation of chick lens crystallin mRNA during normal development and transdifferentiation of neural retina," *Develop. Biol.*, **65**, 372-382 (1978).

Thornton, V. F., and P. J. Evernett, "Endocrine control of oocyte maturation and oviducal jelly release in the toad *Bufo*

bufo," *Gen. Comp. Endocrinol.,* **13,** 268-274 (1969).

Timofeeva, M. Ya, "Synthesis of nucleic acids in early embryonic development," *Author's Abstract of Candidate's Dissertation,* Moscow (1967).

Timofeeva, M. Ya, G. L. Eisner, and N. S. Kupriyanova, "A comparative study of repetitive nucleotide sequences in DNA of differentiated tissues and during malignancy formation", *Mol. Biol.,* **9,** 126-133 (1975).

Timofeeva, M. Ya., T. A. Ivanchik, and A. A. Neyfakh, "Alteration in the activity of synthesis of high-molecular-weight RNA during early development of embryos of *Strongylocentrotus nudus,*" *Dokl. Akad. Nauk SSSR,* **184,** 1014-1016 (1969).

Timofeeva, M. Ya., and K. A. Kafiani, "Nucleic acids in unfertilized eggs and developing embryos of the loach," *Biokhimiya,* **29,** 110-115 (1964).

Timofeeva, M. Ya., and K. A. Kafiani, "Heterogeneity of messenger nucleic acids synthesized at early stages of embryogenesis," *Dokl. Akad. Nauk SSSR,* **164,** 1183-1186 (1965).

Timofeeva, M. Ya., K. A. Kafiani, and A. A. Neyfakh, "Changing of the RNA-synthesizing function of the nuclei at early stages of embryonic development," in: *The Structure and Functions of the Cell Nucleus* [in Russian], Nauka, Moscow (1967), pp. 201-206.

Timofeeva, M. Ya., N. S. Kupriyanova, T. I. Serenkova, I. A. Solovyova, and N. T. Shostak, "5S ribosomal RNA of loach *Misgurnus fossilis* L.," in: *Macromolecules in the Functioning Cell* /English translation/ (A. A. Bayev, ed.), Nauka, Foreign Publications, Moscow (1982).

Timofeeva, M. Ya., T. I. Serenkova, and N. T. Shostak, "The electrophoretic characteristics of low-molecular-weight RNA of unfertilized eggs, embryos, and somatic tissues of the loach," *Mol. Biol.,* **16,** 230-239 (1982).

Timofeeva, M. Ya., and I. A. Solovyova, "Transfer RNA synthesis in early embryogenesis," *FEBS Letters,* **33,** 327-330 (1973).

Timofeeva, M. Ya., I. A. Solovyova, and T. I. Serenkova, "Characteristics of low-molecular-weight RNA synthesized in early embryogenesis," *Mol. Biol.,* **11,** 449-456 (1977).

Timourian, H., "Protein synthesis in sea urchin eggs. I. Fertilization-induced changes in subcellular fractions," *Develop. Biol.,* **16,** 594-611 (1967).

Timourian, H., and G. Watchmaker, "Protein synthesis in sea urchin eggs. II. Changes in amino acid uptake and incorporation at fertilization," *Develop. Biol.,* **23,** 478-491 (1970).

Tobler, H., K. D. Smith, and H. Ursprung, "Molecular aspects of chromatin elimination in *Ascaris lumbricoides,*" *Develop. Biol.,* **27,** 190-203 (1972).

Tocchini-Valentini, G. P., and M. Crippa, "On the mechanism of gene amplification," in: *Colloquia on Biology and Medicine,* Vol. 2 (L. Silvestry, ed.), North Holland, Amsterdam (1971), p. 237.

Tocchini-Valentini, G. P., V. Mahdavi, R. Brown, and M. Crippa, "The synthesis of amplified ribosomal DNA," *Cold Spring Harbor Symp. Quant. Biol.,* **38,** 551-558 (1973).

Trams, E. G., C. J. Lauter, G. J. Koval, S. Ruzdijič, and V. Glišin, "Plasma membrane marker enzymes in developing sea urchin embryos," *Proc. Soc. Exp. Biol. Med.,* **147,** 171-176 (1974).

Treigyte, G., and A. Gineitis, "Specific changes in the biosynthesis and acetylation of nucleosomal histones in the early stages of embryogenesis of sea urchin," *Exp. Cell Res.,* **121,** 127-134 (1979).

Tschudy, D. P., H. S. Marver, and A. Collins, "A model for calculating messenger RNA half-life: short lived messenger RNA in the induction of mammalian δ-aminolevulinic acid synthetase," *Biochem. Biophys. Res. Commun.,* **21,** 480-487 (1965).

Tsien, H. C., and J. M. Wattiaux, "Effect of maternal age on DNA and RNA content of *Drosophila* eggs," *Nature New Biol.*, **230**, 147-148 (1971).

Tucker, E. B., and G. A. Schultz, "Two-dimensional electrophoretic analysis of proteins synthesized during early cleavage period in rabbit embryos," *Exp. Cell Res.*, **114**, 438-443 (1978).

Tufaro, F., and B. P. Brandhorst, "Similarity of protein synthesized by isolated blastomeres of early sea urchin embryos," *Develop. Biol.*, **72**, 390-397 (1979).

Tung, T. C., and Y. F. Y. Tung, "The development of egg fragments, isolated blastomeres and fused eggs in the goldfish," *Proc. Zool. Soc.*, London, **114**, 46-64 (1944).

Tupper, J. T., "Potassium exchangeability, potassium permeability and membrane potential: some observations in relation to protein synthesis in the early echinoderm embryo," *Develop. Biol.*, **32**, 140-154 (1973).

Tupper, J. T., "Inhibition of increased potassium permeability following fertilization of the echinoderm embryo; its relationship to the initiation of protein synthesis and potassium exchangeability," *Develop. Biol.*, **38**, 332-345 (1974).

Turkington, R. W., and G. C. Majumder, "Gene activation during spermatogenesis," *J. Cell. Physiol.*, **85**, 495-508 (1974).

Turner, D. C., R. Gmür, H. G. Lebherz, M. Siegrist, T. Wallimann, and H. M. Eppenberger, "Differentiation in cultures derived from embryonic chicken muscle. II. Phosphorylase histochemistry and fluorescent antibody staining for creatine kinase and aldolase," *Develop. Biol.*, **48**, 284-307 (1976).

Tyler, A., "Masked messenger RNA and cytoplasmic DNA in relation to protein synthesis and processes of fertilization and determination in embryonic development," *Develop. Biol.*, Suppl. 1, 170-226 (1967).

Tyler, A., J. Piatigorsky, and H. Ozaki, "Influence of individual amino acid on uptake and incorporation of valine, glutamic acid and arginine by unfertilized sea urchin eggs," *Biol. Bull.*, **131**, 204-217 (1966).

Vacquier, V. D., "The appearance of β-1,3-glucanohydrolase activity during the differentiation of the gut of sand dollar plutei," *Develop. Biol.*, **26**, 1-10 (1971).

Vacquier, V. D., L. J. Lorh, and D. Epel, "The appearance of α-amylase activity during gut differentiation in sand dollar plutei," *Develop. Biol.*, **26**, 393-399 (1971).

Vacquier, V. D., M. J. Tegner, and D. Epel, "Protease released from sea urchin eggs at fertilization alters the vitelline layer and aids in preventing polyspermy," *Exp. Cell Res.*, **80**, 111-119 (1973).

Van Blerkom, J., and C. Manes, "Development of preimplantation rabbit embryos *in vivo* and *in vitro*. II. A comparison of qualitative aspects of protein synthesis," *Develop. Biol.*, **40**, 40-51 (1974).

Van Blerkom, J., and R. W. McGaughey, "Molecular differentiation of the rabbit ovum. II. During the preimplantation development of in *vivo* and *in vitro* matured oocytes," *Develop. Biol.*, **63**, 151-164 (1978).

Van de Kerckhove, D., "Content of deoxyribonucleic acid content of germinal vesicle of the primary oocyte in the rabbit," *Nature*, **183**, 329-331 (1959).

Vandekerkhove, J., and K. Weber, "At least six different actins are expressed in a higher mammal: an analysis based on the amino acid sequence of the amino-terminal tryptic peptide," *J. Mol. Biol.*, **126**, 783-802 (1978).

Van den Biggelaar, J. A. M., "RNA synthesis during cleavage of the *Lymnaea* egg," *Exp. Cell Res.*, **67**, 207-210 (1971).

Van Gansen, P., C. Thomas, and A. Schram, "Nucleolar activity and RNA metabolism in previtellogenic and vitellogenic oocytes of *Xenopus laevis*," *Exp. Cell Res.*, **98**, 111-119 (1976).

Van Ommen, G. L. B., G. S. P. Groot, and L. A. Grivell, "Transcription maps of mtDNAs of two strains of *Saccharomyces:* transcription of strain-specific insertions; complex RNA maturation and splicing," *Cell,* **18,** 511-523 (1979).

Vanyushin, B. F., and M. D. Kirnos, "The nucleotide composition and pyrimidine clusters in DNA from beef heart mitochondria," *FEBS Letters,* **39,** 195-199 (1974).

Vaughn, L. C., and L. A. Thomson, "A kinetic study of DNA and basic protein metabolism during spermatogenesis in the sand crab, *Emerita analoga,*" *J. Cell Biol.,* **52,** 322-337 (1972).

Vendrely, C., and R. Vendrely, "Sur la teneur individuelle en acide désoxyribonucléique des gametes d'oursins *Arbacia* et *Paracentrotus,*" *Compt. Rend. Soc. Biol.,* **143,** 1386-1387 (1949).

Venkstern T. V., "The processing of tRNA," *Mol. Biol.,* **15,** 5-26 (1980).

Vertel, B. M., and D. A. Fischman, "Myosin accumulation in mononucleated cells of chick muscle cultures," *Develop. Biol.,* **48,** 438-446 (1976).

Vincent, W. S., H. O. Halvorson, H. R. Chen, and D. Shin, "Ribosomal RNA cistrons in single and multinucleolate oocytes," *Biol. Bull.,* **135,** 441 (1968).

Vincent, W. S., H. O. Halvorson, H. R. Chen, and D. Shin, "A comparison of ribosomal gene amplification in uni- and multinucleolate oocytes," *Exp. Cell Res.,* **57,** 240-251 (1969).

Vlasik, T. N., L. P. Ovchinnikov, Kh. Radjabov, and A. S. Spirin, "Translation factors of the wheat embryo extract are RNA-binding proteins," *FEBS Letters,* **88,** 18-20 (1978).

Volfson, V. G., "Changes in transcription characteristics in early embryogenesis of the sea urchin," *Author's Abstract of Candidate's Dissertation,* Institute of Cytology, Leningrad (1972).

Volfson, V. G., and V. I. Vorobyov, "A comparison of the nuclear RNA in sea urchin embryos with RNA obtained *in vitro* by template synthesis on chromatin," *Scient. Commun. of the Inst. of Marine Biology, the Far East Scient. Center, USSR Acad. Sci.,* No. **2,** 48-51 (1971).

Volfson, V. G., and V. I. Vorobyov, "Changes in transcription characteristics in early embryogenesis of the sea urchin," *Ontogenez,* **3,** 498-509 (1972).

Volfson, V. G., V. I. Vorobyov, and L. N. Ananyeva, "Hybridization analysis of RNAs synthesized on chromatin from sea urchin embryos," *Scient. Commun of the Inst. of Marine Biology, the Far East Scient. Center, USSR Acad. Sci.,* No. 2, 45-47 (1971).

Wahrmann, J. P., D. Luzzati, and R. Winand, "Changes in adenyl cyclase specific activity during differentiation of an established myogenic cell line," *Biochem. Biophys. Res. Commun.,* **52,** 576-581 (1973).

Wallace, H., J. Morray, and H. R. Langridge, "Alternative model for gene amplification," *Nature New Biol.,* **230,** 201-203 (1971).

Wallace, R. A., "Enzymatic pattern in the developing frog embryo," *Develop. Biol.,* **3,** 486-515 (1961).

Wallace, R. A., and E. W. Bergink, "Amphibian vitellogenin: properties, hormonal regulation of hepatic synthesis and ovarian uptake and conversion to yolk proteins," *Amer. Zool.,* **14,** 1159 1176 (1974).

Wallace, R A., and J. N. Dumont, "The induced synthesis and transport of yolk proteins and their accumulation by the oocyte in *Xenopus laevis.,*" *J. Cell. Physiol.,* **72,** Suppl. 1, 73-90 (1968).

Wallace, R. A., and D. W. Jared, "Studies on amphibian yolk. VII. Serum phosphoprotein synthesis by vitellogenic females and estrogen-treated males of *Xenopus laevis,*" *Canad. J. Biochem.,* **46,** 953-959 (1968).

Wallace, R. A., and D. W. Jared, "Protein incorporation by isolated amphibian oocytes. V. Specificity for vitellogenin incorporation," *J. Cell Biol.*, **69**, 345-351 (1976).

Wallenfels, B., "Developmental and mutational changes of glycoproteins in the mouse neuronal retina: studies with bovine galactosyltransferase," *Proc. Nat. Acad. Sci. USA*, **76**, 3223-3227 (1979).

Wang, K.-M., "Comparative study of the development enzymes involved in carbohydrate and amino acid metabolism from brain, heart, liver and kidney of chick embryo," *Comp. Biochem. Physiol.*, **27**, 33-50 (1968).

Wang, K., J. McClure, and A. Fu, "Titin: major myofibrillar components of striated muscle," *Proc. Nat. Acad. Sci. USA*, **76**, 3698-3702 (1979).

Ward, R. T., "The origin of protein and fatty yolk in *Rana pipiens*. II. Electron microscopical and cytochemical observations of young and mature oocytes," *J. Cell. Biol.*, **14**, 309-341 (1962).

Waring, G. L., and A. P. Mahowald, "Identification and time of synthesis of chorion proteins in *Drosophila melanogaster*," *Cell*, **16**, 599-607 (1979).

Warren, T. G., M. D. Brennan, and A. P. Mahowald, "Two processing steps in maturation of vitellogenin polypeptides in *Drosophila melanogaster*," *Proc. Nat. Acad. Sci. USA*, **76**, 2848-2852 (1979).

Watson-Coggins, L., and J. G. Gall, "The timing of meiosis and DNA synthesis during early oogenesis in the toad, *Xenopus laevis*," *J. Cell Biol.*, **52**, 569-576 (1972).

Weatherall, D. J., and J. B. Clegg, "Recent developments in the molecular genetics of human hemoglobin," *Cell*, **16**, 467-479 (1979).

Webb, A. C., "An autoradiographic study of tritiated uridine incorporation into the larval ovary of *Xenopus laevis*," *Anat. Rec.*, **184**, 285-300 (1976).

Webb, A., and C. J. Camp, "Mitochondrial DNA synthesis in oocytes of *Xenopus laevis*," *Exp. Cell Res.*, **119**, 414-418 (1979).

Webb, A. C., M. J. La Marca, and L. D. Smith, "Synthesis of mitochondrial RNA by full-grown and maturing oocytes of *Rana pipiens* and *Xenopus laevis*," *Develop. Biol.*, **45**, 44 (1975).

Webb, A., and L. D. Smith, "Accumulation of mitochondrial DNA during oogenesis in *Xenopus laevis*," *Develop. Biol.*, **56**, 219-225 (1977).

Weber, C. S., and W. J. Rutter, "Differential synthesis of aldolases A and B during embryological development," *Fed. Proc. Amer. Soc. Exp. Biol.*, **23**, 487 (1964).

Wegnez, M., and H. Denis, "Recherches biochimiques sur l'oogénèse. 4. Absence d'amplification des gènes organisateurs du RNA 5S et du tRNA dans les petits oocytes de *Xenopus laevis*," *Biochimie*, **54**, 1069-1072 (1972).

Wegnez, M., and H. Denis, "Recherches biochimiques sur l'oogénèse. 6. Propriétés du RNA 5S présent dans les differents compartiments cellulaires des oocytes de *Xenopus laevis*," *Biochimie*, **55**, 1129-1135 (1973).

Wegnez, M., and H. Denis, "Biochemical research on oogenesis. Transfer RNA is fully charged in the 42-S storage particles of *Xenopus laevis* oocytes," *Europ. J. Biochem.*, **98**, 67-75 (1979).

Wegnez, M., H. Denis, A. Mazabraud, and J. C. Clerot, "Biochemical research on oogenesis. RNA accumulation during oogenesis of the dogfish *Scyliorhinus caniculus*," *Develop. Biol.*, **62**, 99-111 (1978).

Wegnez, M., R. Monier, and H. Denis, "Sequence heterogeneity of 5S RNA in *Xenopus laevis*," *FEBS Letters*, **25**, 13-19 (1972).

Weinberg, R. A., and S. Penman, "Small molecular weight monodisperse nuclear RNA," *J. Mol. Biol.*, **38**, 289 (1968).

Weinhold, P. A, and V. B. Rethy, "Ethanolamine phosphokinase: activity and properties during liver development," *Biochim. Biophys. Acta*, **276**, 143-154 (1972).

Weissbach, A, "Vertebrate DNA polymerases," *Cell,* **5,** 101-108 (1975).

Wellauer, P. K., and R. H. Reeder, "A comparison of the structural organization of amplified ribosomal DNA from *X. mulleri* and *X. laevis. J. Mol. Biol.,* **94,** 151-161 (1975).

Wensink, P., and D. D. Brown, "Denaturation map of the ribosomal DNA of *Xenopus laevis,*" *J. Mol. Biol.,* **60,** 235-247 (1971).

Wessells, N. K., and J. H. Cohen, "Early pancreas organogenesis: morphogenesis, tissue interactions and mass effects," *Develop. Biol.,* **15,** 237-270 (1967).

Westin, M., "Effect of actinomycin D on antigen synthesis during sea urchin development," *J. Exp. Zool.,* **171,** 297-304 (1969).

Westin, M., "The occurrence of stage-specific antigens during early sea urchin development," *J. Exp. Zool.,* **179,** 207-214 (1972).

Westin, M., "Phosphatase-active antigens in sea urchin eggs and embryos. I. Substrate specificity, pH-optima and inhibitors. II. A comparison between the activity in unfertilized eggs and plutei," *J. Exp. Zool.,* **192,** 307-322 (1975).

Westin, M., H. Perlmann, and P. Perlmann, "Immunological studies of protein synthesis during sea urchin development," *J. Exp. Zool.,* **166,** 331-345 (1967).

Westin, M., and P. Perlmann, "Immunofluorescence studies of developmental changes in sea urchin eggs and embryos," *Exp. Cell Res.,* **72,** 233-239 (1972).

White, B. N., G. M. Tener, J. Holden, and D. T. Suzuki, "Analysis of tRNAs during the development of *Drosophila,*" *Develop. Biol.,* **33,** 185-195 (1973).

Whiteley, A. H., B. J. McCarthy, and H. R. Whiteley, "Changing populations of messenger RNA during sea urchin development," *Proc. Nat. Acad. Sci. USA,* **55,** 519-525 (1966).

Whiteley, H. R., and A. H. Whiteley, "Changing populations of reiterated DNA transcripts during early echinoderm development," *Current Topics Develop. Biol.,* **9,** 39-88 (1975).

Whittaker, J. R., "Segregation during cleavage of a factor determining endodermal alkaline phosphatase development in ascidian embryos," *J. Exp. Zool.,* **202,** 139-154 (1977).

Wilson, E. B., *The Cell in Development and Heredity,* Macmillan, New York (1928).

Wilt, F. H., "Ribonucleic acid synthesis during sea urchin embryogenesis," *Develop. Biol.,* **9,** 299-313 (1964).

Wilt, F. H., "The control of embryonic hemoglobin synthesis," *Adv. Morphogen.,* **6,** 89-125 (1967).

Wilt, F. H., "The acceleration of RNA synthesis in cleaving sea urchin embryos," *Develop. Biol.,* **23,** 444-455 (1970).

Wilt, F. H., "Polyadenylation of maternal RNA of sea urchin eggs after fertilization," *Proc. Nat. Acad. Sci. USA,* **70,** 2345-2349 (1973).

Wilt, F. H., "The dynamics of maternal poly(A)-containing mRNA in fertilized sea urchin eggs," *Cell,* **11,** 673-682 (1977).

Wilt, F. H., and T. Hultin, "Stimulation of phenylalanine incorporation by polyuridilic acid (poly U) in homogenates of sea urchin eggs," *Biochem. Biophys. Res. Commun.,* **9,** 313-317 (1962).

Wilt, F. H., and D. Mazia, "The stimulation of cytoplasmic polyadenylation in sea urchin eggs by ammonia," *Develop. Biol.,* **37,** 422-424 (1974).

Winesdorfer, J. I., "Demonstration of polyribosomal aggregates and amino acid incorporation in post-mitochondrial supernatants from unfertilized eggs of *Sabellaria cementarium,*" *Amer. Zool.,* **5,** 635 (1965).

Wittliff, J. L., and F. T. Kenney, "Regulation of yolk protein synthesis in amphibian liver. I. Induction of lipovitellin synthesis by estrogen," *Biochim. Biophys. Acta,* **269,** 485-492 (1972a).

Wittliff, J. L., and F. T. Kenney, "Regulation of yolk protein synthesis in amphibian liver. II. Elevation of RNA synthesis by estrogen," *Biochim. Biophys. Acta,* **269,** 493-509 (1972b).

Wolfson, J., D. Dressler, and M. Magazin, "Bacteriophage T7 DNA replication: a linear replicating intermediate," *Proc. Nat. Acad. Sci. USA,* **69,** 499-504 (1972).

Wolgemuth, D. J., G. M. Jagiello, and A. S. Henderson, "Quantitation of ribosomal RNA genes in fetal human oocyte nuclei using rRNA: DNA hybridization *in situ.* Evidence for increased multiplicity," *Exp. Cell Res.,* **118,** 181-190 (1979).

Wolgemuth, D. J., G. M. Jagiello, and A. S. Henderson, "Baboon late diplotene oocytes contain micronucleoli and a low level of extra rDNA templates," *Develop. Biol.* **78,** 598-604 (1980).

Wolstenholme, D. R., and I. B. Dawid, "A size difference between mitochondrial DNA molecules of urodela and anura amphibia," *J. Cell Biol.,* **39,** 222-228 (1968).

Woodland, H. R., "The modification of stored histones H3 and H4 during the oogenesis and early development of *Xenopus laevis,*" *Develop. Biol.,* **68,** 360-370 (1979).

Woodland, H. R., and E. D. Adamson, "The synthesis and storage of histones during the oogenesis of *Xenopus laevis,*" *Develop. Biol.,* **57,** 118-135 (1977).

Woodland, H. R., J. M. Flynn, and A. J. Wyllie, "Utilization of stored mRNA in *Xenopus* embryos and its replacement by newly synthesized transcripts: histone H1 synthesis using interspecies hybrids," *Cell,* **18,** 165-171 (1979).

Woodland, H. R., and C. F. Graham, "RNA synthesis during early development of the mouse," *Nature,* **221,** 327-332 (1969).

Woodland, H. R., and J. B. Gurdon, "RNA synthesis in an amphibian nuclear-transplant hybrid," *Develop. Biol.,* **20,** 89-104 (1969).

Woodland, H. R., J. B. Gurdon, and J. B. Lingrel, "The translation of mammalian globin mRNA injected into fertilized eggs of *Xenopus laevis.* Distribution of globin synthesis in different tissues," *Develop. Biol.,* **39,** 134-140 (1974).

Woodland, H. R., and F. H. Wilt, "The functional stability of sea urchin histone mRNA injected into oocytes of *Xenopus laevis,*" *Develop. Biol.,* **75,** 199-213 (1980a).

Woodland, H. R., and F. H. Wilt, "The stability and translation of sea urchin histone messenger RNA molecules injected into *Xenopus laevis* eggs and developing embryos," *Develop. Biol.,* **75,** 214-221 (1980b).

Woods, P. S., and J. H. Taylor, "Nucleic acid synthesis in dividing cells," *Proc. 10th Int. Congress on Genetics,* **2,** 320 (1958).

Wright, D. A., and F. H. Moyer, "Parental influences on LDH in the early development of hybrid frogs in the genus *Rana,*" *J. Exp. Zool.,* **163,** 215-230 (1966).

Wright, D. A., and C. R. Shaw, "Time of expression of genes controlling specific enzymes in *Drosophila* embryos," *Biochem. Genet.,* **4,** 385-394 (1970).

Wright, D. A., and S. Subtelny, "Nuclear and cytoplasmic contributions to dehydrogenase phenotypes in hybrid frog embryos," *Develop. Biol.,* **24,** 119-140 (1971).

Yamada, T., "Differentiation of lens cells," *Exp. Biol. Med.,* **1,** 77-89 (1967).

Young, B. D., G. D. Birnie, and J. Paul, "Complexity and specificity of polysomal poly(A+) RNA in mouse tissues," *Biochemistry,* **15,** 2823-2829 (1976).

Young, E. M., and R. A. Raff, "Messenger ribonucleoprotein particles in developing sea urchin embryos," *Develop. Biol.,* **72,** 24-40 (1979).

Young, P. G., and A. M. Zimmerman, "Synthesis of mitochondrial RNA in disaggregated embryos of *Xenopus laevis*," *Develop. Biol.*, **33**, 196-205 (1973).

Young, V. R., "The role of skeletal and cardiac muscle in the regulation of protein metabolism," in: *Mammalian Protein in Metabolism*, Vol. 4 (H. Munro. ed.), Acad. Press, New Yorк (1970), pp. 586-660.

Yoshida, K., and A. Katoh, "Crystallin synthesis by chicken lens. III. mRNA stabilization under *in vitro* culture conditions," *Exp. Cell Res.*, **71**, 361-371 (1972).

Yudin, A. L., and A. A. Neyfakh, "Migration of newly-synthesized RNA during mitosis," *Exp. Cell. Res.*, **82**, 210-214 (1973).

Zalokar, M., "Autoradiographic study of protein and RNA formation during early development of *Drosophila* eggs," *Develop. Biol.*, **49**, 425-437 (1976).

Zamboni L., and B. Gondos, "Intercellular bridges and synchronization of germ cell differentiation during oogenesis of rabbit," *J. Cell Biol.*, **36**, 276-82 (1968).

Zampetti-Bosseler, F., G. Huez, and J. Brachet, "Effects of several inhibitors of macromolecule synthesis upon maturation of marine invertebrate oocytes," *Exp. Cell Res.*, **78**, 383-393 (1973).

Zapisek, W. F., and J. Papaconstantinou, "Degradation of ribosomal RNA during chicken lens development," *Biochim. Biophys. Acta*, **299**, 603-611 (1973).

Zeikus, J. G., M. W. Taylor, and C. A. Buck, "Transfer RNA changes associated with early development and differentiation of the sea urchin, *Strongylocentrotus purpuratus*," *Exp. Cell Res.*, **57**, 74-78 (1969).

Zeitz, L., R. Ferguson, and E. Garfinkel, "Incorporation of ^{3}H-TdR and ^{3}H-BudR into sperm DNA of *Arbacia punctulata* after coelomic injection," *Exp. Cell Res.*, **56**, 158-163 (1969).

Zeitz, L., E. Garfinkel, and R. Ferguson, "Determination of the first S-period in sea urchin eggs by the uptake of labeled precursors. Differences found with different precursors," *Biophys. Soc. Abstr. Ann. Meet.*, 14a (1970).

Zeuthen, E., "Segmentation, nuclear growth and cytoplasmic storage in eggs of echinoderms and Amphibia," *Pabbl. Staz. Zool. Napoli Suppl.*, **23**, 47-72 (1951).

Zevin-Sonkin, D., and D. Jaffe, "Accumulation of muscle-specific RNA sequences during myogenesis," *Develop. Biol.*, **74**, 326-334 (1980).

Zimmerman, A. M. "Incorporation of H^3-thymidine in the eggs of *Arbacia punctulata*," *Exp. Cell Res.*, **31**, 39-45 (1963).

Zwaan, J., "Lens-specific antigens and cytodifferentiation in the developing lens," *J. Cell. Physiol.*, **72**, Suppl. 1, 47-72 (1968).

Zylber, E., and S. Penman, "Mitochondrial associated 4S RNA synthesis inhibition by ethidium bromide," *J. Mol. Biol.*, **46**, 201-206 (1969).

SUBJECT INDEX

This book is the first of a two-volume survey of the latest advances in molecular biology of development, the second volume, *Molecular Biology of Development. Problems of Regulation*, is at the printer's. It deals with mechanism of regulation of RNA, DNA and protein synthesis in embryogenesis during differentiation, with the wide range of factors influencing replication, transcription and translation, and with morphogenetic function of nuclei in development.

MIX
Papier aus verantwortungsvollen Quellen
Paper from responsible sources
FSC® C105338

If you have any concerns about our products,
you can contact us on
ProductSafety@springernature.com

In case Publisher is established outside the EU,
the EU authorized representative is:
Springer Nature Customer Service Center GmbH
Europaplatz 3, 69115 Heidelberg, Germany

Printed by Libri Plureos GmbH
in Hamburg, Germany